Vol. 29. The Analytical Chemistry of Sulfur and Its Comp
Karchmer
Vol. 30. Ultramicro Elemental Analysis. By Günther Tölg
Vol. 31. Photometric Organic Analysis *(in two parts)*. By Eugene Sawicki
Vol. 32. Determination of Organic Compounds: Methods and Procedures. By Frederick T. Weiss
Vol. 33. Masking and Demasking of Chemical Reactions. By D. D. Perrin
Vol. 34. Neutron Activation Analysis. By D. De Soete. R. Gijbels. and J. Hoste
Vol. 35. Laser Raman Spectroscopy. By Marvin C. Tobin
Vol. 36. Emission Spectrochemical Analysis. By Morris Slavin
Vol. 37. Analytical Chemistry of Phosphorus Compounds. Edited by M. Halmann
Vol. 38. Luminescence Spectrometry in Analytical Chemistry. By J. D. Winedordner. S. G. Schulman, and T. C. O'Haver
Vol. 39. Activation Analysis with Neutron Generators. By Sam S. Nargolwalla and Edwin P. Przybylowicz
Vol. 40. Determination of Gaseous Elements in Metals. Edited by Lynn L. Lewis, Laben M. Melnick, and Ben D. Holt
Vol. 41. Analysis of Silicones. Edited by A. Lee Smith
Vol. 42. Foundations of Ultracentrifugal Analysis. By H. Fujita
Vol. 43. Chemical Infrared Fourier Transform Spectroscopy. By Peter R. Griffiths
Vol. 44. Microscale Manipulations in Chemistry. By T. S. Ma and V. Horak
Vol. 45. Thermometric Titrations. By J. Barthel
Vol. 46. Trace Analysis: Spectroscopic Methods for Elements. Edited by J. D. Winefordner
Vol. 47. Contamination Control in Trace Element Analysis. By Morris Zief and James W. Mitchell
Vol. 48. Analytical Applications of NMR. By D. E. Leyden and R. H. Cox
Vol. 49. Measurement of Dissolved Oxygen. By Michael L. Hitchman
Vol. 50. Analytical Laser Spectroscopy. Edited by Nicolo Omenetto
Voll. 51. Trace Element Analysis of Geological Materials. By Roger D. Reeves and Robert R. Brooks
Vol. 52. Chemical Analysis by Microwave Rotational Spectroscopy. By Ravi Varma and Lawrence W. Hrubesh
Vol. 53. Information Theory As Applied to Chemical Analysis. By Karel Eckschlager and Vladimir Štěpánek
Vol. 54. Applied Infrared Spectroscopy: Fundamentals, Techniques, and Analytical Problem-solving. By A. Lee Smith
Vol. 55. Archaeological Chemistry. By Zvi Goffer
Vol. 56. Immobilized Enzymes in Analytical and Clinical Chemistry. By P. W. Carr and L. D. Bowers
Vol. 57. Photoacoustics and Photoacoustic Spectroscopy. By Allan Rosencwaig
Vol. 58. Analysis of Pesticide Residues. Edited by H. Anson Moye
Vol. 59. Affinity Chromatography. By William H. Scouten
Vol. 60. Quality Control in Analytical Chemistry. By. G. Kateman and F. W. Pijpers
Vol. 61. Direct Characterization of Fineparticles. By Brian H. Kaye
Vol. 62. Flow Injection Analysis. By J. Ruzicka and E. H. Hansen

Introduction to Photoelectron Spectroscopy

QD
96
.P5
G48
1983

102352

Ghosh, Pradip K.
Introduction to photo-
electron spectroscopy

**East Texas Baptist College
Library**
Marshall, Texas

CHEMICAL ANALYSIS

A SERIES OF MONOGRAPHS ON ANALYTICAL CHEMISTRY AND ITS APPLICATIONS

Editors
P. J. ELVING, J. D. WINEFORDNER
Editor Emeritus: **I.M. KOLTHOFF**

Advisory Board

Fred W. Billmeyer, Jr. Victor G. Mossotti
Eli Grushka   A. Lee Smith
Barry L. Karger   Bernard Tremillon
Viliam Krivan   T. S. West

VOLUME 67

A WILEY-INTERSCIENCE PUBLICATION

JOHN WILEY & SONS
New York / Chichester / Brisbane / Toronto / Singapore

Introduction to Photoelectron Spectroscopy

PRADIP K. GHOSH

Department of Chemistry
Indian Institute of Technology Kanpur

Mamie Jarrett Memorial Library
East Texas Baptist College
Marshall, Texas

A WILEY-INTERSCIENCE PUBLICATION

JOHN WILEY & SONS

New York / Chichester / Brisbane / Toronto / Singapore

Copyright © 1983 by John Wiley & Sons, Inc.

All rights reserved. Published simultaneously in Canada.

Reproduction or translation of any part of this work beyond that permitted by Section 107 or 108 of the 1976 United States Copyright Act without the permission of the copyright owner is unlawful. Requests for permission or further information should be addressed to the Permissions Department, John Wiley & Sons, Inc.

Library of Congress Cataloging in Publication Data:
Ghosh, Pradip K.
   Introduction to photoelectron spectroscopy.

   (Chemical analysis, ISSN 0069-2883; v. 67)
   Rev. ed. of: A whiff of photoelectron spectroscopy. 1978.
   "A Wiley-Interscience publication."
   Includes bibliographical references and index.
   1. Photoelectron spectroscopy. I. Title. II. Series.

QD96.P5G48   1983   543'.0858   82-17374
ISBN 0-471-06427-0

Printed in the United States of America

10 9 8 7 6 5 4 3 2 1

PREFACE

The applications of photoelectron spectroscopy now pervade several important fields of science. This text is intended to be useful to anyone who wants to have an idea about the subject. In particular, it should be helpful to senior undergraduate and beginning graduate students of physics, chemistry, and materials science as a concise review of the subject to date. It should be of help also to practicing scientists in these areas who wish to have an introduction to the principles and scope of applications of the photoelectron spectroscopic technique. Although the chief purpose of this text is to provide a general impression of the field, the appended guide to the literature on the subject is reasonably detailed, representative, and currrent. The book can also be used as a text in a course on electron spectroscopy or on modern physical methods of analysis.

A slimmer version of this book was published in India as *A Whiff of Photoelectron Spectroscopy*. The range of coverage in this expanded edition has remained essentially unchanged. Significant extensions include a more detailed treatment of organic and inorganic molecules (Chapter 4), a number of additional examples on solids (Chapters 5 and 7), and a fairly comprehensive account of the status of quantitative analysis using photoelectron spectroscopy (Chapter 6). The number of references has more than doubled, a reflection of the explosive growth of the field. The notes that accompany the references have been reorganized, in some cases rewritten, and new ones have been added. The case studies included in the text illustrate applications to several fields, and indicate the method followed in making use of spectra from a range of photoelectron spectrometers that vary in degree of sophistication. Both the case studies and the references are chosen with emphasis on the diversity of the applications. Since the book is written as a general reader, reference numbers are avoided in the body of the text. The reference numbers that accompany the figures and the tables serve first to acknowledge the sources of the illustrations and the numerical data; they are also to be taken as markers to major sources on which the text is based. They are supplemented by additional references, comments, and notes in the section of References and Notes.

My primary and overwhelming debt in preparing this volume is to the research workers in this frontier field whose original work I have abstracted here. I am grateful to them and to the publishers of their work for their permission to use in this book material from their research. Some of them helped further by providing clarification of their work, furnishing preprints, and even making available raw data. Yet, despite much encouragement from many research workers and readers of the Indian edition, preparation of the manuscript for the present work ran into difficulties. I owe much to Dr. Kalyan Banerjee and Dr. Virander Chauhan for helping to get it going again and also for reading sections of the manuscript. Crucial assistance came from my students Ashraf Ali, C. S. Sreekanth, and particularly Harish C. Srivastava, who contributed much of their time for months to see the work through. I am most grateful to Gauri Singh for his care and skill in producing the illustrations. I extend my gratitude to my Wiley editors for their extraordinary patience.

The book is dedicated with respect and affection to Professor J. L. Franklin, in whose inspiring association I first felt that charged particles, really, are quite attractive.

Pradip K. Ghosh

IIT Kanpur
July 1982

CONTENTS

CHAPTER 1. INTRODUCTION   1

CHAPTER 2. EXPERIMENTAL TECHNIQUES   23
Resolution and Sensitivity, 23
Cylindrical Grid Analyzer, 26
Spherical Grid Analyzer, 27
Cylindrical Condenser Analyzer, 31
Spherical Condenser Analyzer, 32
Cylindrical Mirror Analyzer, 34
Bessel Box Analyzer, 35
Analyzers for Angular Distribution Measurements, 37
Electron Detectors, 39
Spectrometer Using Monochromatized X-Rays, 40
Spectrometer for Liquid Samples, 42
Photoelectron Spectroscopy in a Strong Magnetic Field, 43
Photoelectron Spectroscopy With Supersonic Molecular Beams, 45
Resonance Light Sources, 46
Spectrometer Using Synchrotron Radiation, 47
Calibration of Electron Energy Analyzers, 49

CHAPTER 3. CORE ELECTRON SPECTRA   53
Core Binding Energies and Chemical Shifts, 55
Charge Potential Model, 60
Valence Potential Model, 63
Group Shifts, 64
Thermodynamic Estimates of Chemical Shifts, 66
XPS, NMR, and Mössbauer Chemical Shifts, 68

Applications of Chemical Shifts, 75
Shake-Up, Shake-Off Processes, 81
Vibrational Broadening of Core Lines, 83
Collective Resonances, 86

CHAPTER 4. VALENCE ELECTRON SPECTRA   89

Spectra of Small Molecules, 89
Spectral Assignment of Large Molecules, 93
Role of Molecular Orbital Models, 94
Vibrational Structure, Lone-Pair Peaks, and Perfluoro Effect, 96
Sum Rule, 97
Other Additive Properties: s-Type and p-Type Bands, 99
Assignment of Benzene Photoelectron Spectrum, 102
Correlation Diagram of Acenes, 109
Correlations in Benzenoid Hydrocarbons, 111
Correlation in Substituted π Systems, 115
Norbornadiene, 117
π-Ionization Energies and Dihedral Angles in Biphenyls, 119
Arylcyclopropanes, 122
Biomolecules and Absolute Donor Capabilities, 125
Study of Tautomeric Equilibria, 126
Other Correlations, 128
Spectra of Hexafluoroacetylacetone Complexes, 131
Spectra of Se_2 and Te_2, 135

CHAPTER 5. SPECTRA FROM SOLIDS AND SURFACES   139

Valence Band Photoemission, 140
Valence Band Spectra, 143
Surface States, 151
Core Level Spectra, 152
Plasmon Peaks, 154
Photoelectron Spectroscopic Determination of Fermi Level, 155
Temperature-Dependent Photoemission, 157
Surface and Bulk Intensities, Depth Profiling, 160

Initial Oxidation of Polycrystalline Zinc, 163
Adsorption Studies on Polycrystalline Nickel, 166
Adsorption of Organic Molecules on Platinum Single Crystal, 171
Oxidation of GaAs (110) Surface, 174
Carbon Monoxide Adsorption on Platinum, 177

CHAPTER 6. PHOTOIONIZATION CROSS SECTIONS AND QUANTITATIVE ANALYTICAL CHEMISTRY APPLICATIONS 179

Experimental Determination of Photoionization Cross Sections, 181
Theoretical Calculations of Photoionization Cross Sections, 185
Hydrogenic Systems, 186
Quantum Defect Method, 187
Central Field Calculations, 188
Hartree–Fock Calculations and Other Methods, 193
Cross Sections for Molecules, 197
Cross Section Characteristics and Nodal Properties of Orbitals, 199
A Model for Approximate Photoionization Cross Sections, 200
Applications to Analytical Chemistry, 202
A Model for Quantitative Applications, 204
Mean Free Path Estimates, 209
An Alternative Approach to Quantitative XPS, 214
Trace Metal Analysis by Extraction onto Solid Surfaces, 226
Trace Analysis by Volatilization Technique, 229
Matrix Dilution Technique, 229
Analysis of Organic Polymers, 230
A Status Summary and Comparison of XPS with Other Analytical Methods, 230

CHAPTER 7. ANGULAR DISTRIBUTION OF PHOTOELECTRONS 233

Angular Distribution from Gaseous Samples: Asymmetry Parameter β, 234
β from Theory: One-Electron System, 243
β for Multielectron Atoms, 246

CONTENTS

β in Autoionization, 250

Anisotropic Electron–Ion Interaction Effects in Open Shell Atoms, 255

Differential Cross Section and β for Molecules, 256

Differential Cross Section of Vibrational Levels, 260

Angular Distribution Data and Molecular Orbital Parameters, 266

Mean Free Path Estimates from Angle-Resolved Photoemission from Solids, 270

Surface Sensitivity Enhancement at Grazing Electron Exit Angles and Grazing X-Ray Incidence Angles, 273

Symmetry Selection Rules and Single Crystal Effects in Angle-Resolved Photoemission, 274

Chemisorption, Surface States, and Angle-Resolved Photoemission, 277

Angle-Resolved Photoemission from Oriented Molecules on Single Crystals, 279

Parity Considerations and Angular Distribution, 281

Core Level Photoemission and Photoelectron Diffraction, 282

REFERENCES AND NOTES   285

INDEX   371

Introduction to Photoelectron Spectroscopy

CHAPTER

1

INTRODUCTION

Until about the mid-1960s the principal techniques for studying atomic and molecular structure employed optical spectroscopy, various diffraction methods, magnetic resonance spectroscopy, and mass spectrometry. The optical spectroscopic techniques use emission and absorption characteristics of the sample. The diffraction methods depend on characteristic diffraction patterns of the material under study. Magnetic resonance spectroscopy consists of studying absorption of electromagnetic radiation by the sample under the influence of a magnetic field. In mass spectrometry, the primary step involves ionization of the sample followed by a mass-to-charge ratio analysis of the resulting ions.

Electron impact ionization is known to affect primarily the outer shells of atoms and molecules. When the projectile electron energy is high, multiple ionization is frequently observed. In contrast, photoionization – ionization caused by energetic photons – is less constrained, and ionization of inner-shell electrons takes place with comparable facility, and, in the photoionization process, ordinarily only a single electron is ejected. In electron impact ionization, it has been possible to derive the energies of vibrational levels, as well as those of excited electronic states of the molecular ion, from plots of ion intensities versus electron energies, that is, from ionization efficiency curves. This method, however, has not become popular because of experimental difficulties with obtaining intense monoenergetic beams of electrons over a large energy range. Whereas a similar technique is possible in principle for the photoionization process, a somewhat different aspect of the latter has emerged as a powerful tool in chemical analysis and in the determination of orbital energies of both neutral and ionic species. The method is applicable not only to samples in the free gaseous state but also to those in condensed states: liquids and solids. It utilizes the characteristic kinetic energy distribution of the photoelectrons that are ejected in the photoionization process.

The basic photoelectron spectroscopic technique involves ionization of the sample atom or molecule M by a beam of monoenergetic photons, in which process the sample loses an electron:

$$M + h\nu \to M^+(E_{int}) + e \tag{1.1}$$

M^+ is the resulting ion formed in the state with internal energy E_{int}, and e is the product photoelectron. E_{int} includes electronic, vibrational, and rotational energy of the ions; $E_{int} = 0$ means that the ion is formed in its ground state. In order that photoionization may occur, it is essential that the incident photon possess an energy higher than the lowest ionization potential I_p of the sample atom or molecule. It follows that the excess energy available after ionization, $h\nu - I_p - E_{int}$, must appear as translational energy of the products, that is, the ion and the electron (Fig. 1.1).

Conservation of linear momentum requires that this excess energy be partitioned in inverse proportion to the masses of the products. This means that the electron, which is about two thousand times lighter than the lightest atom, carries practically all the excess energy. The error involved in assuming that absolutely all the excess energy appears as translational energy of the electrons is, in the worst case, of the order of 1 part in 10^4, which is negligible for ordinary applications. For ionization of heavier atoms and molecules, this error is even less. Thus, if monoenergetic photons are used for ionization and the photon energy $h\nu$ is known, a simple determination of the kinetic energy of the photoelectrons, which is equal to $h\nu - (I_p + E_{int})$, directly provides the ionization potential of the sample to form its own ion in a certain internal energy state. In the lowest internal energy state of the ion, $E_{int} = 0$ and the photoelectrons

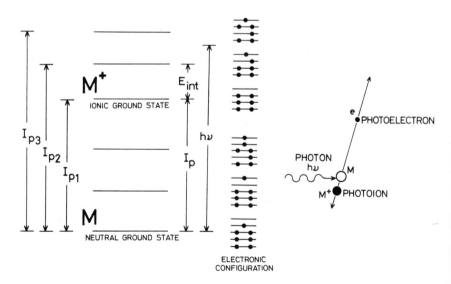

Fig. 1.1. The basic photoionization process.

INTRODUCTION

correspond to the lowest ionization potential of the target particle. Since ionization potentials are characteristic properties of atoms and molecules, the method therefore provides a direct means for chemical analysis. The basic experimental requirements are simple; one needs to have a monoenergetic photon source and an electron energy analyzer.

But the ionizing photon can do more than remove the most loosely bound electron; if the photon energy is sufficiently high, it can also cause the removal of tightly bound electrons. Removal of the most loosely bound electron of the neutral ground state molecule results in an ion with the remaining electrons in their lowest energy states; this constitutes the lowest energy state of the ion. If a more tightly bound electron is removed, however, the resulting ion can be viewed as having a hole in the lower energy state, with the electron from the hole occupying an excited energy level; this overall assembly of the electrons is an excited state of the ion located above the ground ionic state. Thus, with energetic photons different electrons are removed, one at a time, and corresponding photoelectrons with energies $h\nu - I_{p1}$, $h\nu - I_{p2}$, and so on, are ejected. A photoelectron kinetic energy analysis of such a situation yields a series of ionization potentials I_{p1}, I_{p2}, and so on. Since the ionization potential of an electron from a given state is also a close measure of the binding energy of the electron in that state, these two terms are sometimes used synonymously. For molecules, the energy required to populate the ground vibrational state of the ground electronic state of the molecular ion is called the *adiabatic ionization potential*. The *vertical ionization potential*, which is higher than the adiabatic ionization potential, involves ionization to higher vibrational states of the ground electronic state of the ion. The transition probability of this process, like ordinary electronic transitions, is governed by the Franck–Condon principle.

The depth to which atomic and molecular levels can be probed with this technique depends on the upper limit of the photon energy used. The most common low-energy photon sources use helium with HeI resonance radiation ($2\,^1P \rightarrow 1\,^1S$) of wavelength 58.4 nm, which has an energy of 21.22 eV. With this radiation source, energy states having binding energies up to 21.22 eV can be investigated. The use of resonance radiation from some other species leads to somewhat higher energies (HeII 40.8 eV, $n = 2 \rightarrow n = 1$). But if orders of magnitude increase in photon energy is sought, then one has to resort to X-ray sources with energies in keVs, such as Al K_α (1487 eV) or Mg K_α (1254 eV). With the availability of high-energy photon sources, one can measure the binding energies of the orbital electrons that are close to the nucleus, the *core electrons*. In general, for measuring binding energies of the *valence electrons*, the electrons that are distant from the nucleus and participate in bonding, ultraviolet radiation sources like that from HeI are used, and for core electrons, X-ray sources are used. Although a uniform convention on nomenclature is yet to emerge, analysis

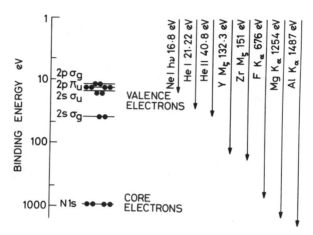

Fig. 1.2. Energy ranges of some radiation sources [30].

using low-energy photon sources has been simply called photoelectron spectroscopy or *UV-PES*, and when X-ray sources are employed, the term used is *X-PES*, or alternatively *ESCA*, which stands for electron spectroscopy for chemical analysis. The basic physical process involved in both techniques is the same. The terms *PESOS*, photoelectron spectroscopy for outer shells, and *PESIS*, photoelectron spectroscopy for inner shells, have also been used for UV-PES and X-PES respectively. The trend, however, is to use the term *UPS* for ultraviolet and *XPS* for X-ray photoelectron spectroscopy. The ranges of some photon sources in relation to molecular nitrogen core and valence orbitals are shown in Fig. 1.2.

Two other processes may occur simultaneously with photoionization, and they can also be used to derive information of the type available from photoelectron spectroscopy. Both of these processes occur when an inner-shell electron is removed by photoionization. In one process, the hole created by the loss of an electron owing to ionization is filled by an electron from one of the outer shells; thus a radiation equivalent to the energy difference between the levels is emitted, usually in the X-ray region. This process is the *X-ray fluorescence,* and analysis of this radiant energy leads to information about the energy levels of the sample. In the second process, the energy that would otherwise have been released as radiation concomitant with the electron transition from the outer to the inner shell is used in ejecting another electron whose ionization potential is lower than the available energy from the electron relaxation process. This is known as the *Auger process*, and an energy analysis of the Auger electron also leads to information about the energy levels of the sample. Both X-ray fluorescence and the Auger process are two-step processes, unlike simple photoionization, which is a one-step process. They are shown schematically in Fig. 1.3.

INTRODUCTION 5

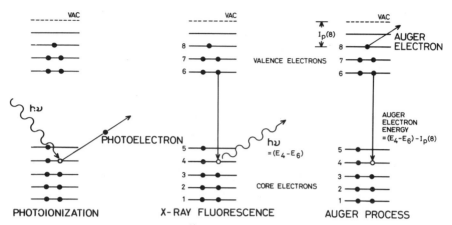

Fig. 1.3. X-ray fluorescence and Auger processes.

Since the phenomena of X-ray fluorescence and the Auger process are independent of the mechanism of the primary ionization step, they can be initiated by electron impact ionization as well as by photon impact. And whereas the kinetic energy of the photoelectron resulting from direct photoionization *increases* with incident photon energy, the Auger electron kinetic energy is *independent* of the energy of the projectile responsible for primary photoionization. This difference is used to distinguish between photoelectrons and Auger electrons.

Auger electron spectroscopy is often used in conjunction with or complementary to photoelectron spectroscopy. The steps in Auger electron emission are shown schematically in some detail in Fig. 1.4. After the ionization of an inner electron by either photons or electrons, energetic ions or even neutrals, a hole in one of the inner levels of the atoms is created. This situation is energetically unstable, and relaxation takes place with the hole being filled by an

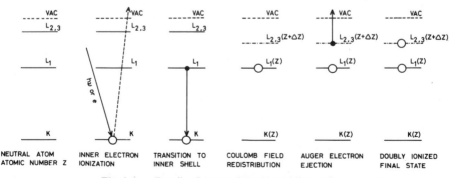

Fig. 1.4. Details of Auger electron emission [32].

electron from a higher state. The available energy $E_K - E_{L_1}$, where E's are binding energies, may then be used to release a second electron, the Auger electron. The latter may originate from the same level or from one having lower binding energy. If the Auger electron has sufficient excess kinetic energy, then it gets ejected into the vacuum where it is amenable to a kinetic energy measurement.

The Auger electron may be expected to have a kinetic energy of $E_K - E_{L_1} - E_{L_{2,3}}$; the term $L_{2,3}$ represents the level from which the second electron is lost. But after the primary ionization, a redistribution of the coulomb field takes place and the atomic energy levels shift. This means that a better approximation to the kinetic energy of the Auger electron can be obtained by considering binding energies for excited atoms. An expression that gives sufficiently accurate values for the energies of Auger transitions is

$$E_{ABC}(Z) = E_A(Z) - \tfrac{1}{2}[E_B(Z) + E_B(Z+1)] - \tfrac{1}{2}[E_C(Z) + E_C(Z+1)] \quad (1.2)$$

where $E_{ABC}(Z)$ is the expected kinetic energy of Auger transition ABC in an atom having atomic number Z, A is the level initially ionized, and B and C are the levels in which the two electrons (the electron that fills the inner hole and the Auger electron) originate. $E_B(Z)$, $E_B(Z+1)$, $E_C(Z)$, and $E_C(Z+1)$ are the binding energies of B and C levels in atoms of atomic number Z and $Z+1$ respectively. Implicit in the term E_{ABC} is the way nomenclatures are made in Auger transitions. The level initially ionized is written first (A), followed by the level from which the electron relaxes (B), and finally the level from which the Auger electron is lost (C). Thus, the transition shown in Fig. 1.4 would be labeled $KL_1L_{2,3}$. In solids, however, electrons may originate in the valence band. Then KL_1V or KVV would represent suitable Auger transitions. The scheme of Fig. 1.4 is strictly valid for isolated atoms having discrete energy levels. Auger transitions also occur in solids and large valence band populations lead to intense Auger electron intensities. Between the X-ray fluorescence and the Auger process, the efficiency of X-ray fluorescence at low primary energies is not much more than 0.05 and the Auger process dominates. Only at transition energies greater than 10 keV do rates of radiative transitions become comparable to Auger transitions.

Returning to photoelectron spectroscopy, let us now look into the nature of the information one can obtain about free atoms and molecules, that is, about those in the gaseous state. Since in the gaseous state perturbations due to neighboring sample atoms and molecules are negligible, widths of the energy levels are solely their natural widths determined by the lifetimes in the levels. Whether all the energy levels will be distinctly observed through electron energy analysis depends on the linewidth of the monochromatic radiation and the resolution of the electron energy analyzer employed.

Experimentally, a photoelectron spectrum is obtained by radiating the sample with a beam of monochromatic photons and energy-analyzing the ejected photoelectrons. After the electron energy analysis, the photoelectron spectrum shows a number of types of photoelectrons that differ in kinetic energy and also in flux. Knowledge of the initial photon energy is necessary to convert this into binding energy information, since the photoelectron energy is the difference between the incident photon energy and the binding energy of the electron. Thus, with a given sample, depending on the photon source employed, the numerical values in the abscissa scale of the plot of the electron intensity and electron kinetic energy may be different, but they all reduce to an identical binding energy spectrum once information about the incident photon energy is introduced.

Figure 1.5 shows the photoelectron spectrum of argon with clearly resolved spin—orbit coupled states $^2P_{3/2}$ and $^2P_{1/2}$. This spectrum was obtained using HeI 21.22 eV radiation. The upper scale on the abscissa represents the kinetic energy of the photoelectrons and the lower scale represents the binding energy, obtained by subtracting the photoelectron kinetic energies from 21.22 eV.

The basic principles that apply to atomic photoelectron spectra are also applicable to photoelectron spectra of molecules. This is illustrated by using the N_2 and N_2^+ energy states. A few low-lying potential energy curves of nitrogen are shown in Fig. 1.6. We see that with HeI radiation of 21.22 eV energy, levels up to the $B^2\Sigma_u^+$ molecular ionic state can be populated. The population distribution in the various ionic states depends on the photoionization cross sections to these states at 21.22 eV energy. The fine structure of distribution into vibrational levels of the various electronic states of the ion ordinarily depends on

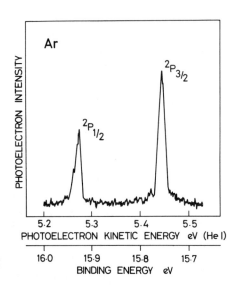

Fig. 1.5. Photoelectron spectrum of argon [5].

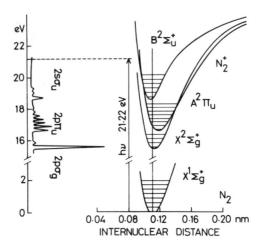

Fig. 1.6. Photoelectron spectrum of N_2 [30].

the shapes and locations of the potential energy curves of the ionic electronic states with respect to the neutral ground state and hence on the respective Franck–Condon factors. Figures 1.2 and 1.6 show that the HeI photon energy is adequate only for ionizing up to $2s\sigma_u$ electrons (to the state $B^2\Sigma_u^+$); the higher binding energy state $2s\sigma_g$, lying closer to the core, remains unaffected.

The relative location of the molecular and ionic potential energy curves, which determine the Franck–Condon factors, depends on the nature of the electron orbital from which the electron is lost, that is, on whether it is bonding, antibonding, or nonbonding. Depending on the nature of the orbital involved, this effect causes changes in the vibrational frequency of the ionic states and hence in the spacing of the vibrational energy levels. For example, the loss of a bonding electron in ionization reduces the charge distribution between the nuclei and decreases the force constant of the molecular ion. Since the vibrational frequency ν and the force constant k are related by a formula of the type $\nu = (1/2\pi)\sqrt{(k/\mu)}$ where μ is the reduced mass, the effect of decreasing the force constant is also to lower the energy spacing between the vibrational levels. The effect of the loss of an antibonding electron is just the opposite – it strengthens the bond, decreases the internuclear distance, increases the force constant, and hence increases the energy spacing between the levels. Nonbonding electron removal has little effect on the internuclear distance and the force constant, and the spacing between the vibrational energy levels of the molecular ion remains unaffected. This criterion of change in vibrational spacing can be applied to find out whether the molecular orbital involved in ionization has bonding, nonbonding, or antibonding characteristics. For example, according to this analysis, photoelectron spectral data indicate that the $2p\sigma_g$ orbital is near nonbonding (weakly bonding) because the vibrational spacing decreases on ionization from 2345 cm^{-1} (vibrational spacing of the ground state of N_2) to 2191 cm^{-1}, that the $2p\pi_u$ orbital is strongly bonding because in the band assigned to it the vibrational spacing decreases drastically from 2345 to 1850 cm^{-1}, and that the $2s\sigma_u$ orbital is essentially nonbonding (weakly antibonding) because the vibrational spacing due to this band increases from 2345 to 2397 cm^{-1}. Although knowledge of the nature of an orbital can be used to predict shifts in vibrational frequencies in spectra, a more important application is the reverse study of the changes in vibrational energy level spacings in order to derive information about the nature of the orbital involved in ionization.

In addition to obtaining information about the bonding nature of orbitals involved in ionization, vibrational structure in photoelectron spectroscopy is a valuable tool for determining vibrational energy levels of those molecular ions where absorption or emission spectroscopic work has proved to be very difficult. An example is the molecular ion H_2^+ or HBr^+. In many such molecules only scant data on vibrational-energy-level spacings of molecular ions can be obtained with the help of emission spectroscopic work, and the question of absorption spectroscopy almost does not arise, as only a handful of gaseous molecular ions have so far been observed in absorption and that too under rather special experimental conditions. Photoelectron spectroscopy, however, easily gives the energy spectrum of the molecular ionic vibrational levels, and in many cases it constitutes a powerful alternative to absorption spectroscopy of gaseous ions. The HeI UPS of HBr^+ is presented in Fig. 1.7, which shows a vibrational progression up to $v' = 6$, wherefrom the vibrational frequencies and the vibrational constants of the ionic state can be easily derived.

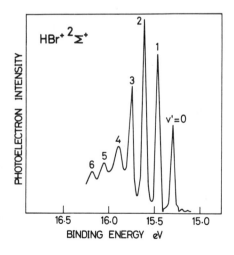

Fig. 1.7. Photoelectron spectrum of HBr [112].

We can appreciate from the above discussion that the scope of application of photoelectron spectroscopy goes far beyond the determination of valence energy levels. Binding energies of valence electrons can be found out through UPS, but several types of additional information can be obtained by using higher energy photons. The main difference between the effects of low-energy photons and X-rays is that the latter are able to reach the core electrons and eject them. One of the important applications of XPS is the determination of binding energies of core electrons in various atoms. Since the magnitude of the core electron binding energy is expected to strongly reflect the attractive forces of the nucleus on the core electrons, this should depend on the charge of the nucleus of the atom chosen, and this makes the core binding energy a characteristic property of the atom considered. For this reason, photoelectron spectroscopy finds application in chemical analysis, and out of this originates its acronym ESCA, mentioned earlier. Core binding energies of $1s$ electrons of the second period elements Li, Be, B, C, N, O, and F derived from XPS are shown in Fig. 1.8. The ordinate represents photoelectron count rate, which is equivalent to photoelectron intensity. Since the attractive force of the nucleus on the $1s$ electrons increases with the atomic number, it is natural to expect that the $1s$ binding energy for these atoms will be the lowest for lithium and the highest for fluorine. This is found to be correct ($1s$ binding energy for free-atom Li = 58 eV, Be = 115 eV, B = 192 eV, C = 288 eV, N = 403 eV, O = 538 eV, and F = 694 eV), which shows that photoelectrons from XPS can be used for chemical identification of samples. In Fig. 1.8, the intensities of photoelectron peaks from various elements are different; this is because even at a photon energy adequate for ionization, the ionization cross sections are dif-

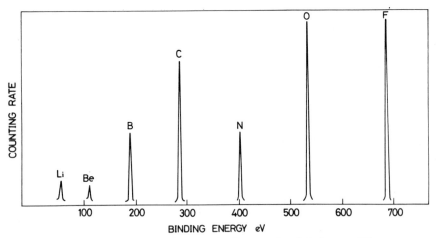

Fig. 1.8. Core ($1s$) binding energies for second period elements [1].

ferent for each species. In fact, from a knowledge of such experimental peak intensities, the magnitude of incident photon flux, and sample density, it is possible to determine absolute values of photoionization cross sections. For solid samples, mean free paths of photoelectrons generated in the solids also play a role in experimentally observed photoelectron intensities.

But applications of the characteristic core binding energies extend even further. Whereas the binding energy of a certain type of core electron is quite characteristic of a given atom, if an XPS is taken of a molecule in which an atom of a particular kind is present, the core photoelectron peak of the atom does not always appear at precisely the same characteristic energy; there is a slight shift depending on the molecule chosen. This happens because although the core electron binding energy is principally determined by the nuclear charge, there is some subsidiary effect due to the valence electrons influencing it. The more pronounced the effect of the valence electrons, the greater is the shift in the binding energy. Since the valence electron distribution varies, depending on the nature of bonding, it is reasonable to expect that the core binding energy would reflect, through *chemical shifts,* changes in valence charge distributions of the molecule. This is observed in practice, and for this reason the chemical shift of core electron binding energies holds great potential for use as a probe in investigating the nature of bonding in molecules. A simple case of this chemical shift in core binding energy can be seen in the behavior of carbon monoxide and carbon dioxide. Since oxygen is more electronegative than carbon, we can expect the carbon valence electrons in carbon monoxide to be somewhat shifted towards the oxygen atom. This would cause the influence of carbon nucleus on the C $1s$ electron to increase, and its effect would be to bind the C $1s$ electron more strongly to the carbon nucleus and hence increase the C $1s$ core binding energy. In carbon dioxide, the presence of two oxygen atoms ought to augment this effect, and the core binding energy should be even higher. This is indeed what is experimentally observed. Carbon $1s$ core binding energy is found to be 296 eV in carbon monoxide and 297.5 eV in carbon dioxide, compared with 288 eV for the free atom case. In the oxygen atom the effect is just the reverse. Availability of the carbon electron decreases the oxygen $1s$ binding energy in carbon monoxide, but the same effect is somewhat more in carbon dioxide as there is more shift of bonding electrons. Thus the core binding energy for oxygen $1s$ is lower in carbon dioxide than in carbon monoxide.

Applying these chemical shift principles to larger molecules reinforces this point and demonstrates its scope of applications. Taking the case of ethyl chloroformate, $ClCO \cdot OCH_2CH_3$, one observes that there are three carbon atoms, one in the carbonyl group, one in its somewhat adjacent methylene group, and the third in the farthest methyl group. All these carbon $1s$ peaks must be close to 288 eV, the core binding energy of carbon $1s$ in the free atom, but one should also expect some appropriate chemical shifts depending on the particular electronic environment of each of the carbon atoms. As explained earlier, the carbonyl carbon, because of its proximity to oxygen as well as chlorine, must have the highest binding energy, the methylene carbon somewhat lower, and the methyl carbon C $1s$ the lowest binding energy. The spectrum of ethyl chloroformate is shown in Fig. 1.9, and the binding energies of the three carbon $1s$ peaks confirms this analysis. This also shows how it may provide an entirely new technique in organic and inorganic chemistry for determination of charge distribution and hence may constitute a useful method in structural investigations. The spectrum of ethyl trifluoroacetate, taken at much higher resolution, is also shown in Fig. 1.9, in which the chemically shifted peaks manifest much more clearly. The chemical shifts of the peaks are shown with respect to the C $1s$ binding energy of CH_3 carbon, which has a binding energy of 291.2 eV. The trifluoromethyl carbon C $1s$, with three highly electronegative fluorine atoms, has the highest binding energy.

The process of photoionization described earlier may become quite complicated owing to various additional factors, for example, the occurrence of *autoionization*. In autoionization, as shown schematically in Fig. 1.10, the primary process is photoabsorption from the ground state X to populate an excited bound state such as e_3, located above the ground electronic state i of the ion of the neutral molecule. This is followed by a radiationless transition to the state f, without energy change, to form an ion and an electron. The kinetic energy of the electron thus formed is equal to the energy difference between the states f and i. There may be radiative transitions from the discrete excited

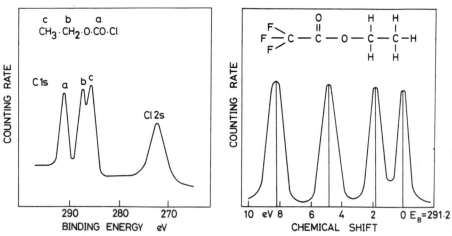

Fig. 1.9. C $1s$ chemical shifts in ethyl chloroformate and ethyl trifluoroacetate [30, 145, 253].

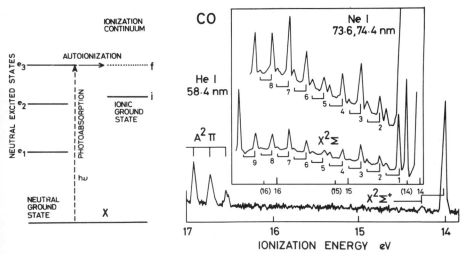

Fig. 1.10. The autoionization process and the effect of autoionization in CO photoelectron spectra [119].

state e_3 to lower excited states located below the ionization limit i, but because of autoionization, the emission intensity from such transitions may be expected to be quite low. Electronic autoionization is a very fast process, involving a time period of the order of $10^{-15}-10^{-11}$ sec compared with allowed radiative transitions, which are in the range of $10^{-9}-10^{-7}$ sec. When autoionizing levels such as e_3 exist, autoionization is by far the dominant process and one observes a sharp drop in emission intensity from the autoionizing levels. Further, since photoabsorption to a discrete excited state is a resonance excitation process — that is, its transition probability to the excited state peaks when the photon energy is equal to the excitation energy and is negligible when it is not — transitions to autoionization states also peak at photon energies corresponding to their respective excitation energies. When a series of such autoionizing levels exist, population to these states as a function of photon energy therefore may be expected to show collectively a series of peaks. Since all these states autoionize and produce free electrons, the resulting photoelectron spectrum also may be expected to show a series of resonance maxima and off-resonance minima.

In molecules, two types of autoionization may be recognized. The autoionization state e_3 may have a Rydberg character, that is, an outer electron distributed around an inner core. In the first type of molecular autoionization, the molecular ion that is produced as a result of autoionization is in the same state as the inner core of the autoionizing species. The autoionization state may be vibrationally or rotationally excited such that its total energy is above the ionization limit. The excess vibrational or rotational energy may then relax into the energy of the Rydberg electron, which is then ejected. This process is termed *vibrational or rotational autoionization.* The rates of such autoionization processes decrease rapidly if the transitions involve more than one quantum of vibrational energy transfer. In the second type of molecular autoionization, termed *electronic autoionization,* the resulting molecular ion is not the same as the core of the autoionizing state. This is always the case in atomic autoionization, but it is also quite frequently observed in molecules. Since the electronic autoionization lifetime is very short, decay of autoionization excited states by radiative fluorescence is unlikely. Such radiative transitions are more probable for vibrational autoionization, which is a much slower process.

The part played by autoionization in photoelectron spectroscopy depends strongly on the wavelength of radiation used to effect photoionization. When radiation of a fixed energy $h\nu$ is used, every ionic state in the interval $h\nu - I$, where I is the lowest ionization potential in the system, contributes to the true wave function of the final state reached on photoabsorption. The contribution of a particular autoionization state is determined by the proximity of the autoionization state to $h\nu$ and the strength of mixing of the excited neutral with the ion state. A strong mixing implies a large continuum character for the autoionization state and, as indicated earlier, a dominant oscillator strength at $h\nu$. When an autoionization state contributes significantly to the true wave function, the photoelectron spectral intensities may show considerably distorted structure, such as intensity oscillations as a function of incident photon energy. For example, in the ionization of xenon leading to the formation of Xe^+ $^2P_{3/2}$ and Xe^+ $^2P_{1/2}$ ions,

Photon wavelength at threshold of ionization

$$Xe(5s^2 5p^6\ ^1S_0) + h\nu \longrightarrow Xe^+(5s^2 5p^5\ ^2P_{3/2}) + e \quad 102.21\,\text{nm} \quad (1.3)$$
$$(12.13\,\text{eV})$$

$$Xe(5s^2 5p^6\ ^1S_0) + h\nu \longrightarrow Xe^+(5s^2 5p^5\ ^2P_{1/2}) + e \quad 92.28\,\text{nm} \quad (1.4)$$
$$(13.43\,\text{eV})$$

the ratio R of the two cross sections $\sigma_{3/2}$ and $\sigma_{1/2}$ to form respectively the $^2P_{3/2}$ and $^2P_{1/2}$ ionic states, which also reflects the photoelectron intensity ratio, has a value of 1.60 measured at a large number of wavelengths in the range 92.28–46.0 nm. But near the absorption resonances in this region, R changes significantly. At 55.86 nm, R has a value of 1.46; at 54.39 nm, R is 0.77; and at 52.18 nm, R is 1.49. Such excursions of R can be traced to variations of cross sections as a function of the incident photon energy. Cross section calculations show that whereas $\sigma_{3/2}$ decreases at these wavelengths, $\sigma_{1/2}$ remains unchanged at 55.86 nm and 52.18 nm, but increases at 54.39 nm.

Autoionization also affects the photoelectron spectral intensities of molecules. Direct photoionization from the ground state of a molecule populates the vibrational levels of the molecular ion that are reachable in vertical transitions, and the intensities are dictated by Franck–Condon factors. If the incident photon energy, however, happens to coincide with the energy of a neutral excited state located higher than the lowest molecular ionic state, then the neutral state is populated by resonance absorption and can autoionize into the lower-lying ionic states. Vibrational levels of the molecular ion populated in this manner may involve drastically different levels from those that are directly populated from the ground state, and the overall population densities therefore may not follow Franck–Condon factors. For example, in the case of photoionization of O_2 using HeI radiation, the spectra show bands due to the ionic states $X^2\Pi_g$ at 12.06 eV, $a^4\Pi_u$ at 16.1 eV, $A^2\Pi_u$ at 16.8 eV, $b^4\Sigma_g^-$ at 18 eV, and $B^2\Sigma_g^-$ at 20 eV. The photoelectron spectrum of O_2 with NeI ($h\nu = 16.85$ eV, which is close to a series of Rydberg levels whose limit coincides with the $A^2\Pi_u$ state of O_2^+) shows a dramatic development of the $X^2\Pi_g$ vibrational structure, with the vibrational progression extending up to at least $v' = 20$. There is also a hundredfold enhancement of the $a^4\Pi_u$ vibrational states.

Such participation of autoionization states has also been detected in H_2. Experiments with HeI radiation yield ionic vibrational level populations that reflect the Franck–Condon factors. However, experiments with variable energy photon sources to observe threshold photoelectron spectra show, when normalized to the intensity of $v' = 3$, anomalous non-Franck–Condon distribution for the vibrational states $v' = 0, 1, 2$, in contrast to regular Franck–Condon distribution for the states $v' = 4, 5, 6$. This indicates that autoionizing Rydberg states influence ionization into $v' = 0, 1, 2$ states, and $v' = 3, 4, 5, 6$ states are formed essentially by direct photoionization.

Figure 1.10 also shows the effect of participation of autoionization states in the photoelectron spectrum of CO. The use of HeI 58.4 nm radiation gives only two vibrational peaks due to $X^2\Sigma^+$ and three peaks due to the first excited ionic state $A^2\Pi$. Photoionization by NeI doublet (73.6, 74.4 nm) on the other hand, shows a remarkable development of vibrational peaks due to $v' = 0-9$ of the $X^2\Sigma^+$ state. These are in two different progressions caused by the two exciting NeI lines. The relative intensities of the lines in the two progressions are different and are best explained by contributions from autoionization transitions.

When the photoelectron leaves the sample atom or molecule, if no significant energy changes of the other electrons present take place then the experimental ionization energy may be taken to be equal to the orbital energy of the departing electron. This is the substance of *Koopmans's theorem*, which states that the negative of the one-electron orbital energies and the ionization potentials can be taken to be the same; the remaining electrons are assumed to remain frozen in their orbitals and do not undergo any change while ionization occurs. The energy level diagrams that can be constructed from the ionization potentials obtained from photoelectron spectral data are thus really the energy level diagrams of the resulting ions.

Table 1.1. Atomic Relaxation Energies for Ionization from Various Subshells [117]

Atom	1s	2s	2p	3s	3p	3d	4s
Nonrelativistic Calculations							
H	0.0						
He	1.5						
Li	3.8	0.0					
Be	7.0	0.7					
B	10.6	1.6	0.7				
C	13.7	2.4	1.6				
N	16.6	3.0	2.4				
O	19.3	3.6	3.2				
F	22.0	4.1	3.8				
Relativistic Calculations							
F	22.2	4.1	3.9				
Ne	24.8	4.8	4.7				
Na	23.3	4.1	4.7	0.3			
Mg	24.6	5.2	6.0				
Al	26.1	6.1	7.1	1.0	0.2		
Si	27.1	7.0	8.0				
P	28.3	7.8	8.8				
S	29.5	8.5	9.6	1.4	0.9		
Cl	30.7	9.3	10.4				
Ar	31.8	9.9	11.1	1.8	1.4		
K	31.2	9.1	10.5				
Ca	32.0	9.6	11.1				
Sc	33.8	11.5	12.9				
Ti	35.4	13.0	14.4	3.6	3.4	2.0	0.3
V	37.0	14.5	16.0				
Cr	38.6	15.9	17.4				
Mn	40.1	17.2	18.8			3.6	0.4
Fe	41.6	18.5	20.2	5.7	5.3		
Co	43.2	19.8	21.6			4.1	0.0
Ni	44.7	21.1	22.9	6.7	6.3		
Cu	49.2	23.7	25.7	7.7	7.2	5.3	0.3

Koopmans's theorem can be used in a limited way to obtain an ordering of neutral orbitals in terms of energy. The zero of the energy scheme can be taken as the level that signifies where a given orbital electron is detached from the atomic core. The various atomic orbitals can then be located relative to this base by an amount that is negative by their respective ionization energies. The true values of the orbital energies of the neutral, however, would be somewhat different from the values obtained from the use of photoelectron spectral ionization energies and Koopmans's theorem. The differences are due to the energies of reorganization or relaxation of the remaining electrons, the relaxation occurring concomitant with the ionization process. In order to have some idea about the magnitude of these *relaxation energies,* theoretically obtained values of relaxation energies of a few atomic states using Hartree–Fock–Slater wave functions are presented in Table 1.1. Note that the absolute values of the relaxation energies are low in valence-shell ionization, $\sim 1-3$ eV, but are quite high for ionization of core electrons, $\sim 10-50$ eV.

The one-electron model, the model in which the total wave function of an n-electron system is an antisymmetrized product of n-one-electron wave functions — that is, Hartree–Fock (HF) orbitals — is quite helpful, but it is not the only way to construct an independent particle model. Strictly speaking, the process of photoionization causes a transition from the initial system of the neutral species described by a many-body wave function to a final system containing the ion and the electron by another many-body wave function. The one-electron HF orbitals constitute a helpful description because they happen to concentrate most of the many-body effects, such as those involved in photoionization, into one orbital. It can well happen, however, that what is simply described by a one-electron process (using canonical HF orbitals) actually involves several orbitals (the many-body effect) in another type of orbital representation, in which case description of ionization from one orbital would lose meaning. This limitation of one-electron models should be borne in mind.

Theoretical methods used in calculations of binding energy vary greatly, from semiempirical to rigorous, each method having its characteristic strengths and limitations. A comparison can be made between binding energies obtainable from eigenvalues of orbitals using nonrelativistic as well as relativistic HF wave functions and binding energies from differences between the total electronic energies of the neutral atom and its ion, and a comparison of both of these can be made with experimental ionization potentials. The difference between the energies from the eigenvalue method and the experimental ionization energies, although large in absolute values, is small in percentage, and the total-energy method does not always lead to better agreement with experiment. For calculations of the binding energies of inner shells, it is essential that relativistic effects be considered; for the outer shells, contributions from electron correlation become very important. Although concern has been expressed about the validity of Koopmans's theorem and the consequent uncertainties in orbital order reached through its assumption, there is also the view that it is only one of several uncertainties and not necessarily the dominant one. In practice, at the present state of development of the subject, Koopmans's theorem is extensively applied in the interpretation of photoelectron spectra.

Photoelectron spectra are often quite complex because of various interactions that lead to more than one final state. When multiple final states occur, the photoionization process can populate each of them. Every such transition results in a photoelectron line, and thus multiple lines are observed. There are several kinds of multiple or split lines; one of much practical importance is that in which an ionic state is formed by the removal of an electron from a degenerate subshell in a closed-shell molecule. As a result of this removal, a coupling between the spin and orbital angular momenta of the ion takes place, the degeneracy is lifted, and multiple levels manifest separately. Photoionization transition to these levels causes a splitting, the *spin–orbit splitting*, of the photoelectron lines. The energy of separation of the levels depends on the magnitude of the orbital and spin angular-momentum vectors, their orientation, and the proportionality constant — that is, the spin–orbit coupling constant — that is characteristic of the subshell.

Spin–orbit separation exists also in the initial states, and it is well characterized for the inner shells. There is no spin–orbit splitting for s subshells. For a closed shell with $l > 0$, $l + \frac{1}{2}$ and $l - \frac{1}{2}$ states arise with a population of $2l + 2$ and $2l$ respectively. The $P_{1/2}-P_{3/2}$ level splittings for the rare gas atoms are as follows: Ne 0.12 eV, Ar 0.18 eV, Kr 0.73 eV, Xe 1.41 eV, and Rn 4.07 eV. The extent of this doublet splitting, which reflects the magnitude of the spin–orbit coupling constant, progressively increases with atomic number. For a given atom, the relative splitting in each of the subshells with $l > 0$ is of the same magnitude. In the valence shell, the magnitude of spin–orbit splitting is affected by electron correlation, and additional coupling can occur if the shell is partly occupied.

Using experimental coupling constants, spin–orbit splitting can be calculated from atomic spectroscopy and from molecular orbital calculations. Some semiempirical models have been applied to halide molecules of the type PX_3, and these models have been extended to incorporate nd valence electrons. Spin–orbit splittings for each of the I_2 molecular orbitals have been studied, as well as spin–orbit splitting in alkyl halides and its interaction with conjugate systems.

Multiplet splitting arises when interaction takes place between an unpaired electron formed by photoelectron ejection and any unpaired electron already existing in the atom or molecule in any of its incomplete shells. There are a large number of species with unpaired electrons, among them NO and O_2, the transition metals that possess partially filled d orbitals, and rare earths and actinides with partially filled f shells. For example, in Fe_2O_3 there are five unpaired electrons in the $3d$ shell. Photoionization of Fe $3s$ results in two possible final states. And a $3d$ electron ejection from Mn^{2+} to form Mn^{3+} leads to two states ^5S or ^7S, the first one a pentuplet and the second one a septet. The first state is due to the remaining $3s$ electron having an antiparallel spin and the second state a parallel spin with respect to the five unpaired d electrons. The energy separation ΔE in multiplet splitting varies with the environment of the atom concerned, and this can be used to derive chemically useful information. Although the core binding energy chemical shifts provide similar information, one advantage to multiplet splitting is that calibration in terms of absolute binding energies is not critical. In the atomic case, the intensity of such multiplet lines is proportional to the multiplicity expression $(2L + 1)$ $(2S + 1)$ where L and S represent total orbital and spin angular momenta. When a vacancy is formed in an inner shell having an angular momentum greater than zero, the number of multiplet states greatly increases. With configuration mixing allowed, the number of possible states increases still further. The extent of multiplet splitting can be used to estimate the bonding character of the unfilled valence shell. It has been shown that for a given ligand, the value of multiplet splitting follows the number of unpaired spins of the metal atom.

Yet another kind of splitting arises when a high degree of symmetry possessed by a molecule is destroyed by photoelectron ejection. As a consequence, the energy state of the positive ion undergoes a splitting known as *Jahn–Teller splitting*. For a linear molecule, the corresponding splitting is called the *Renner–Teller splitting*. Photoelectron spectra of CH_4, SiH_4, GeH_4, CBr_4, and similar molecules show Jahn–Teller splitting. The overall Jahn–Teller broadening decreases with a decrease in the force constant; thus the broadening is in the order $CH_4 > SiH_4 > GeH_4$. The effect turns out to be small when heavier atoms are substituted for hydrogen; it increases when the central atom is heavy, such as in the molecule $Pb(CH_3)_4$. The spin–orbit splittings of various degenerate states of a given polyatomic molecule may differ drastically, whereas Jahn–Teller splittings are usually uniform in size. In the assignment of photoelectron bands, theoretical calculations on the magnitude of the shift can provide considerable guidance to the type of interaction that is operative.

We noted earlier that a count of the number of electrons resulting from photoionization, along with information on a few other experimental parameters, can be used to derive photoionization cross sections to various ionic states. An important aspect of such electron flux measurements in photoelectron spectroscopy involves the measurement of the number of electrons ejected as a function of the angle between the direction of the photon beam and the direction of the electron ejection. For a dipole transition involving an unpolarized beam of photons and randomly oriented molecules, as is usually the experimental case, the intensity of a given photoelectron peak is given by

$$I(\alpha) \propto 1 + \frac{\beta}{2}(\tfrac{3}{2}\sin^2\alpha - 1) \tag{1.5}$$

where α is the angle between the direction of the photon beam and the ejected photoelectron, and β is called the *angular parameter* or *asymmetry parameter*. The β depends on the photoelectron energy and also on the nature of the atomic or molecular orbbital from which the photoelectron is ejected. It may take values from $+2$ to -1. The above expression undergoes a change if one uses a polarized beam of photons or a beam with an arbitrary degree of polarization. Equations have been derived for such cases from which β can be evaluated. The importance of β measurements is that the magnitude of β can be related to the nature of the orbital from which the electron is ejected. The $I(\alpha)$ measurements need to be carried out at various photon energies before such correlations are attempted, but even then the orbital–β correlations are not exactly straightforward.

Like photoelectron spectroscopic studies of gaseous molecules, there are now extensive applications to systems in condensed states: in solids, liquids, and solutions. The liquids and solutions require special techniques in sample handling. Examples of applications of photoelectron spectroscopy in these areas though are fewer in number they are rapidly increasing. In solids, which constitute an important area of application of photoelectron spectroscopy, the characteristic core level photoemission can be used for qualitative and quantitative analysis of the sample, and the chemical shift of the core photoelectron peaks can be used for characterization of the chemical nature of the material. Photoelectron spectroscopy can also be used to study valence band structure using both UPS and XPS. The advantage of UPS in studying band structure is that owing to the higher monochromaticity of the incident radiation UPS leads to higher resolution spectra. Usually spectra are taken for a series of incident photon energies as the band structure of the final state strongly influences the photoelectron energy distribution curve. In XPS, where the final state is located at a much higher energy, a spectrum using one photon energy is adequate. Both XPS and high-energy UPS can routinely be used for studying deep-lying regions of valence bands. Such photoelectron spectroscopic measurements have been used to study valence band for a large number of materials and relating it to theoretically derived bulk band structure.

Besides studies on bulk electronic structure, an important field of application of photoelectron spectroscopy is the study of surface properties. When the ionizing radiation strikes a solid surface, the emerging photoelectrons are mostly limited to a certain depth because of the mean free path limitations of the electrons inside the solid. It turns out that most of the electrons that do emerge from the surface originate from a depth not greater than about 2 nm. This means that photoelectron spectra can be used very effectively to study surface properties, including adsorption phenomena. Details of the origin of catalytic activity in a material also can be studied directly. Adsorption processes have been studied in detail using XPS, UPS, and Auger spectroscopy in situ, while adsorption actually takes place.

Besides application to organic and inorganic chemistry, there have been numerous applications of photoelectron spectroscopy in biochemistry. In biochemical applications, it has been used in conjunction with surface etching techniques (such as those using oxygen plasma) for investigating the surface and subsurface composition of cells. The method involves alternate scans of photoelectron spectra and the etching process, by which means atomic distributions as a function of depth from the surface can be measured. In such experiments, atom ratios in a sample are determined from line intensities and elemental sensitivity factors. *E. coli* cells, for example, have been studied in this manner. Phosphorus core level signals have been used to determine the distribution of teichoic acids through the cell wall of two bacillus species. Sodium distribution was found to be heavier on the surface and potassium concentration was found to increase with depth.

That there is water of hydration in chlorophyll *a* and that it disappears completely at 250°C was first directly observed by following C$1s$ and O$1s$ signals from chlorophyll in a temperature-dependent XPS study on chlorophyll *a*. XPS has also been a very useful technique for studying metalloproteins, using the knowledge that oxidation states can be identified from chemical shifts of core photoelectron signals.

Metalloporphyrins have been studied in a similar manner by following N $1s$, C $1s$, and metal $2p$ core electron peaks. Such studies show that the N's in the free base have higher electron densities than in the complexed ligand, and the electron densities on the N's decrease with increasing electronegativity of the metal. The method has also been applied to some metal phthalocyanine complexes such as Fe, Co, and Mg, and to chemisorption studies of HCO_2H, pyridine, and H_2O. In pyridine adsorption, it has been found that Fe $2p$ binding energies do not shift, which indicates that there is no charge transfer, whereas HCO_2H and FePc show strong chemical shifts. UPS of phthalocyanine (H_2Pc) and the Mg, Fe, Co, Ni, Cu, and Zn complexes all show the first ionization potential as ~ 6.4 eV, apparently independent of the metal ion, indicating that the highest occupied molecular orbital has a predominantly ligand character.

Chemical equivalence where there is more than one kind of atom can be studied by XPS. From UPS–XPS studies, drug activities have been correlated with electron donating ability. In liquid samples, XPS techniques have been used to obtain spectra of biologically active molecules, such as salazopyrin in ethylene glycol solution.

There also has been a series of photoelectron spectroscopic studies of psychotropic drugs. Interactions of the aromatic rings and amino function of these molecules are believed to occur at specific receptor sites in the central nervous system. It has been found that ionization potentials of these molecules can be related to their donor capabilities. Good correlations have been found to exist between π-ionization potential and the ability of a drug to displace bound d-LSD from rat brain homogenates. Psychotropic activities correlate well with ionization potentials if lipophilicities are considered.

By now we should have some idea about the basic principles that guide photoelectron spectroscopy as well as the general domain of its applications. Briefly, the technique is applicable to all three states of matter, and it not only permits in principle both qualitative and quantitative analysis of all the elements, it is perhaps the only technique that provides a direct access to orbital energetics. A wide range of systems have already been studied, very large organic molecules and polymers as well as small transient species. Besides the traditional domains of atomic and molecular structure determination and solid state studies, there have been applications to space science, geology, and biochemistry. In the chapters that follow, we consider a few of the applications of photoelectron spectroscopy in some detail, using core electron spectra as well as valence electron spectra. But before we take up these applications, let us consider the experimental techniques.

CHAPTER

2

EXPERIMENTAL TECHNIQUES

The essential constituents of a photoelectron spectrometer are (1) an intense monochromatic light source, (2) an energy analyzer for the ejected photoelectrons, and (3) a detector for the energy analyzed electrons. Each of these components is available in quite a large number of types varying in degree of sophistication, and various combinations of the three are possible to make a complete instrument. In this chapter, the elements of several types of photoelectron spectrometers are presented, from the simplest to some of the most advanced.

RESOLUTION AND SENSITIVITY

In any photoelectron spectrometer it is desirable that both the spectral *resolution* and the *sensitivity* of photoelectron detection be as high as possible. Unfortunately, these are conflicting requirements; they cannot be increased simultaneously, and compromises are necessary. The energy resolution is limited by the width of the photoelectron lines and is determined by several factors: the inherent width of the ionizing radiation, the intrinsic width of the level that is responsible for photoelectron emission, and the finite resolving power of the electron energy analyzer used. The resolution of an electron energy analyzer is usually measured as the *full width at half maximum height* (abbreviated FWHM) of the electron line divided by the electron energy, that is, ΔE (FWHM)/E_0 where E_0 is the energy at which the photoelectron line appears. Spectrometer base resolutions are usually expressed as the energy width at the base. The inverse of the resolution is the *resolving power*.

Electron energy analyzers are of two types, the *retarding potential analyzers* and the *dispersive analyzers*. In the first type, the photoelectrons are subjected to a variable retarding potential and a photoelectron spectrum results from differences in the relative transmission of electrons of different energies at a given retarding field. In the second type, there is spatial resolution of photoelectrons of different energies. In the dispersive instruments now much in use, the base resolution depends on the *dispersion* as well as the *trace width* of a monoenergetic electron beam. The dispersion, which provides information about the spatial expanse of the spectrum for a given energy range, is defined as $D = \Delta x/(\Delta E/E_0) = E_0 x/E$, where the direction x lies on the focal surface. Contributing to the trace width are the entrance slit width S_1, which is magnified M times at the focus, the exit slit width S_2, and the maximum values x_i and x_j of both spatial and angular coordinates. If terms of the third and higher orders are neglected, the base resolution is given by

$$R = \frac{|M|S_1 + S_2 + \sum_{i,j} C_{ij} x_i x_j}{D} \qquad (2.1)$$

where C_{ij}'s are appropriate coefficients to $x_i x_j$.

The number of photoelectrons detected depends on the product of several factors: (1) the photoionization cross section, (2) the intensity of the ionizing radiation at the sample, (3) the useful area of the entrance slit, and (4) the useful solid angle of the electron analyser. Where grids are used transmission factor of the grids must also be considered, and if several detectors can be used simultaneously, then the number of such detectors must be considered. The *transmission* of a spectrometer is defined as the fraction of electrons reaching the detector from a point source that emits electrons isotropically, and it is given by the product of the useful solid angle of the instrument and the transmission factor. The integral of the point source transmission over the entrance slit area is called *luminosity*, and the integral of the solid angle over the slit area is called *étendue*. Usually the ratio of the slit width to some appropriate length unit in the instrument, such as the radius of the main electron trajectory in the analyzer, is what determines the contribution to the resolution. As a result, the useful slit area, and hence the luminosity, is generally proportional to the square of the linear dimension of the instrument. For a comparison of different types of instruments, therefore, the luminosity should be normalized by dividing by the square of the relevant length unit.

In instruments having a well-defined focal surface on which energy analyzed photoelectrons are focused, one may use a multidetector array. This consists of an ordered arrangement on the focal surface of a large number of individual electron detectors. With such an array, simultaneous detection of photoelectrons of more than one value of energy is possible. The gain in such arrangements can be understood by considering the information acquisition rate as a measure of sensitivity. If in an individual detector and in a component of multidetectors the collection areas are the same, then the information acquisition rate in a multidetector is multiplied by the number of individual detectors. Enhancement of sensitivity also takes place when satellite lines associated with primary photoionizing radiation and background contribution from bremstrahlung are contained. For this reason, monochromatization of X-rays helps increase sensitivity as well as resolution. With low-energy radiation sources, both the energy width of the source and the energy distribution of the sample species affect the resolution.

With regard to the linewidth of the ionizing radiation, when continuum sources are used, monochromatization is necessary, and the energy spread of the monochromatized radiation depends on the resolving power of the monochromator and other instrumental factors. In line sources, pressure broadening, nonthermal recoil of the emitting atoms, and Doppler broadening due to thermal motion of the emitters all contribute to the linewidth. Stark broadening due to interaction of the emitters with electrostatic fields of charged particles in the electric discharge radiation source, van der Waals broadening due to collisions with impurity gases, and resonance broadening as a result of collisions of the emitters with atoms of its own kind contribute to pressure broadening. With regard to the recoil of emitters, the recoil energy E_r of photon emission, given by $E_r = (h\nu)^2/2Mc^2$, is absorbed by the emitting atom, M being the mass of an isolated atom and ν the emitted frequency. This recoil causes a loss in photon energy, that is, it results in a red shift $\Delta \nu_r$ in the frequency of the emitted photon. For a photon energy of $\sim 10\,\text{eV}$, $\Delta \nu_r/\nu$ is $\sim 10^{-8}$ for an emitter having an atomic weight of unity; in the ultraviolet region, therefore, the red shift is negligible. The Doppler half-width $\Delta \nu_D$ of the radiation due to the thermal motion of the emitters is given by $\Delta \nu_D = 7.16 \times 10^{-7} \bar{\nu}\, (T/M)^{1/2}\,\text{cm}^{-1}$, where $\bar{\nu}$, in cm^{-1}, is the radiation frequency at peak emission, M is the atomic weight of the emitter in amu, and T is the temperature of radiating atoms in degrees Kelvin. For HeI 58.4 nm radiation at 300°K, $\Delta \nu_D \simeq 1.04\,\text{cm}^{-1}$, which is equivalent to an energy spread of $\sim 6.54 \times 10^{-5}\,\text{eV}$.

For resonance line sources, the resonance broadening is usually more important than the van der Waals broadening. At 1 torr, the half-width due to resonance broadening is only about $3 \times 10^{-7}\,\text{eV}$, which is much smaller than the Doppler broadening. The precise extent of Stark broadening due to the effect of ions or electrons in a discharge is difficult to estimate in lamps employed in photoelectron spectroscopy; but in the electron density range $10^{10} - 10^{14}\,\text{cm}^{-3}$ and electron temperatures 1–15 eV, which represent the ranges of these parameters in most discharge sources, the effect of Stark broadening is negligible.

Photoelectron line shift also occurs because of a nonthermal recoil of photoions; maximum recoil takes place when the recoil is in the direction of the incident photon. The maximum fraction of photon energy carried by a photoion is approximately $(m/M)(\frac{1}{2}mv^2/h\nu)$, where m is the electronic and M the photoionic mass, v is the velocity of the ejected photoelectron, and $h\nu$ is the photon energy. For ionization of a hydrogen molecule by HeI 58.4 nm radiation, this fraction is about 5×10^{-3}, and the maximum value of the ion recoil energy is about 1.5×10^{-4} eV. This is negligible compared with an energy resolution of 10^{-2} eV, which is that of most of the electron energy analyzers in use. A distribution of translational energies of the sample molecules that undergo photoionization thus contributes to the photoelectron kinetic energy width; this energy broadening ΔE_t is given by $\Delta E_t = (2m^2 v^2 T/M)^{1/2}$, the T representing the temperature of the species. Considering ionization by HeI radiation and a minimum ionization potential of 7 eV, ΔE_t is less than 10^{-2} eV for $M \geqslant 80$. If rotational structure in photoelectron spectra is to be resolved, ΔE_t therefore needs to be reduced as much as possible, primarily by working at low temperatures and using molecular beams.

For UPS, the light source requirement is almost invariably met by employing a dc, ac, or microwave discharge and then using the resonance radiation from the discharged gas, which is usually He or Ne. Since intensities of the lines other than the resonance lines are usually insignificant, further monochromatization of the radiation emanating from the discharge is ordinarily not necessary. The vacuum UV radiation from the discharge tube is collimated and guided to the sample chamber through a windowless region. These discharges usually take place in torr range pressures, and if the sample is also present in gas phase, then in order to avoid undesirable photoelectron neutral collisions the photoionization region pressure must be well below 10^{-3} torr. Differential pumping between the source and the sample region is essential in order to maintain this large pressure difference. In the analyzer and detector regions, the lower the pressure the better, and pressures in the range $10^{-5} - 10^{-10}$ torr are used. Ultrahigh vacuum conditions are essential in the sample chamber when solid surfaces are studied and contamination must be avoided. The HeI resonance line at 58.4 nm, which has an energy of 21.22 eV, is the most common source of low energy monochromatic radiation; for somewhat higher energies the HeII line at 30.4 nm with 40.8 eV energy can be employed. A simple discharge tube and photoionization chamber, used in one of the first photoelectron spectrometers, are shown in Fig. 2.1.

CYLINDRICAL GRID ANALYZER

Photoionization takes place in the region P, and after energy analysis through grids G_1 and G_2, the photoelectrons are detected at the collector C. The collector signal is then suitably recorded and displayed. The electron energy spectrum is scanned by applying a retarding potential difference voltage between the grids G_1 and G_2. The collector current $\sim 10^{-12}$ A is taken to the input of a sensitive electrometer with an input resistor $\sim 10^{10}$ Ω. The voltage due to

SPHERICAL GRID ANALYZER

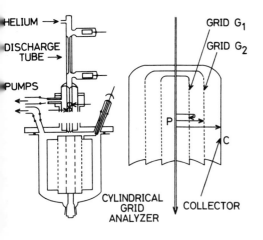

Fig. 2.1. Cylindrical grid analyzer [134].

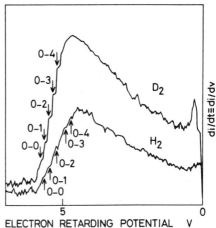

Fig. 2.2. Photoelectron spectra of hydrogen and deuterium using cylindrical grid analyzer [134].

the signal is small, which keeps the collector practically at zero potential. For HeI radiation, the grid G_2 is held at about -20 V, while G_1 changes from this negative value (no retarding potential between G_1 and G_2) to about zero that is, to a retarding potential of 20 volts. This gradually inhibits the transmission of the high-energy electrons. The G_2-C voltage difference is always kept greater than the G_1-C voltage difference, thus ensuring that the positive ions do not reach the collector. The retarding potential spectrum, which is the intensity of the collected electrons as a function of the increasing retarding potential — also called the photoelectron stopping curve — can be displayed as such or in a differentiated form. Photoelectron spectra of molecular hydrogen and deuterium taken in this instrument and displayed in the differentiated form are shown in Fig. 2.2.

SPHERICAL GRID ANALYZER

Although vibrational structure can be discerned in photoelectron spectra obtained from cylindrical grid analyzers, as the various bands indicate in Fig. 2.2, the resolution is hardly adequate, and accurate measurements of vibrational constants or transitional probabilities are difficult. The photoelectron intensity has a $\sin^2\alpha$ dependence in its spatial distribution, where α is the angle between the incident unpolarized photon direction and the photoelectron. The cylindrical grids affect the radial component of electron velocity, so the electrons

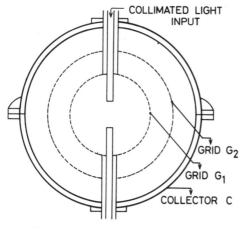

Fig. 2.3. Spherical grid analyzer [135].

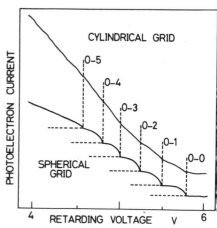

Fig. 2.4. Photoelectron spectrum of hydrogen using spherical grid analyzer [135].

ejected at a wide range of angles cause a smoothing out of the retarding potential spectrum. This problem can be largely corrected by using a set of grids of spherical shape around the photoionization region, as shown in Fig. 2.3. In the photoionization region the photoelectron ejection is always normal to the retarding field, and in such a grid configuration a constant potential difference of 3 volts between the two grids is quite adequate to prevent the positive ions from entering the electron retarding region between the collector and grid G_2. The electron retarding voltage is applied between grid G_1 and the collector. The two grids at any particular retarding voltage permit the passage of only those electrons that have higher than a particular kinetic energy, and therefore the grids may be said to act as a high-pass energy filter. The higher the retarding potential, the fewer are the number of electrons that can pass. Highest in kinetic energy are photoelectrons generated in the ionization process in which the ground vibrational state of the ground electronic state of the molecular ion is populated. These photoelectrons are stopped only at the highest retarding potential, as marked by the 0–0 point in Fig. 2.4. When the retarding potential is decreased sufficiently such that the next lower energy group of photoelectrons can also pass — in this case, those which involve population of the first vibrationally excited state of the molecular ions — the collector current increases by a step, as seen at point 0–1, and so on with the other steps. The improvement in resolution achieved by the spherical grid analyzer can be seen in the hydrogen photoelectron retarding potential curve shown in Fig. 2.4.

In the analyzer configuration shown in Fig. 2.3, the signal-to-noise ratio in the spectrum is not very satisfactory. Several techniques have been devised to improve it. At each step of the retarding potential curve, the electrons that cause the steps have energies very close to zero, the fast electrons contributing to the background. A solution may therefore be found if these low-energy electrons can somehow be separated from the accompanying high-energy electrons. In one technique, a second grid pair turned the opposite way separates the low-energy electrons. In another, a post-monochromator with a negative potential performs the same function by focusing only the low-energy electrons. This is shown in Fig. 2.5, where the photoelectrons originate from a solid sample. The central stop is to prevent the fast electrons from reaching the collector, and thus to reduce the background. The resolution is about 0.05% at an étendue of 0.50 cm² sr, although in comparison with dispersive analyzers the background still remains quite high.

In another instrument that uses energy filters, the electrons are retarded in a planar field; those electrons which just manage to pass through the filter are trapped in a cage and detected. The high-energy ones go through the cage and are collected elsewhere. If we consider a two-dimensional section across the planar field, and an electron having an energy E that enters the field at an angle α with respect to the normal and is retarded by a potential difference V_0, then the angle θ with respect to the normal at which the electron leaves the field is related to E, V_0 and α by the relation

$$E = \frac{eV_0}{\cos^2\alpha - \sin^2\alpha \cot^2\theta} \tag{2.2}$$

When the three-dimensional situation is considered, if the respective half-angles of electron entry and exit with respect to the normal are α_0 and θ_0, which are

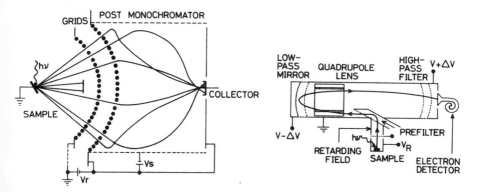

Fig. 2.5. Improved spherical grid analyzer [136].

Fig. 2.6. Energy analyzer using low-pass mirror and high-pass filter [138].

constrained within the limits $0 < \alpha_0 < 0.1$ rad and $0 < \cos^2\theta_0 < 1$, then the full base resolution width is given by

$$\Delta E \simeq eV_0 \alpha_0^2 (1 + \cot^2\theta_0) \qquad (2.3)$$

Since the electrons are preretarded, the relative resolution of the analyzer does not have to be high, and the instrument can be operated with a constant retarding field. This analyzer has a high luminosity and good background rejection.

In an improved version of this type of analyzer, shown in Fig. 2.6, the electrons pass through a variable retarding potential V_R and also through a prefilter, following which they are subjected to another retarding potential $V - \Delta V$, which reflects the low-energy electrons and thus acts as a low-pass mirror. The electron beam is then subjected to another retarding potential $V + \Delta V$, which acts as a high-pass filter, and the detector receives electrons in the kinetic energy range $e(V - \Delta V)$ and $e(V + \Delta V)$; e and $V < 0$, $\Delta V > 0$. The quadrupole lens is used for alignment of the beam.

These instruments are not particularly useful for monochromatized X-ray sources. The one shown in Fig. 2.5 operates at large solid angles and that in Fig. 2.6 has a very narrow beam structure. As a consequence in both devices, although the resolution is high the relative intensity is low. Further, since they have large étendues, gaseous samples are less suitable, as rather large differential pumping systems are required. These instruments, however, have a rather large tolerance of stray magnetic fields, as much as two orders of magnitude larger than that permissible in dispersive instruments.

Some of these improved retarding potential analyzers, although convenient to use and having reasonably good resolution, generally suffer from the disadvantage that the primary spectrum is obtained in integral form where photoelectrons of different energy groups are not entirely sorted out. Recently analyzers have been designed with resolution in the range of 15 to 25 meV (ΔE FWHM), where the energy spectrum is obtained directly by modulating the retarding potential or by a combination of the retarding field and an energy selective electron optical lens. As a result, it transmits only very slow electrons which just overcome the retarding field, and the modulation yields a differential spectrum.

Superior resolution with significantly less background is obtained with dispersive electron energy analyzers. Typically, in such instruments, photoionization is carried out in a separate ionization chamber. The photoelectrons are extracted out of this chamber and injected into a magnetic or electrostatic electron energy analyzer where, depending on the applied magnetic or electric field strength, electrons of a specific energy are focused at an exit point. Among the dispersive electron energy analyzers are the magnetic spectrometers. The resolving powers of magnetic spectrometers can be very high, but luminosity values are necessarily low. A typical figure may be 0.04% energy resolution at a luminosity of about 2×10^{-5} cm^2 sr for an instrument with about 30 cm radius. Since such instruments have a well-defined focal plane, multidetector arrays can be conveniently used. The most serious disadvantage of the magnetic analyzers is that they are very sensitive to stray magnetic fields, and for a resolution on the order of 0.01% the stray field needs to be reduced to about 0.1 milligauss over a rather large volume. This can be achieved by using large compensating Helmholtz coils, but they make the analyzer space-consuming and immobile. Furthermore, constant monitoring is required for the stray field. Often, for enhanced efficiency in electron energy analysis, the analyzer field is set in such a way that only electrons of a predetermined energy are focused at the detector. A preretardation is applied to the photoelectrons emanating from the sample so that all photoelectrons in turn have the predetermined energy. A display of the detector output against the preretardation field strength then gives the photoelectron spectrum. In magnetic analyzers application of any preretardation technique is difficult, since the elements that are needed for preretardation are situated within the magnetic field.

CYLINDRICAL CONDENSER ANALYZER

The electron optical properties of a radial electric field established between the plates of a cylindrical condenser, a cross section of which is shown in Fig. 2.7, can be used for photoelectron energy analysis. Electrons injected at the entrance slit describe circular orbits of radius r when the relationship $r = mv^2/eE_r$ is satisfied, E_r being the radial field and $\frac{1}{2}mv^2$ the electron kinetic energy.

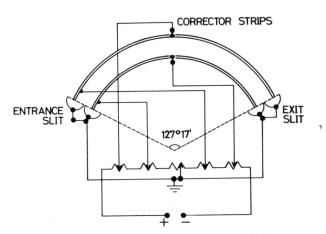

Fig. 2.7. Cylindrical condenser analyzer [142].

The field therefore deflects the electrons according to their kinetic energy. This means that for every value of electron kinetic energy there exists a field E_r at which the trajectory radius is equal to that of the cylindrical condenser and the electrons are focused at the exit slit. A detector placed at the exit point gives, if the radial field E_r is scanned, an energy spectrum of the injected electrons. Direction focusing is also achieved because, at a deflection angle of 127° 17′ ($\pi/\sqrt{2}$ radians), electrons placed in one field boundary are focused at the other. A resolution better than 0.2% is easily achieved, the ultimate resolution achievable being of the order of 0.05%.

SPHERICAL CONDENSER ANALYZER

As an improvement on the cylindrical 127° 17′ analyzer, the concentric double hemispherical analyzers or double spherical sectors have advantages in both resolution and transmission. Spherical sector analyzers offer resolution better than 0.02%. At the same value of radius, the resolving power of the hemispherical analyzer is about twice that of a cylindrical sector, and it can employ an operating voltage nearly twice that of the latter. Two-dimensional focusing is intrinsic to the spherical fields, and a much finer beam can be obtained at the exit point of the analyzer. An additional second stage can be used to even further improve the energy resolution. One such analyzer, which uses a section of the hemispheres, is shown in Fig. 2.8. The peripheral equipotential lines can be simulated by metal wires with appropriate potentials. This reduces adverse effects on electron trajectories due to fringe fields.

In such spherical fields the charged particles in general describe elliptic paths, and only those particles that have suitable energy and angle of injection give rise to circular paths. The relationship between the electron energy E and

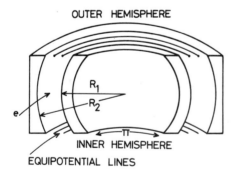

Fig. 2.8. Double hemispherical analyzer.

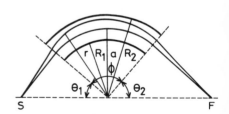

Fig. 2.9. Electron trajectories in a spherical sector condenser [1].

the field E_s for which a circular path is obtained in a sector spherical condenser, as shown in Fig. 2.9, is $E_s = (E/e)(R_2/R_1 - R_1/R_2)$, and for point focus, the condition $\tan \phi = -\tan(\theta_1 + \theta_2)$ needs to be satisfied. The base-width energy resolution $\Delta E/E$ for this type of analyzer is given by

$$\frac{\Delta E}{E} = \left| \frac{s}{2a} \right| + |C_1 \alpha_r^2| + |C_2 \alpha_z^2| + \left| C_3 \left(\frac{s}{2a}\right)^2 \right| + \left| C_4 \left(\frac{h}{2a}\right)^2 \right| + \left| C_5 \alpha_r \frac{s}{2a} \right| + \left| C_6 \alpha_z \frac{h}{2a} \right| + \left| k \frac{\alpha_r}{a} \right| \qquad (2.4)$$

where s is the width and h the height of the effective source, a is the radius of the optic circle, α_r and α_z are the radial and azimuthal entrance angles relative to the central ray, and the last term with k is for relativistic correction. The C_1 through C_6 are the aberration coefficients; they are dependent on the sector angle and can be calculated from detailed ion optical considerations of the sector analyzer. With a knowledge of the magnitude of these aberration coefficients, such as shown in Fig. 2.10, the resolution $\Delta E/E$ can be optimized. For an instrument using a sector angle ϕ of 157.5°, an optic circle radius of 36 cm, interelectrode distance of 8 cm, azimuthal sector angle of the electrodes of 60°, and with $k \leqslant 2\%$ of C_1 at energies < 5 keV, the theoretical resolution is $\sim 5 \times 10^{-4}$ for an angle-defining aperture of 1×60 mm^2. (An analyzer with these dimensions has been used in the photoelectron spectrometer discussed below under the heading "Spectrometer using Monochromatized X-rays" and shown in the accompanying Fig. 2.17. The spherical condenser analyzer and the

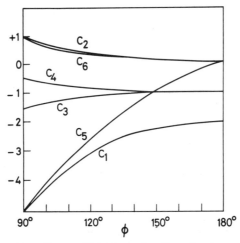

Fig. 2.10. Aberration coefficients as a function of sector angle [145].

cylindrical mirror analyzer, described below, are the two commonly used electron energy analyzers in present-day high resolution photoelectron spectrometers.)

The cylindrical and spherical fields are actually special cases of toroidal fields. Some other toroidal fields have been found to be superior to the hemispherical field for both electron spectrometers and monochromators. Calculations show that a gain in luminosity of about a factor of 3 for the spectrometer and at least a factor of 6 for the monochromator can be expected. The major drawback is the large sector angle needed in the spectrometer ($\sim 219°$) and the large angle between the beam and the image planes.

CYLINDRICAL MIRROR ANALYZER

An electron energy analyzer that uses two coaxial cylinders and has the properties of space focusing is shown in Fig. 2.11. It is simple to construct. Electrons with a kinetic energy eV_e leaving the point S located on the coaxis of the two cylinders at an angle θ can converge again on the same axis at a point I after being reflected by a potential $-V_p$ applied to the outer cylinder. First-order focusing can be obtained at any θ between 0 and $\pi/2$. Second-order focusing is achieved when θ is equal to $42°18.5'$, when the distance D between the image and the object is $6.12 r_1$ (r_1 being the radius of the inner cylinder). Base resolution of the analyzer is given by $\Delta E/E = 21 (\sin \theta \, \Delta\theta)^3$. For a finite-sized object, $\Delta E/E$ is somewhat higher. The V_e/V_p ratio required for a second-order focusing is $1.3098/\log(r_2/r_1)$ where r_2 is the radius of the outer cylinder. Another attractive feature of this analyzer is its high luminosity, as electrons from the entire 2π angle around the cylindrical axis are collected by the detector. A hydrogen photoelectron spectrum taken with a similar cylindrical mirror analyzer is shown in Fig. 2.12.

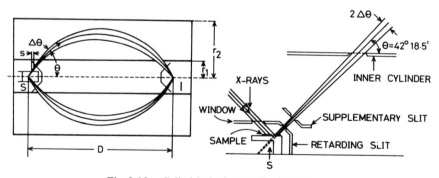

Fig. 2.11. Cylindrical mirror analyzer [150].

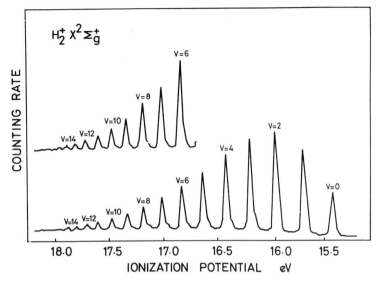

Fig. 2.12. Hydrogen spectrum using cylindrical mirror analyzer [592].

BESSEL BOX ANALYZER

Another electron energy analyzer, recently reported, that also has certain elements of simplicity in design is shown in Fig. 2.13. The analyzer consists

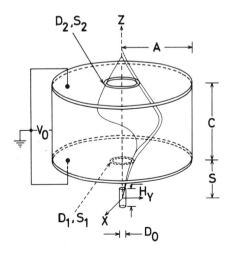

Fig. 2.13. Bessel box analyzer [153, 154].

of a cylindrical chamber of radius A and length C. The bottom and top endplanes of the cylinder are isolated from the lateral surface, and they act as mounts for the circular slits S_1 and S_2 at diameters D_1 and D_2 for the entrance and exit of electrons. S is the distance between the source and the entrance slit S_1. The expression for potentials inside the analyzer region involves Bessel functions, and thus the analyzer is also called the Bessel box. The potential inside the cylindrical region is given by

$$\phi(r, z) = \frac{4V_0}{\pi} \sum_{\substack{k \\ \text{odd}}} \left[\frac{1}{k} \sin \frac{k\pi(z-S)}{C} \right] \frac{I_0(k\pi r/C)}{I_0(k\pi A/C)} \tag{2.5}$$

where V_0 is the potential on the lateral surface of the cylinder, the top and bottom surfaces having $\phi = 0$, and I_0 is the modified Bessel function of order zero. The field components $E_r(r, z) = -[\partial\phi(r, z)/\partial r]$ and $E_z(r, z) = -[\partial\phi(r, z)/\partial z]$ can be found out from the above equation. Equations of motion can be obtained by equating the force experienced by a particle of charge e in E_r and E_z fields to mass-acceleration products in r and z directions. After integration of the r and z equations and elimination of the time parameter, trajectories inside the Bessel box can be determined. Computational studies using some practical values of parameters yield magnitudes of attainable resolution. With $A = 7.50$ cm, $C = 7.50$ cm, $S = 2.00$ cm, $D_0 = 0.10$ cm, $H = 0.10$ cm, $D_1 = 1.00$ cm, $S_1 = 0.01$ cm, $V_0 = -10$ V, $D_2 = 1.66$ cm, $S_2 = 0.015$ cm, $(E/\Delta E)_{\text{calc}} = 296$, which has been realized in the fabricated instrument as $(E/\Delta E) \simeq 303$.

The potential distribution inside the analyzer permits a number of trajectory classes. All have identical initial and terminal radii on the end-planes, but they differ in the number of times they cross the axis. The higher modes have very high dispersion but poor focusing properties. Since the energies of the modes differ, the fundamental mode, which has the highest kinetic energy electrons, can be filtered from the others by placing a retarding element in front of the detector. Alternatively, since the exit angles differ for the various modes having a common exit radius, mode separation can be achieved by a pinhole baffle placed between the electron detector and the annular exit slit. A comparison of the performances of the Bessel box analyzer and the cylindrical mirror shows that for the same axial source-to-focus distance, identical acceptance solid angle, and source disk diameter, the resolving power in the Bessel box analyzer is about 300 compared with 158 expected in cylindrical mirror analyzer. But this does not mean absolute superiority of the former. With sources of finite size, the Bessel box is superior; with point sources, the cylindrical mirror analyzer is superior. One expects an $E/\Delta E \simeq 3000$ with a Bessel box analyzer of double the size mentioned above, and with essentially the same solid angle of acceptance.

ANALYZERS FOR ANGULAR DISTRIBUTION MEASUREMENTS

In angular distribution measurements of photoelectrons, simultaneous determination of photoelectron intensities at two different angles is often needed. Sometimes the same electron energy analyzer can be used for this purpose, as shown in Fig. 2.14. In this experimental scheme, two rather diminutive channel electron multipliers are used as detectors, and the slits are so arranged that the two multipliers record photoelectron current for two specific angles. In order to have a somewhat higher angular resolution, a small accelerating potential in the ionizing region is provided in the same instrument, so that the solid angle of electron injection at the analyzer increases.

For angle-resolved photoelectron spectroscopy seeking information at a wide range of emission angles, it is of advantage to be able to move the entire analyzer around the emitting system. Miniaturization of the analyzer is of much help, but design simplifications may be necessary, including some sacrifice in performance characteristics. Such miniature analyzers have been made; the one shown in Fig. 2.15 involves a plane-parallel-plate electron energy analyzer, also called the plane mirror analyzer. The parallel plates have a separation of distance d, and the potential difference between them is V_a. The electrons that are injected at an angle θ and brought to a focus on the plane satisfy the relationship

$$x = (h+y) \cot \theta \left(1 + \frac{1}{\cos 2\theta}\right) \tag{2.6}$$

In particular, only those electrons are brought to a focus at point I of the plate

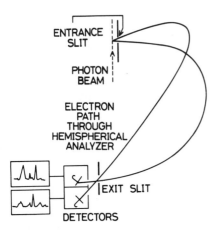

Fig. 2.14. Angular distribution measurements at two angles using a single-electron energy analyzer [157].

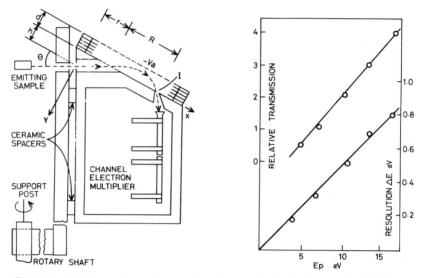

Fig. 2.15. Miniature plane mirror analyzer for angular distribution studies [162].

that have an electron energy E_k given by

$$E_k = \frac{h(eV_a/4d)}{\cos 2\theta \sin^2\theta} \qquad (2.7)$$

where h is the perpendicular distance from the entrance slit to the base plate and e is the electronic charge. A photoelectron kinetic energy spectrum, therefore, can be obtained by scanning the voltage V_a. For $\theta = 30°$, second-order focusing condition occurs, and then $R = 2r = 2\sqrt{3}h$. E_k at this condition may be considered to be the pass energy E_p, given by $2heV_a/d$. The energy resolution is given by $\Delta E/E_p = [(s_1 + s_2)/2R] + 4.6\,(\Delta\theta)^3$. The first term is due to slit width contribution, s_1 and s_2 being respectively the effective widths of the entrance and exit slits. The second term is due to angular aberration, $\Delta\theta$ representing the maximum angular deviation from 30°. In order to keep transmission and resolution of the analyzer fixed, the analyzer is operated either with preretardation or with preacceleration, keeping V_a fixed. It has been found that $E_p = 7$ eV represents a good compromise between the conflicting requirements of good resolution and adequate transmission, and at this E_p, $\Delta E = 0.3$ eV. With this value of E_p, the range of photoelectron kinetic energy for which more than 50% transmission is expected is $1\text{ eV} \leqslant E_k \leqslant 23\text{ eV}$.

For complete mapping of the entire space, two such analysers may be employed, one capable of movement on the horizontal plane and the other on the vertical, as shown in Fig. 2.16. Results obtained on photoionization of GaSe using this arrangement of analyzers are also shown in the same figure. The

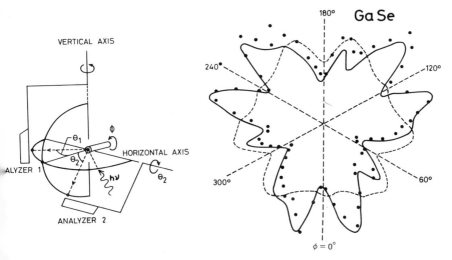

Fig. 2.16. Angular distribution measurements using two analyzers, and angular distribution results of GaSe [162].

photoelectron energy (6 eV, corresponding to the most intense peak at photon energy 17 eV) and the angle of emission with respect to the surface normal were held constant. In obtaining data the sample was rotated about the surface normal, changing the angle ϕ. The full circles shown in the figure were obtained with analyzer 2, using an angle of emission $\theta_2 = 60°$. The total curve was obtained by a threefold symmetrization of the full-circle data. Each point was averaged with the two points $+120°$ on each side and a smooth curve drawn through the results. This uses the fact that the rotational symmetry of the GaSe crystal is at least threefold. The dashed curve represents threefold symmetrized results of analyzer 1, at angle of emission $\theta_1 = 65°$. This curve has been rotated $90°$ and both curves thus relate to the same crystal axis. Although the angles of emission θ_1 and θ_2 are not quite equal, the main difference in the curves are attributable to polarization of the synchrotron radiation, which was the source of the photons. Unlike the full curve, the dashed curve displays mirror symmetry. Asymmetry is expected for photoelectrons not in the plane of incidence. Considerable additional information can be obtained by having such analyzers movable in mutually perpendicular planes.

ELECTRON DETECTORS

Detectors used in photoelectron spectroscopy vary a great deal in degree of sophistication. The simplest consists of a plain charge collector called a Faraday cup, but this requires a sensitive electrometer since it has to measure rather small currents. Furthermore, since the current is very small, to develop even a small input voltage, high input resistances are required in the electrometer. As a result, the response time is very long as the latter is strongly dependent on the input resistance. For fast response, electron multipliers can be used which have gains of the order of 10^6. With these, from an input of 10^{-12} A of electron current, output current of 10^{-6} A can be obtained that can be used with almost any recording instrument, and a fast scan can be achieved. But the gains of the electron multipliers are sensitive to aging and other operating conditions, and frequent calibration is needed if quantitative measurements are also to be made. This problem can be circumvented by using the multiplier in the pulse counting mode. In this mode, the arrival of each individual electron at the input of an electron multiplier results in a small output pulse; the number of such pulses are then counted. For dispersive analyzers with well defined focal planes, multi-detector arrays can be used.

SPECTROMETER USING MONOCHROMATIZED X-RAYS

For photoelectron spectrometers using X-ray sources, the electron energy analyzer and detector requirements are by and large similar to those used along with the vacuum ultraviolet sources, except that relativistic corrections on electron trajectories need to be made when considering high-energy electrons generated by X-rays. There are, however, considerable complications in the technology of generating X-rays, in their monochromatization and in reducing the effect of inherent linewidth, the latter determining materially the limit of attainable resolution. X-rays are generated by allowing an intense beam of electrons to strike an appropriate target material, such as Mg or Al, which results in emission of characteristic X-radiation. Nowadays the degrading influences of X-ray linewidth and bremstrahlung on observed spectra have mostly been eliminated. One of the recent high-resolution XPS–UPS instruments, shown in Fig. 2.17, uses a finely focused electron beam from a 6 kilowatt (500 mA, 12–20 kV) electron gun impinging on the anode to generate the X-rays. A rotating anode is used to keep the target cool and thus prevent evaporation of the target material. The resulting X-rays are then monochromatized and focused simultaneously by a spherically bent quartz crystal placed on a Rowland circle. The full width at half maximum (FWHM) intensity of the monochromatic beam is about 0.21 eV. The X-rays pass through a thin Al foil and enter a gold-plated sample compartment. Solids to be studied, which can have any shape, are set such that the illuminated surface faces the photoelectron exit slit. A spherical 157° sector electrostatic analyzer, which has been described earlier, is used for energy analysis of the photoelectrons. The analyzed electrons

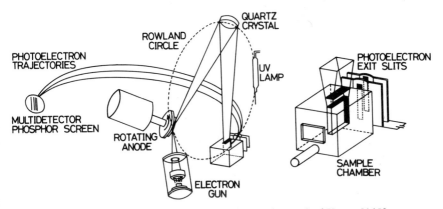

Fig. 2.17. XPS–UPS instrument using monochromatized X-rays [145].

fall on a multidetector system where two wafers in tandem, each having many small holes, are placed where the energy analyzed electrons impinge. The wafers are made of a material that gives good secondary electron emission, yielding an electron current gain of about 10^4 per wafer. Each primary electron thus produces a pulse of about 10^8 secondary electrons, which are then accelerated to 4 kV and allowed to hit a phosphor screen. The light images thus produced on the screen are scanned by a television camera and quantitative results brought out using a multichannel analyzer and an on-line computer. This instrumental arrangement is typical of the high-resolution XPS instruments now in common use. A carbon 1s core electron spectrum of CH_4 obtained from this instrument is shown in Fig. 2.18. The asymmetry in the peak is believed to be due to

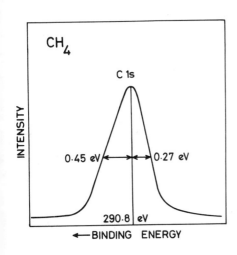

Fig. 2.18. C 1s core electron spectrum from the instrument of Fig. 2.17 [145].

excitation of symmetric vibrations in the molecular ion after photoionization of the core electron.

SPECTROMETER FOR LIQUID SAMPLES

The initial applications of photoelectron spectroscopy were on solids and gases. Recently, photoelectron spectroscopy has been applied to liquids as well. The liquid sample in the form of a beam is passed through the photoionization region, where the ionizing monochromatic radiation is shone on the slit side of the electron spectrometer, as illustrated in Fig. 2.19. The liquid to be studied is continuously circulated in the system. To minimize interference of the vapor phase spectrum with the liquid phase, rapid removal of the vapor phase sample from the ionization chamber by fast differential pumping is required. This may be assisted by having a liquid beam as narrow as possible and precooling the liquid to reduce its vapor pressure. With fast pumping, the maximum vapor pressure change that has been achieved is of the order $1/r$, the r being the distance from the beam; this has been realized in a water beam with a diameter of about 0.2 mm. The reservoir pump prevents backstreaming of the liquid into the ionization region and removes residual gases from the liquid sample. The two parallel channels following the reservoir are used to provide a variable beam flow speed. Typical values of the liquid beam flow are 0.5–1 ml

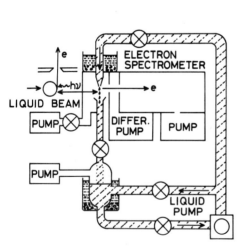

Fig. 2.19. Photoelectron spectrometer for liquid sample [166].

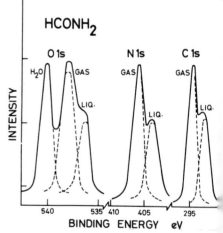

Fig. 2.20. Photoelectron spectra of HCONH$_2$ in liquid and vapor phases [166].

per second. The O 1s, N 1s, and C 1s lines from $HCONH_2$ are shown in Fig. 2.20. The presence of traces of water in the sample causes an O 1s peak to appear.

The difference in core level binding energies between the liquid and vapor phases is of the order of 1.6 eV. Part of the explanation of the observed low-energy shift of the liquid peaks with respect to the corresponding peaks of the vapor phase species is believed to lie in the relaxation effects associated with the ionization process. When a molecule in the liquid is ionized, besides the electronic relaxation in the orbitals of the molecule, polarization of the surrounding molecules occurs. The magnitude of the electronic relaxation is around 15–20 eV for ionization of 1s orbitals in the second row elements. The extramolecular relaxation associated with polarization, which causes a low-energy shift of the binding energy, is estimated to be up to about 10% of the total relaxation energy. It is observed that the shift between the vapor signal and the liquid signal is approximately the same for all the levels of the molecule, including the valence levels. This indicates that regardless of the nature of the hole created in ionization, localized core hole or diffuse valence hole, the liquid behaves isotropically as regards the extramolecular relaxation.

Quick-frozen solutions have also been the subject of photoelectron spectroscopic studies. Certain investigations have shown that the immediate surroundings of the solute remains analogous with that in the original liquid solution if the latter can be rapidly frozen. Some recent work in photoelectron spectroscopy has made use of this technique to investigate interactions between cations and anions in aqueous solution.

PHOTOELECTRON SPECTROSCOPY IN A STRONG MAGNETIC FIELD

Recently a new spectrometer has been described that holds potential for obtaining photoelectron spectra of atomic and molecular ions. Figure 2.21 shows an exploded view of this apparatus. Its principle of operation is as follows.

Primary ions are generated either by photoionization or by electron impact ionization; the resulting ionic species are then photoionized by HeI radiation. The photoelectrons emanating from the ionization cell move towards the retarding plate. The presence of a strong magnetic field constrains the electrons to move along the magnetic axis, and they are energy analyzed first by retardation and then by a crossed electric field. The photoelectron spectrum thus obtained is due to the original neutrals as well as to any ions formed in the cell in the primary ionization. The contribution to the spectrum due to the ions is easily sorted out by their mass selective removal by the ion cyclotron resonance technique. The latter is especially convenient to carry out in the system since a strong magnetic field is already present.

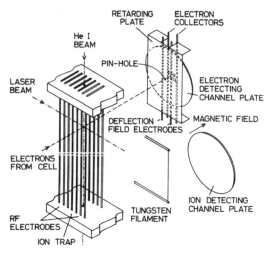

Fig. 2.21. Photoelectron spectrometer in a strong magnetic field [179].

The tungsten filament that provides electrons for primary ionization is positioned parallel to the trapping well and outside the trapping region. It is so placed that electrons from the filament cross the axis of the light beam. The ion trap is normally held at + 0.6 V. The trapping potentials concentrate the ions in the HeI beam region and the photoelectron detection system responds to electrons emanating from this zone. There is also a provision for introduction of a tuned dye laser beam into this region to allow excitation of ionic as well as neutral species. A crossed electric field is applied between the two deflecting field electrodes located behind the retarding plate. This electric field, in conjunction with the magnetic field at right angles to it, deflects the electrons vertically, the extent depending on the electron kinetic energy. The electron collectors behind the electron detecting channel plate are so located that one of them detects the undeflected electrons and the other the deflected ones.

The oscillating rf potentials used for ion cyclotron resonance removal of ions have a mean value of 0 V. These rf potentials are applied in antiphase to the outer portions of the cage structure. Accelerated ions are detected by deflection out of the cage along the magnetic axis onto an off-axis ion detecting channel plate.

With an apparatus of this type, it should be possible to obtain second ionization potentials for molecular and fragment ions. This possibility includes molecular ions that are formed directly, ions produced in ion–molecule collisions, negative ions, and also those ions that are first excited to metastable states. An important application of this magnetically collimated photoelectron energy analyzer is in the new technique of photoelectron spectromicroscopy, microscopy using energy-selected photoelectrons.

PHOTOELECTRON SPECTROSCOPY WITH SUPERSONIC MOLECULAR BEAMS

Another recent improvement of the photoelectron spectroscopic technique is employment of supersonic molecular beams. If sample species are injected into the photoionization region in the form of a molecular beam at right angles to the photoelectron sampling direction, then, since there would be no molecular motion in the beam with respect to the electron sampling direction, Doppler broadening of the photoelectron lines is eliminated and much higher resolution spectra result. In the past, molecular beam sample injection has been used for photoelectron spectral study of those molecules that have reasonable vapor pressures only at high temperatures. This has involved mostly studies of inorganic halides. Molecular beams used in those instruments consisted of free jet expansion of the sample. Recently an apparatus has been reported that constitutes probably the first use of a differentially pumped supersonic beam source for photoelectron spectroscopy. In this instrument a much larger nozzle throughput may be used, so that for a given background pressure in the main chamber a higher number density in the beam is obtained. Ratio of the beam to background gas signals is typically 12:1 for a background pressure of 2×10^{-5} torr. A section of this apparatus is shown in Fig. 2.22.

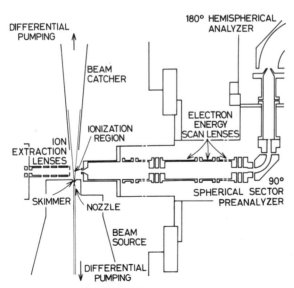

Fig. 2.22. Photoelectron spectroscopy using supersonic nozzle beam [187].

A molecular beam having a beam divergence of 10° is formed by expanding the high pressure sample gas (1000 torr) through a 70 μ diameter nozzle and then extracting the central portion of the beam by using a 1.1 mm diameter skimmer. The ionizing photon beam is introduced perpendicular to both the nozzle beam and the photoelectron sampling direction. The region where photoionization takes place has a size of about 2 mm in diameter, and it is located 3.2 cm from the electron energy analyzer aperture of 1.8 mm in diameter. Photoelectrons from the ionization zone are drawn into the electron energy analyzer, which contains the electron energy scan lenses used for retardation, and, in tandem, a 90° spherical sector preanalyzer and a 180° hemispherical analyzer. The photoions formed in the process of photoionization are concurrently drawn to the left through a set of ion extraction lenses and to a quadrupole mass spectrometer, where a mass analysis identifies the species photoionized. The background pressure in the nozzle chamber is 2×10^{-4} torr, and the pressure where photoionization takes place is 2×10^{-5} torr. Pressures in the mass analyzer chamber is 1×10^{-7} torr, in the front part of the electron energy analyzer 5×10^{-8} torr, and 3×10^{-9} torr in the hemispherical analyzer region. Spectral resolution measured is 10 meV (FWHM). The apparatus has been used for high resolution experiments on small molecules, such as rotational relaxation studies on H_2. In a recent experiment, spectra of rotationally cold ethylene molecules have been obtained at an instrumental resolution of 11 meV FWHM, and much of the vibrational structure has been resolved for the first time.

RESONANCE LIGHT SOURCES

There are constant improvements and innovations on the various constituents of photoelectron spectrometers. In one involving light source properties, capillary-constrained dc discharge lamp operates with water-cooled aluminum electrodes at relatively high discharge currents, and a variation of the discharge pressure is used to control preferential emission of one or the other helium resonance lines. The mechanical design as well as the performance characteristics of the lamp are shown in Fig. 2.23. Typical operating currents are in the range 50–100 mA at 500 V, with starting voltages about 5 kV. Another lamp improvement, shown in Fig. 2.24, employs a cylindrical cathode with a pair of wire anodes inside that are symmetrically located with respect to the axis of the cathode. Electrons follow long oscillatory paths between the anodes, and this results in an increase in the number of ions produced by a single electron. This also leads to an electron density increase in the discharge plasma and results in the requirement of a very small discharge current for a given emission intensity. This particular lamp operates satisfactorily as a photoelectron spectrometer light source at pressures as low as 3×10^{-4} torr and a discharge current

SPECTROMETER USING SYNCHROTRON RADIATION

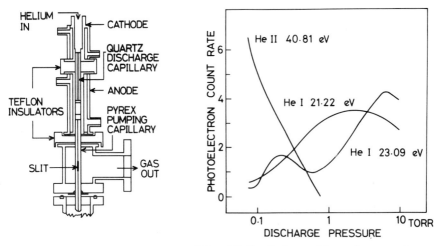

Fig. 2.23. Light source having intense HeI and HeII radiation, and its performance characteristics [190].

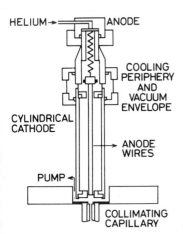

Fig. 2.24. Light source for operation at low pressure and power [191].

of 3 mA, and it provides somewhat more intense radiation of the HeII first resonance line compared with that of HeI.

SPECTROMETER USING SYNCHROTRON RADIATION

There are practically no simple and useful radiation sources between the rare gas resonance lines and the characteristic X-ray lines of Mg, Al, and similar elements. In this regard, *synchrotron radiation* is the most promising emission source in the intervening wavelength region for photoelectron spectroscopic studies. Synchrotron radiation is obtained from very high energy electrons and positrons moving in circular orbits as, for example, in accelerators and storage rings. At relativistic energies, the radiation is emitted at a very small angular cone around the instantaneous direction of flight of the charged particles. The intensity of radiation filling the accelerator plane remains constant till the cutoff energy ϵ_c, after which the intensity drops rapidly. ϵ_c in keV is given by

$$\epsilon_c = 2.218 E^3 R^{-1} \tag{2.8}$$

where E is the particle energy in GeV and R is the radius of the orbit in meters. The angular width ψ is given by mc^2/E where m is the particle mass and c is the velocity of light. The photon intensity against photon energy as well as the angular divergence of synchrotron radiation from a couple of accelerators are shown in Fig. 2.25a and Fig. 2.25b. The time structure of synchrotron light sources follows that of the orbiting particles. Emission of radiation takes place in bunches of 0.1 to 1 msec pulses depending on the accelerator or storage ring, and this results in peak intensities $\sim 10^4$ times the average photon density. The pulses may have a periodic structure of 2 nsec to 1 μsec due to repetition frequency of bunches, 30 nsec to 1 μsec due to period of revolution, or 20 msec, which is the period of injection into accelerators. The radiation is characterized by a high degree of polarization. Radiation off the plane is elliptically polarized, which can be decomposed into left and right circularly polarized radiation with

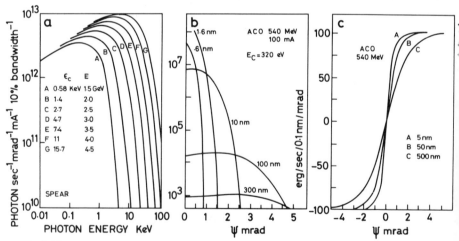

Fig. 2.25. Characteristics of synchrotron radiation. Adapted from C. Kunz [46] (copyright 1974, Pergamon Press Ltd), and from P. Pianetta and I. Lindau [205].

a degree of circular polarization as shown in Fig. 2.25c. With synchrotron radiation, the entire range between the rare gas resonance line sources and the X-rays can be covered. Since the source of synchrotron radiation is chemically very pure, solid samples can be maintained at pure surface condition without much additional pumping between the source and the sample chamber. Furthermore, being strongly polarized in the electron orbital plane makes the radiation very useful for the study of momentum distribution of valence electrons in a crystal. Its major limitation is that its use is limited by the availability of an appropriate synchrotron.

A synchrotron radiation source photoelectron spectrometer is shown schematically in Fig. 2.26. In this particular instrument a radiation from a 1.3 GeV synchrotron is used. The radiation is dispersed vertically by a grazing incidence spectrometer, and the horizontal focus line that is formed by a toroidal mirror acts as an incident slit of the spectrometer. The dispersed image is focused by a concave grating into a point of the sample position, and a two-stage cylindrical mirror type of analyzer is used for electron energy analysis. The GaSe results shown in Fig. 2.16, it may be recalled, were obtained by using synchrotron radiation.

CALIBRATION OF ELECTRON ENERGY ANALYZERS

As regards calibration of electron energy analyzers and photoelectron spectra, two aspects need to be considered: calibration of spectral energies and calibration of spectral intensities.

Calibration of binding energies for gaseous samples is the simplest. A common procedure is to mix the sample gas, before the spectrum is run, with a reference gas whose binding energy is known. In UPS, the reference gas is usually a rare

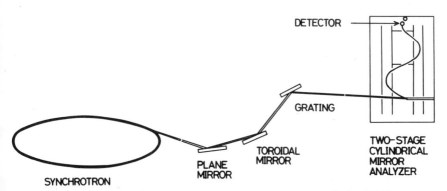

Fig. 2.26. Photoelectron spectrometer using synchrotron radiation [203].

gas such as Ar, Kr, and Xe, which provides a calibration of the instrumental energy scale. As binding energy values of several reference samples are now well known, such calibration is possible at both valence electron levels and core electron levels.

The energy scale of a photoelectron spectrometer may be calibrated by taking one or more metallic samples where binding energy differences of photoelectron peaks are precisely known. On the basis of the experimentally observed spectra, instrumental parameters can then be adjusted. Some such photoelectron lines that are used for calibration are

Cu	$2p_{3/2}$ and $3s$		CF_4	F $1s$
Ag	$3p_{5/2}$ and $3d_{5/2}$		CO_2	O $1s$ and C $1s$
Au	$4f_{5/2}$ and $4f_{7/2}$		N_2	N $1s$
Graphite	$1s$		Ar	$2p_{3/2}$ and $3p$
Ne	$1s$, $2s$, and $2p$		Kr	$3p_{3/2}$ and $3d_{5/2}$

The larger the binding energy difference between the calibrant peaks the better, because for a given absolute energy uncertainty of the lines, the calibration per unit energy has less uncertainty if the calibrant lines are widely spaced. Besides this method of using a single source of photons, a set of photon sources of accurately known energies, such as Al K_α, Mg K_α, He I, and He II, may be used for calibration. For photoionization of a given electron in the sample, the use of two photon sources yields two photoelectron peaks of two different kinetic energies, the difference being equal to the difference in photon energies, and this can be used for calibration. For solids, the use of core lines is desirable since valence band maxima may vary with the photon energy used.

With a metal, the energy difference between the kinetic energy of the fastest electrons, those ejected from the Fermi level, and the cutoff energy of secondary electrons — zero kinetic energy — is equal to the difference between the photon energy and the work function of the metal sample. Therefore, the energy scale can be calibrated if the incident photon energy and the work function of the metal sample are both accurately known. Polycrystalline tungsten is a good calibrant with a work function of 4.54 ± 0.07 eV. For solid conductors, the binding energy of an electron with respect to the Fermi level can also be written as $E_B^F = h\nu - E - \phi_{sp}$ where E is the kinetic energy of the photoelectron and ϕ_{sp} is the work function of the spectrometer material. If the spectrometer is calibrated, then the binding energy is obtained from the incident photon energy, the experimentally determined kinetic energy, and ϕ_{sp}. Otherwise, since two metals in contact equilibrate and have the same Fermi level, for a sample in contact with a foil of a reference metal one may write $E_{B,\text{sample}}^F - E_{B,\text{ref}}^F = E_{\text{ref}} - E_{\text{sample}}$. Therefore, if the kinetic energy difference of the two peaks is measured, and $E_{B,\text{ref}}^F$ is known, then the $E_{B,\text{sample}}^F$ can be determined. And if the work function ϕ_s of the clean material is known, the binding energy with respect to vacuum level may be obtained, since $E_B^{vac} = E_B^F + \phi_s$.

In a nonconducting sample, sample charging takes place as the electrons lost due to photoionization are not compensated for rapidly enough in the sample. As a result, the positively charged sample decelerates the emitted photoelectrons and shifts the spectrum to lower kinetic energies. By applying a bias potential to the sample, the rate of neutralization of holes can be checked and charging can be detected. Flooding the sample with low-energy thermionic electrons is another method of checking; peaks then move to higher apparent kinetic energies, which is an indication of charging. In such situations, provided the substrate is not reactive, vacuum deposition of a small quantity of a calibrant material such as gold can be used. Gold equilibrates with nonconducting sample, and Au 4f peaks may be used as calibrating lines for the sample. For liquids, the problem is still more complex, and some physically admixed solids such as graphite have been used as calibrants. There is no definitive method of expressing the binding energy values for gas molecules as adsorbates; a convenient one is to refer them to the Fermi level of the substrate. For comparison with gas phase binding energy values of these molecules, an appropriate correction would be to use the work function of the surface after adsorption.

For quantitative measurement of photoelectron peak intensities, it is also essential that the relative luminosity of an electron energy analyzer be known as a function of electron energy. We noted earlier that the luminosity is synonymous with the transmission in that it represents the fraction of electrons detected out of the total that is produced. In such calculations of luminosity, it is important that the finite source size is taken into account since point source calculations may result in considerable error. In order to have suitable calibration of electron energy analyzers, several methods may be followed. One may first measure the total number of photoelectrons formed and then the number detected. For example, in UPS instruments this involves use of a vacuum ultraviolet monochromator, a light source producing a number of lines in the vacuum ultraviolet, and a calibrated photodetector for measuring the intensity of radiation. Knowledge of the energy–brightness relationship of a retarding lens, as is sometimes used in front of an electron energy analyzer, also serves as a calibrating method. Once such measurements are made on some samples in a dependable manner, the relative intensity data, which reflect relative photoionization cross sections, can then be used to calibrate any electron energy analyzer provided a photoelectron spectrum is obtained in the same analyzer using the same sample. The other point of concern is the modulation of intensity caused by the angular distribution. Since most photoelectron energy analyzers sample a cone of angles at right angles to the photon beam, the observed intensity I_{obs} and the true intensity are related as:

$$I_{obs} = I_{true}\left[1 - \cos\gamma + \tfrac{1}{4}\beta\frac{g-2}{g+1}(\cos^3\gamma - \cos\gamma)\right] \quad (2.9)$$

where I_{true} is a fixed fraction of the total number of electrons ejected into a solid angle of 4π steradians, γ is the half angle of the cone of acceptance, β is the asymmetry parameter, and g the degree of polarization of radiation. If the photon beam is incident along the y-axis and the photoelectrons are sampled along the z-axis, then $g = I_x/I_z$ where I_x and I_z are intensities of polarization components in x and z directions. If the cone half-angle is small, it can be shown that

$$I_{obs} = I_{true}\left(1 - \tfrac{1}{2}\beta\frac{g-2}{g+1}\right) \quad (2.10)$$

which, for unpolarized radiation, i.e., when $g = 1$, becomes

$$I_{obs} = I_{true}\left(1 + \frac{\beta}{4}\right) \quad (2.11)$$

When such I_{obs} values and β values are available on some compounds, any energy analyzer can be calibrated conveniently.

In general, relativistic corrections are necessary when electron kinetic energies are high — that is, when incident photon energies are high — as the electrons in the analyzers then no longer converge at the nonrelativistic focus. Relativistic calculations have been carried out for toroidal analyzers such as the spherical sector plate analyzer, and also for parallel plate and cylindrical mirror analyzers, and the results have been applied to the calibration of XPS instruments.

CHAPTER

3

CORE ELECTRON SPECTRA

Photoionization by high-energy X-ray photons causes the ejection of core electrons. An analysis of the consequent photoelectron spectra yields core electron binding energies. The gross core electron binding energies are directly useful for qualitative identification of the atomic constituents of a substance, and the chemical shifts of the core electron binding energies provide useful information about the electronic environment of the constituent atoms. In this chapter we first consider the chemical shifts of the core binding energies and the correlations between atomic charge and chemical shift. We next explore possible relationships between core energy chemical shifts with parallel chemical shifts observed in nuclear magnetic resonance spectra and Mössbauer spectra. A few applications of chemical shifts are then presented. This is followed by consideration of the fine satellite structure associated with core photoelectron peaks, the effects of molecular ion vibrations on spectral linewidths and their possible correlations with Franck–Condon profiles, and, finally, the effects of collective resonance on core electron spectra. Here the discussions are limited to core electron spectra of free molecules; later, spectra of solid samples will be considered.

The principles that determine the magnitude of the core binding energies are simple and have been considered earlier. The deeper the location of a core electron, the greater is the magnitude of the binding energy. And for a given core-shell electron – $1s$, $2s$, $2p$, and so on – the binding energy increases with the atomic number across a period. In this connection one notes that with heavy atoms, the binding energies of the core electrons that lie very deep are so high that in such cases only the higher shell electrons are accessible for photoelectron spectroscopic studies with conventional radiation sources. In contrast, the chemical shifts of binding energies for any given type of electron are very small and are dependent on the nature of neighboring atoms. The adjacent atoms determine the type of bonding and hence the nature of charge distribution in the atom of interest. The charge distribution, in turn, depending on changes in the nature of bonding, causes small changes in the binding energy.

Table 3.1. Binding Energies of Electrons in Some Elements (Free Atom) (eV) [1, 14]

	K	L_1	L_2	L_3		K		L_3		K			K
H	13.60				Na	1075			K	3610	Rb		15203
He	24.59				Cl	2829			Br	13481	I		33176
Li	58	5.392			Ar	3206.3			Kr	14327	Xe		34565
Be	115	9.322				K				L_1		L_2	L_3
B	192	12.93		8.298									
C	288	16.59		11.26	Sc	4494				503		408	403
N	403	20.33		14.5	Fe	7117				851		726	713
O	538	28.48		13.62	Ni	8338				1015		877	860
F	694	37.85		17.42									
Ne	870.1	48.47	21.66	21.56		M_1	M_2	M_3	M_4	M_5			$N_{1,2}$
					Sc	55	33	33					6.540
					Fe	98	61	59		8			7.870
					Ni	117	75	73	10	9			7.635

	K	L_1	L_2	L_3	M_1	M_2	M_3	M_4	M_5	P_1	N_1	N_2	N_3
Au Z = 106	80729 164820	14356 32811	13738 31787	11923 23915	3430 8698	3153 8263	2748 6338	2295 5644	2210 5292		764 2413	645 2202	548 1664

	N_4	N_5	N_6	N_7	O_1	O_2	O_3	O_4	O_5	O_6	O_7	P_1	P_2	P_3	$P_{4,5}$	Q_1
Au Z = 106	357 1330	339 1238	91 797	87 770	114 590	76 499	61 350	12.5 216	11.1 194	39	35 96	9.23	65	39	7	9

NOTE: $K = 1, L = 2, M = 3, N = 4, O = 5, P = 6$ and $Q = 7$. Subscript $1 = s_{1/2}, 2 = p_{1/2}, 3 = p_{3/2}, 4 = d_{3/2}, 5 = d_{5/2}, 6 = f_{5/2}$ and $7 = f_{7/2}$. Thus, P_4 represents the $6d_{3/2}$ subshell.

CORE BINDING ENERGIES AND CHEMICAL SHIFTS

Table 3.2. Carbon 1s Binding Energies (eV)[4]

CH_4	290.7	CS_2	293.1
$\underline{C}H_3CH_2 \cdot CO \cdot OC_2H_5$	290.8	CH_3COCH_3	293.8
$\underline{C}H_3CH_2OH$	290.9	$H \cdot \underline{C}OCH_3$	293.9
$\underline{C}H_3 \cdot CO \cdot CH_3$	291.2	$C_2H_5 \cdot CO \cdot O\underline{C}_2H_5$	294.5
$\underline{C}H_3 \cdot CHO$	291.3	$CH_3 \cdot \underline{C}O \cdot OH$	295.4
$\underline{C}H_3 \cdot COOH$	291.4	CO	295.9
CH_3OH	292.3	CO_2	297.5
$CH_3\underline{C}H_2OH$	292.3	CHF_3	298.8
$CH_3\underline{C}H_2 \cdot CO \cdot OC_2H_5$	292.4	CF_4	301.8

CORE BINDING ENERGIES AND CHEMICAL SHIFTS

Some elemental core electron binding energies as well as the chemically shifted binding energies of some gaseous compounds involving C 1s, N 1s, O 1s, F 1s, S $2p_{3/2}$, and Xe $3d_{5/2}$ electrons are presented in Tables 3.1 through 3.7. Where more than one atom of the same kind is present in the molecule, the binding energy values refer to the atom underlined. F 1s, O 1s, N 1s, C 1s, Cl $2p_{1/2,3/2}$, S $2p_{1/2,3/2}$, Si $3p_{1/2,3/2}$, and Br $3p_{1/2,3/2}$ binding energy shifts for functional groups in polymeric molecules are given in Fig. 3.1.

Theoretical core electron binding energies can be calculated from *self-*

Table 3.3. Nitrogen 1s Binding Energies (eV) [4, 14]

$C_6H_5NH_2$	405.5	NO	410.7
NH_3	405.6	$C_6H_5NO_2$	411.6
N_2H_4	406.1	$N\underline{N}O$	412.5
$\underline{N}NO$	408.6	NO_2	412.9
$\underline{N}_2$	409.9	ONF_3	417.0

Table 3.4. Oxygen 1s Binding Energies (eV) [4]

CH_3CHO	537.6	SO_2	539.6
$C_2H_5 \cdot C\underline{O} \cdot OC_2H_5$	537.6	H_2O	539.7
$CH_3 \cdot C\underline{O} \cdot OH$	538.2	$CH_3 \cdot CO \cdot \underline{O}H$	540.8
C_2H_5OH	538.6	CO_2	540.8
$C_2H_5 \cdot CO \cdot \underline{O}C_2H_5$	538.8	N_2O	541.2
CH_3OH	538.9	NO_2	541.3
CH_3COCH_3	539.0	CO	542.1
SOF_2	539.4	O_2	543.1
		NO	543.3

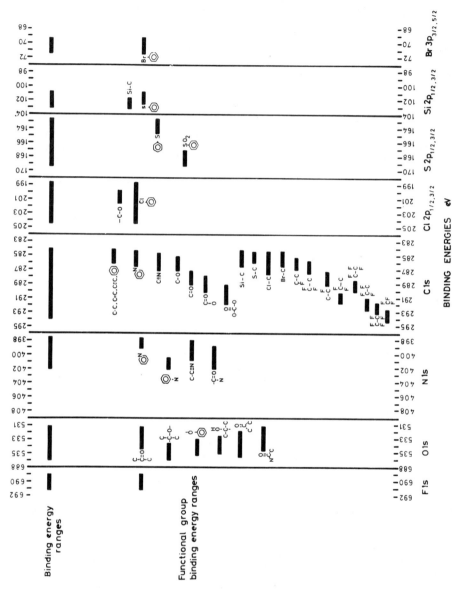

Fig. 3.1. Core level binding energy shifts for functional groups in polymeric molecules. Adapted

Table 3.5. Fluorine 1s Binding Energies (eV) [4, 14]

SOF_2	693.3
CHF_3	694.1
SF_6	694.6
CF_4	695.0

Table 3.6 Sulfur $2p_{3/2}$ Binding Energies (eV) [4, 14]

S	168
CS_2	169.8
H_2S	170.2
SO_2	174.8
SOF_2	176.2
SF_6	180.4

consistent field (SCF) methods *ab initio* and can be expressed as the difference between the total electronic energies of the neutral species and that of the ion. Thus, for an atom A, the binding energy can be expressed as $E_B = E_B(\Delta \text{SCF}) = E_{\text{tot}}(A) - E_{\text{tot}}(A^{+*})$, where the asterisk indicates that a core electron is missing. Such *ab initio* calculations, however, are quite costly to carry out as they require much computer time. Estimates of core binding energies can also be made on the basis of Koopmans's approximation, which assumes that the negative of one-electron orbital energy of the electron that is lost in ionization and the experimental binding energy of the electron are approximately equal, that is, $E_B \simeq E_B^{\text{Koop}} = -\epsilon_i$. In the general case, ϵ_i is the energy of the electron in molecular orbital i moving in the field of all other electrons as well as nuclei of the molecule. Such calculation of E_B using Koopmans's approximation, however, invariably gives higher values of binding energy as it does not consider the effect of electron relaxation. We have already noted that whereas in valence electron ionization the contribution of relaxation is small, the effect of relaxation must be taken into account when calculating the binding energy in core electron ionization.

The relaxation energy in terms of the Koopmans's binding energy and the true binding energy may be written as $E_{\text{relax}} = (-\epsilon_i) - E_B$. Some values of relaxation energies of atoms have already been given in Table 1.1. In molecules, too, when an electron is lost the electron relaxation takes place in the resulting ion. Whereas there is no canonical way by which such relaxation energy can be apportioned to various possible component sources, in one way of viewing it, the core relaxation energy in a molecule may be considered to consist of two parts: one due to orbital contraction around the atom from which the electron is lost, and the other due to redistribution of charge in the overall molecule.

Table 3.7. Xenon $3d_{5/2}$ Binding Energies (eV) [4, 14]

Xe	676.4	$XeOF_4$	683.4
XeF_2	679.4	XeF_6	684.3
XeF_4	681.9		

Table 3.8. Core Electron Binding Energies and Relaxation Energies of a Few Molecules [252]

Molecule		Core Electron Binding Energy (eV)			Relaxation Energy (eV)		
		$E_B^{Koop} = -\epsilon_{1s}$	$E_B(\Delta SCF)$	$E_B(\text{expt.}) = I_p$	E_{relax}	$E_B^{Koop} - E_B(\Delta SCF)$	$E_B^{Koop} - E_B(\text{expt.})$
C1s	CH$_4$	305.2	291.0	290.7	15.87	14.2	14.5
	HCN	307.8	296.7	294.0	15.12	11.1	13.8
	CO	310.7	298.4	296.2	14.01	12.3	14.5
	CH$_3$F	308.0	296.0	294.0	15.81	12.0	14.0
N1s	NH$_3$	422.8	405.7	405.6	18.78	17.1	17.2
O1s	H$_2$O	559.4	539.4	539.7	21.42	20.0	19.7
	CO	563.5	542.1	542.3	20.65	21.4	21.2
F1s	HF	715.2	693.3	694.0	23.4	21.9	21.2
Ne1s	Ne	891.4	868.4	869.1	21.2	23.0	22.3

It turns out that the major contribution to the relaxation is due to the first process. One could, however, alternatively consider the relaxation energy to consist of energies of the following steps: relaxation of any core electrons left behind in the atom following the ionization, reorganization of the valence electrons of the same atom, and reorganization of the electrons of neighboring atoms. In such calculations it has been found that the contribution to the relaxation energy due to the residual core is relatively insignificant and that the major contribution to the relaxation is due to the valence electrons of the same atom.

The total electronic energy of an atom depends on the electron populations in the various shells of the atom as well as on appropriate shielding constants of the shells. Subsequent to ionization, the tendency for both intraatomic and interatomic electron flow is due to the presence of a positively charged core. Since the electronegativity of an atom changes with its electronic environment, in this case of electron flow associated with core ionization it is reasonable to expect the flow to continue till the electronegativities of neighboring atoms are equalized, although it may mean the formation of partially charged atoms in the molecule. The electron population in various atomic shells, the shielding constants, and the partial atomic charges in a molecule are all related quantities, and one needs to minimize the total electronic energy with respect to all of them. The difference between the minimum energy obtained with relaxation and the Koopmans's binding energy E_B^{Koop}, which is the total electronic energy calculated using populations and shielding constants appropriate to an unrelaxed core, is the core relaxation energy, defined earlier. It has been possible to make fairly accurate estimates of the core relaxation energy on this basis. Some values of relaxation energies calculated using the above principles are presented in the fourth column of Table 3.8.

In this table, the first column shows the binding energies according to Koopmans's approximation, the second column $E_B(\Delta\text{SCF})$ is the binding energy obtained by calculating the total electronic energy of both the neutral and the ion by the SCF method and then taking their difference, the third column gives the experimental ionization potentials, the fourth column presents relaxation energies calculated by equalization of electronegativities. The differences $E_B^{\text{Koop}} - E_B(\Delta\text{SCF})$ and $E_B^{\text{Koop}} - E_B(\text{expt})$, presented respectively in the fifth and sixth columns, show that the relaxation energies calculated on the basis of the equalization of electronegativity principle are reasonably satisfactory.

The relaxation energy may be considered essentially independent of the chemical environment. It is also reasonable to assume that the binding energy of a core electron changes because of a change of potential felt at the core electron site. This change in the potential would be primarily due to the valence electrons of the same atom that loses the core electron, with an additional contribution due to the electrons of neighboring atoms. On this basis, it is possible to construct a simple *charge potential model* to explain the chemical shifts in binding energies.

CHARGE POTENTIAL MODEL

Let us assume that the atoms in a molecule can be approximated by hollow nonoverlapping electrostatic spheres and that the core binding energy depends on the potential felt at the core site due to the valence electrons of the same atom as well as electrons of the neighboring atoms. The shift in binding energy E_B of a core electron of an atom A in a molecule M, with respect to the binding energy of the same core electron of atom A in a reference compound, can then be written

$$\Delta E_B^A = \Delta E_v^A + \Delta E_m^A \tag{3.1}$$

where ΔE_v^A represents the contribution to the chemical shift due to the valence electrons of the same atom and ΔE_m^A the contribution of electrons from other atoms in the molecule. More explicitly, the above equation can be written

$$\Delta E_B^A = a\left[\left(\frac{q^A}{R_A}\right)_M - \left(\frac{q^A}{R_A}\right)_{\text{ref}}\right] + a\left[\left(\sum_{C \neq A}\frac{q^C}{R_{AC}}\right)_M - \left(\sum_{C \neq A}\frac{q^C}{R_{AC}}\right)_{\text{ref}}\right] \tag{3.2}$$

where q^A and R_A represent respectively the charge and the radius of the valence shell of A, q^C the charge on a neighboring atom, R_{AC} its distance from atom A, and a stands for a conversion factor. The terms in the first set of square brackets represent the effect due to intraatomic potential, and the terms in the second set of square brackets represent the effect due to interatomic potential, with summations carried over all neighboring atoms. It is possible to rearrange the above equation by first separating out the $\Sigma q^C/R_{AC}$ term for the reference molecule and then writing

$$\Delta E_B^A = k_A' \Delta q^A + a\left(\sum_{C \neq A}\frac{q^C}{R_{AC}}\right)_M - a\left(\sum_{C \neq A}\frac{q^C}{R_{AC}}\right)_{\text{ref}} = k_A' \Delta q^A + V_A + l' \tag{3.3}$$

where Δq^A is the charge difference on the atom A between the molecule it is in and the reference system, and k_A' represents an average interaction between a core electron and a valence electron belonging to atom A. This amounts to assuming an average radius of valence shells of atom A. The second term can be represented by an effective interatomic potential V_A, and the third term, since it depends only on the reference compound, by a constant l'. By separating the contribution of the reference system from the first term and combining it with l', it is possible to express the shift in the binding energy in terms of q^A as

$$\Delta E_B^A = k_A q^A + V_A + l \tag{3.4}$$

where k and l for a particular core ionization can be determined independently if for a set of compounds the charges are known and chemical shift data are also available.

Determination of atomic charge q_P^A using Pauling's method depends first on an evaluation of the partial ionic character I of the bond in which the atom of interest is involved. For this purpose, the equation $I = 1 - [\exp - 0.25(\chi_A - \chi_C)^2]$ can be used where χ_A and χ_C are the electronegativities of the two bonding atoms A and C in Pauling's scale. The charge q_P^A is then given by $q_P^A = Q^A + \Sigma_C I$ where Q^A is the formal charge on the atom. When the structural formula of the molecule is written, if there is no formal charge on the atom, Q naturally is to be taken as zero. Some corrections in electronegativities, however, are needed if a formal charge is present. For example, with a charge of $+1$ (or -1) on the atom, the electronegativity of the atom increases (or decreases) by a factor of $\frac{2}{3}$ the difference between the electronegativity of atom A and that of the next atom with one higher (or lower) atomic number. The summation of the second term containing I needs to be carried over all the atoms which are bonded to the atom A.

A plot of the C 1s shift in bromomethane in Fig. 3.2a shows that the shift is about 5.7 eV per unit charge. The improved linearity of the plot using a higher electronegativity indicates that bromine electronegativity is better explained by a value greater than 2.8. Such studies can also be used to evolve a uniform scale of electronegativity on the basis of chemical shift data. With atomic charge calculated from the semiempirical SCF CNDO/2 method, the correlations obtained between the charge and binding energy shifts for C 1s, N 1s, and S 2p using the charge potential model are shown in Fig. 3.2c.

The C 1s chemical shifts for several types of hybridization (sp^3, sp^2, sp, aromatic) in various molecules are correlated with charge parameter q_P in Fig. 3.2b. As is evident from the figure, a good linear correlation exists. Though the secondary substitutent effect still persists, to a first approximation, the localized bond treatment with the nearest neighbor approximations in charge potential model holds good. This figure also presents the chemical shift obtained from graphite, -0.7 eV.

The potential model given above involves charges and potentials that really refer to the initial states. One might suspect that in such a situation the relaxation energy might yield too high a binding energy. On the other hand, if only final-state properties are used, a low binding energy, due to concentration of negative charge around the core, might result. Based on this analysis, models

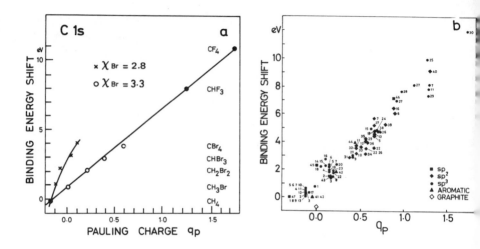
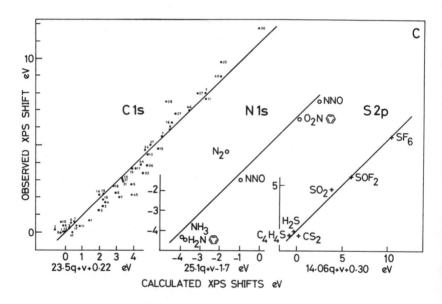

Fig. 3.2(a) Plot of C 1s binding energy shifts against Pauling charge, calculated by using the equation $q_P^A = Q^A + \Sigma I$, where $I = 1 - [\exp - 0.25\,(x_A - x_C)^2]$. The electronegativities of Br are indicated in the figure.
(b) Plot of C 1s chemical shifts for various types of hybridized carbon atoms as a function of the charge parameter q_P.
(c) Plots of experimental binding energy shifts against shifts calculated by the charge potential model, using CNDO charges for C 1s, N 1s, and S 2p electrons [1, 4, 253].

have been constructed where the binding energy is considered to be a function of the average of initial-state and final-state charges. The situation thus resembles a half-ionized core with the binding energy expressed as $E_B = k(q_i + q_f - 1)/2 + V_{av} + l$. Such models have been used with considerable success.

VALENCE POTENTIAL MODEL

The core electron binding energy shifts have also been interpreted in terms of a *valence potential model*. The valence potential ϕ is defined as the potential experienced by a core electron of an atom A due to the valence electron density of the molecule; it is related to the binding energy E_B^A of the core electron by

$$E_B^A = e\phi_{VAL}^A + \text{constant} \qquad (3.5)$$

When this binding energy is compared with that of a corresponding core electron in a reference compound, one obtains

$$\Delta E_B^A = e\Delta\phi_{VAL}^A \qquad (3.6)$$

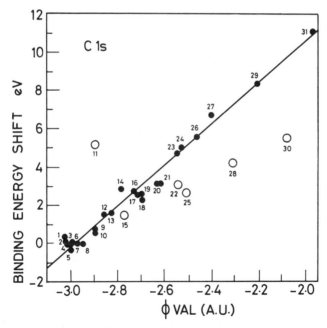

Fig. 3.3. Correlation of C 1s binding energy shifts with valence potential ϕ_{VAL} from EHT calculations [255].

The quantum mechanical average value of the potential is given by

$$\phi_{VAL}^A = -2 \sum_i \left\langle \phi_i \left| \frac{1}{R_A} \right| \phi_i \right\rangle + \sum_{C \neq A} \frac{q_C}{R_{AC}} \tag{3.7}$$

In the first term, which represents the electronic contribution, R_A is the distance of the ith valence electron from the nucleus A, ϕ_i is the wave function of the valence electron, and the summation is over all the doubly occupied molecular orbitals. The second term represents the nuclear potential due to the neighboring atoms, where q_C represents the nuclear charge on atom C and R_{AC} is the internuclear distance. It has been found that the contribution to ϕ_{VAL}^A from 1s core is insensitive to the environment. For the neighboring atoms, there is almost complete screening by the core molecular orbitals of their respective nuclei, so that instead of q_C, a reduced charge q_C^* needs to be used.

Figure 3.3 presents a correlation between the experimental C 1s binding energy shifts with valence potentials (ϕ_{VAL}, atomic units) obtained from extended Hückel theory (EHT) calculations. The calculated binding energy shifts are in good agreement with the experimental ones. The circles below the line are for molecules in which the C atom is attached to the elements of the second row (such as N, O, F) of the periodic table. For such molecules a separate linear correlation can be found. The model predicts binding energy shifts close to experimental shifts for some CNDO (C 1s) and EHT (C 1s, O 1s, S 2p) studies. Other MO schemes such as NDDO, INDO, and MINDO, which assume less drastic differential overlap neglect, can provide better values for ϕ_{VAL}.

GROUP SHIFTS

It is possible to extend the applicability of the potential model by introducing the electroneutrality condition $q^A = \sum_{C \neq A} -q^C$ and making an appropriate choice of a reference compound such that $l = 0$. By introducing these two conditions in the charge potential model, one can write

$$\Delta E^A = k_A q^A + \sum_{C \neq A} \frac{q^C}{R_{AC}} + l = k_A q^A + \sum_{C \neq A} \frac{q^C}{R_{AC}}$$

$$= \sum_{C \neq A} \left(\frac{1}{R_{AC}} - k_A \right) q^C \tag{3.8}$$

The summation on the right side includes all the neighboring atoms that influence A. This summation however, can be subdivided and carried out group by group according to the atoms that are attached to the atom A, and one may write

$$\Delta E^A = \sum_G \left[\sum_{\substack{\text{all C's} \\ \text{in Group } G}} \left(\frac{1}{R_{AC}} - k_A \right) q^C \right] = \sum_G \Delta E_G \qquad (3.9)$$

ΔE_G's then represent the *group shifts*. Some of the group shifts for the carbon atom, along with the group electronegativities, are given below.

	ΔE_G	χ_G		ΔE_G	χ_G
– CH$_3$	– 0.32 eV	2.0	– CF$_3$	0.59 eV	3.1
– CH$_2$OH	– 0.42	2.0	– OH	1.51	3.5
– C(=O)OH	– 0.15	2.3	– OCH$_3$	1.46	3.5
– H	0.01	2.6	– Cl	1.56	3.6
– NH$_2$	0.25	2.9	– F	2.79	4.0

The application of this group shift principle, however, requires an assumption that the presence of one group attached to an atom does not affect the shift of

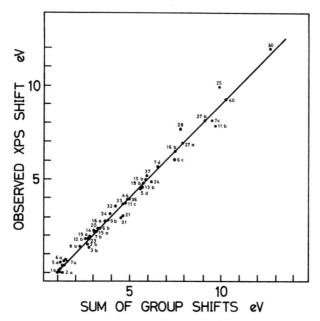

Fig. 3.4. Plot of experimental binding energy shifts against sums of group shifts for C 1s electron [253].

other groups attached to the same atom. This is not strictly true, but the group shift concept is a useful one, as can be seen from the plot of observed XPS shift versus the sum of group shifts presented in Fig. 3.4. Although not all the shifts of molecules conform to the simple potential models presented above or to the group shift model, it is clear nevertheless that the correlations are of great value in predicting XPS shifts, or conversely, in deriving estimates of charges on atoms if the shifts are experimentally known.

THERMODYNAMIC ESTIMATES OF CHEMICAL SHIFTS

In a rather interesting way of looking at core electron binding energies, thermodynamic data have been utilized to predict XPS shifts. The rationale is the assumption that the valence electrons of a molecule remain practically unaffected if one of the atomic cores that lacks an electron is replaced by the complete core of an atom with one unit greater nuclear charge. According to this *equivalent cores approximation*, the energies of the following two reactions, for example, should be identical

$$NH_3^{+*} + O^{6+} \longrightarrow OH_3^+ + N^{6+*}$$
$$N_2^{+*} + O^{6+} \longrightarrow NO^+ + N^{6+*}$$
(3.10)

the asterisk indicating, as before, that a core electron and not a valence electron is missing. Here N^{+*} is replaced by O^+ and O^{6+} is replaced by N^{6+*}. It is not necessary to assume that the energy of such a replacement is zero; it is sufficient to assume that, for atoms of a given element, all such replacements involve the same energy change. On the basis of the above two reactions, one concludes that in any reaction in which an interchange of an incomplete core and an equivalent equally charged complete core takes place, the energy change is zero:

$$NH_3^{+*} + NO^+ \longrightarrow OH_3^+ + N_2^{+*}$$
$$\Delta E = 0$$
(3.11)

where O and N cores interchange. But the core binding energy difference between ammonia and molecular nitrogen is the energy change in the reaction

$$NH_3 + N_2^{+*} \longrightarrow NH_3^{+*} + N_2$$
$$\Delta E = E_B(NH_3) - E_B(N_2)$$
(3.12)

The significance of the reaction (3.11) is that by adding (3.11) and (3.12) one obtains

$$NH_3 + NO^+ \longrightarrow OH_3^+ + N_2$$
$$\Delta E = E_B(NH_3) - E_B(N_2)$$
(3.13)

The energy change of the overall chemical reaction turns out to be the difference in the core binding energies. Since reaction (3.13) involves well characterized chemical species, this means that XPS shifts can easily be derived from simple thermodynamic data of heats of formation. Some ion–molecule reactions that can be used for estimates of chemical shifts of other nitrogen compounds are:

$$(CH_3)_2NH + NO^+ \longrightarrow (CH_3)_2OH^+ + N_2$$
$$CH_3NH_2 + NO^+ \longrightarrow CH_3OH_2^+ + N_2$$
$$HCN + NO^+ \longrightarrow HCO^+ + N_2$$
$$N_2O + NO^+ \longrightarrow NO_2^+ + N_2 \quad (3.14)$$
$$NO + NO^+ \longrightarrow O_2^+ + N_2$$
$$NF_2 + NO^+ \longrightarrow OF_2^+ + N_2$$

A plot of the N 1s XPS binding energies of nitrogen compounds expressed with respect to the N 1s binding energy of N_2 against the thermodynamically derived relative binding energies is shown in Fig. 3.5. This method has been found to be applicable to a wide range of compounds of many elements, including chemical shifts of B 1s, C 1s, O 1s, and so on. A similar correlation for some xenon compounds is shown in Fig. 3.6. This method can also be used to derive thermodynamic information on gaseous ions from data on XPS shifts.

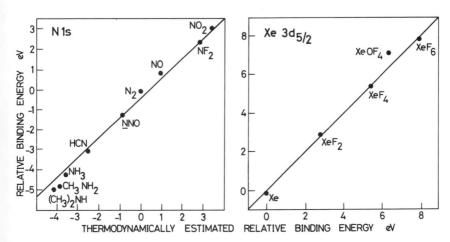

Fig. 3.5. N 1s binding energy shifts from photoelectron spectra versus thermodynamically estimated relative binding energies in some nitrogen compounds [257].

Fig. 3.6. Xe $3d_{5/2}$ binding energy shifts from photoelectron spectra versus thermodynamically estimated relative binding energies in some xenon compounds [257].

XPS, NMR, AND MÖSSBAUER CHEMICAL SHIFTS

Since chemical shifts are observed in other physical methods used in the determination of molecular structure, it is of interest at this point to examine whether there exists any simple correlation of core electron chemical shifts obtained in photoelectron spectroscopy with other well-known chemical shifts, for example, NMR chemical shifts and Mössbauer isomer shifts.

Simple generalizations cannot be applied to NMR chemical shifts, and, therefore, no simple correlation with XPS shifts can be expected. NMR shifts depend on local diamagnetic and local paramagnetic currents about the nucleus and on currents in distant groups. The relative importance of these contributions varies with the structure of the molecule, and no straightforward correlation is known to exist between them. It is reasonable, however, to expect a close relationship between the chemical shift in the diamagnetic screening constant in NMR and the electrostatic potential. Since the electrostatic potential is known to closely predict the core electron chemical shifts, some linear correlations between XPS and NMR shifts may be expected in closely related compounds where the NMR shift changes because of changes in effective nuclear charge rather than magnetic anisotropy effects.

One has to be careful, however, in selecting a group of compounds that may be called closely related. Consider, for example, the shifts that have been observed

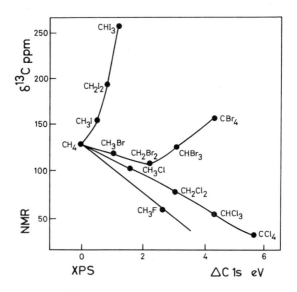

Fig. 3.7. Comparison of ^{13}C NMR chemical shifts and C $1s$ XPS shifts in halomethanes [117, 263].

in a series of halomethanes. ^{13}C NMR shifts and C 1s core electron chemical shifts of this series of compounds are shown in Fig. 3.7. We see no simple correlation between the two types of shift as progressively more hydrogen atoms are substituted by halogens. Actually, for such correlation, a more compatible set of closely related compounds are those that have the same number of halogen substituents, such as CH_3I, CH_3Br, CH_3Cl, CH_3F, or a set like CH_2Br_2, CH_2Cl_2, CH_2F_2. Figure 3.8 shows the relationships that exist between the shifts for these sets of compounds, and a near linear correlation can be seen. In general, such types of correlations may be expected if the rest of the molecule remains essentially unchanged; in this case, the hybridization of the carbon atom remains unaltered. Figure 3.9 shows such a plot for monosubstituted methanes. Because of a large neighboring anisotropic effect, the halogens behave somewhat differently. Figure 3.10 shows the nature of correlations for fluoromethanes using C 1s and F 1s photoelectron lines. Figure 3.11 shows the NMR—XPS (S 2p) correlation plot for a series of organometal phenyl sulfides, X denoting the metal atom. It shows that the core electron shifts of the first atom in aromatic substituents, in this case S, may correlate well with ^{13}C shifts in aromatic rings. Figure 3.12 shows the correlation plot for p-fluorophenyl sulfur compounds. ^{19}F meta shifts, indicated by the dashed line, are quite insensitive compared with para shifts. Rather complex dithiole compounds also give a good correlation with fluorophenyl tags, as shown in Fig. 3.13.

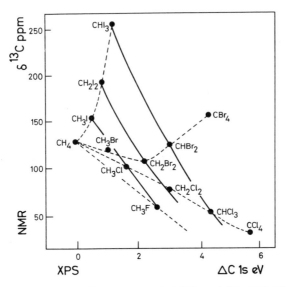

Fig. 3.8. Correlations between ^{13}C NMR chemical shifts and C 1s XPS shifts in halomethanes [262].

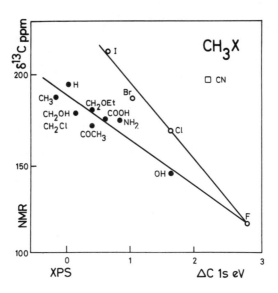

Fig. 3.9. Correlations between ^{13}C NMR and C 1s XPS shifts in monosubstituted methanes [264].

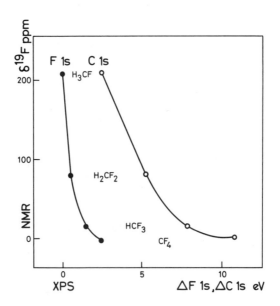

Fig. 3.10. Correlations between ^{19}F NMR and C 1s, F 1s XPS shifts in fluoromethanes [265].

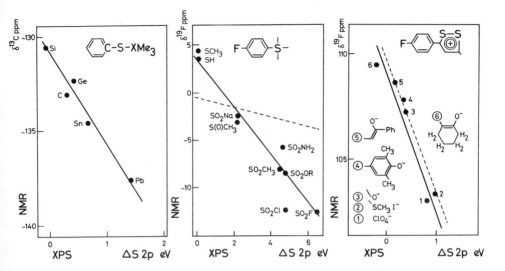

Fig. 3.11. Correlations between ^{13}C NMR and S $2p$ XPS shifts in organometal phenyl sulfides. Adapted from S. Pignataro et al. [266] (copyright 1972, Pergamon Press Ltd).

Fig. 3.12. Correlations between ^{19}F NMR and S $2p$ XPS shifts in parafluorophenyl sulfur compounds. Adapted from J. W. Elmsley and I. Phillips [267] (copyright 1972, Pergamon Press Ltd).

Fig. 3.13. Correlations between ^{19}F NMR and S $2p$ shifts in dithiole compounds. Adapted from B. J. Lindberg et al. [268] (copyright 1974, Pergamon Press Ltd), and from B. J. Lindberg et al. [269].

The solid line in Fig. 3.13 is the least squares regression. The dashed line is the least squares regression from a plot of δ^{19}F for the tagged compounds against ΔS $2p$ for the corresponding nontagged compounds. The shift between these lines gives a measure of the substituent effect of the fluorine tag on ΔS $2p$. Where correlations exist, the lines do have characteristic slopes, for example, in Fig. 3.11 it is 4.6 ppm/eV. Some such characterizations, where good linear correlations exist, are shown in Table 3.9. In general, it may be said that in seeking correlations the core electron chemical shift should preferably be connected with substituent groups exerting a magnetically isotropic effect on the nuclear magnetic resonating atom. Correlations may be expected if the immediate environment of the resonating atom does not vary much and its hybridization remains constant. If the resonating atom is not a terminal one, the variation of only one of the substituents is likely to yield correlations, and in aromatic systems correlations may be expected if variations involve substituents in distant positions.

In Mössbauer spectroscopy, gamma rays that are emitted by the sample in

Table 3.9. Correlations Between NMR and XPS Shifts [262]

Compounds	NMR	XPS	Slope (ppm/eV)	Structure Data	Correlation
$H_{4-x}CHal_x$	$^{13}C\,sp^3$	$\Delta C\,1s$	43	x unrestricted	No general correlation
				x = 1	Good correlation
H_3CX	$^{13}C\,sp^3$	$\Delta C\,1s$		X unrestricted	No general correlation
			43	X = halogen	Good correlation
			24	X = first row element	Correlation
$F-CH_{3-x}F_x$	^{19}F terminal	$\Delta F\,1s$	114	primary subst. C	Smooth, curved
		$\Delta C\,1s$	34	x unrestricted	Correlations
$H-CH_2X$	^1H terminal	$\Delta C\,1s$	1.3	primary subst. C	Slightly curved correlation
⌬C−SX(Me)₃	^{13}C aromatic	$\Delta S\,2p$	4.6	primary subst. S; X = group IV elements	Good correlation
F−⌬−S−	^{19}F terminal aromatic	$\Delta S\,2p$	2.5	para-S subst. of type −SOX, −SO₂X	Correlation
F−⌬−S−S−X⁻⊕	^{19}F terminal aromatic	$\Delta S\,2p$	7.8	para-C subst. X distant variation	Good correlation
F−⌬−C−	^{19}F terminal aromatic	$\Delta C\,1s$	1.6	para-C subst. first order subst. −M subst.	Correlation scattered

transitions from nuclear excited states to the ground state are studied. The sample nuclei are irradiated by a gamma-ray source that contains nuclei of the same kind as that of the sample. Radiation frequency of the gamma-ray source depends on the energy difference E between the nuclear excited and ground states, and it is shifted by both recoil energy E_R and Doppler shift E_D of the nuclei. A relative movement of the radiation source with respect to the sample thus allows a scan of irradiation energy, and any absorption and resonance fluorescence by the sample is detected.

The Coulomb interaction energy between the nucleus and the electron distribution is included in the total electronic and nuclear energy of these excited states. Nuclear transitions are accompanied by small changes in nuclear dimensions, and since the coulomb energy is dependent on the overlap of the nuclear and electron charge densities at the nucleus, this change in nuclear dimension influences the energy of the emitted gamma radiation. Direct valence electron contribution to the electron charge density at the nucleus happens to be much less than that due to the core electrons. But variations of the valence electron population in different molecules cause, through the core electron charge distribution, minute differences in the gamma radiation frequency. These are detectable in high-resolution Mössbauer spectroscopy and constitute what are known as chemical isomer shifts.

The chemical isomer shift, δ, would therefore depend on both the s-electron charge density ($l \neq 0$ electrons have a node and therefore zero charge density at the nucleus) and the relative change in nuclear dimensions, such that the δ between two compounds A and B (for example, the absorber material) can be expressed as

$$\delta_{AB} = K \{|\psi_s(0)_A|^2 - |\psi_s(0)_B|^2\} \frac{\Delta \langle r^2 \rangle}{\langle r^2 \rangle} \qquad (3.15)$$

where $|\psi_s(0)|^2$ represents the s-electron charge density at the nucleus, the nuclear size factor $\Delta \langle r^2 \rangle / \langle r^2 \rangle$ is the relative change in average of the square of the nuclear radius after deexcitation, and K is a proportionality constant. $\Delta \langle r^2 \rangle$ can be either positive or negative and depends on the isotope and the nature of the transition. From *ab initio* calculations on electron densities at nucleus on a series of molecules it is possible to determine, using Eq. (3.15), both the proportionality constant and the nuclear size factor $\Delta \langle r^2 \rangle / \langle r^2 \rangle$. Because of various complications, however, the latter is not known much better than ± 100%, and this puts a major constraint on the applications of Mössbauer isomer shifts to chemical bonding, although the sign of $\Delta \langle r^2 \rangle$ is now known for most common Mössbauer transitions. Yet the relative population of s and other electrons is not easily discernible from these data because $|\psi_s(0)|^2$ may increase either due to an increase in s electron population or a decrease in p, d, or f populations. The latter effect causes deshielding of the s electrons, which in turn makes the s electron orbitals contract, and it is difficult to separate the extent of these two effects. Knowledge of XPS shifts is particularly useful here in evolving a better insight into the nature of bonding.

A combination of Mössbauer and XPS shifts can be used to make inference about population changes, as shown in Table 3.10. In the table, P_A represents the total electron population on atom A; $P_A(s)$ and $P_A(l \neq s)$ denote respectively the s electron and the p, d, or f electron population on A; + and − denote a

Table 3.10. Correlation of Electron Population with Mössbauer and XPS Shifts [270]

Experimental Data		Interpretation		
ΔE_A	$\delta_A \Delta \langle r^2 \rangle$	ΔP_A	$\Delta P_A(s)$	$\Delta P_A(l \neq s)$
+	+	−	+(s), 0	−
+	0	−	−(s)	−
+	−	−	−	+(s), 0, −(s)
0	+	0	+(s)	−(s)
0	0	0	0	0
0	−	0	−(s)	+(s)
−	+	+	+	+(s), 0, −(s)
−	0	+	+(s)	+
−	−	+	0, −(s)	+

significant increase or decrease, +(s) and −(s) stand for either a small increase or a small decrease respectively, and 0 indicates approximately zero effect. Figure 3.14 gives XPS-Mössbauer shift correlation of some Fe(II) low spin complexes. $\Delta \langle r^2 \rangle$ is negative for the Fe^{57} (14.4 keV) transition. The negative slope of the lines in Fig. 3.14 therefore implies a concurrent decrease of $\delta \Delta \langle r^2 \rangle$ and the XPS shift ΔE with increase in the Mössbauer shift δ. This corresponds

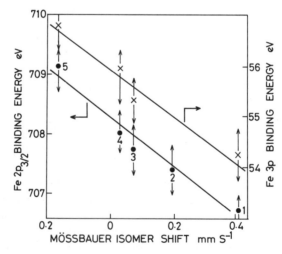

Fig. 3.14. Correlations of Mössbauer shifts with Fe $2p_{3/2}$ and Fe $3p$ XPS shifts. (1) [Fe^{II}(phen)$_3$][ClO$_4$]$_2$. (2) trans-Fe(isocy)$_4$Cl$_2$. (3) trans-Fe(isocy)$_4$(SnCl$_3$)$_2$. (4) [Fe(SnCl$_3$)(isocy)$_5$][ClO$_4$]. (5) Na$_2$Fe(CN)$_5$NO·2H$_2$O (phen = 1,10 phenanthroline, isocy = p-methoxy phenyl isocyanide) [272].

to the case in the last row of Table 3.10 where ΔE as well as the factor $\delta \Delta \langle r^2 \rangle$ are negative, indicating a gradual increase in electron population on the iron atom, contributed primarily by increased π-backbonding from the ligands. The Mössbauer isomer shifts are more amenable to interpretation in terms of ground state electron distribution, on the basis of which linear δ_A versus ΔE_A correlation in a series of compounds should be predictable. It is reasonable to expect that comparisons with XPS shifts would be more meaningful with the availability of more dependable information on nuclear size factors.

APPLICATIONS OF CHEMICAL SHIFTS

We now consider a few applications of core binding energy shifts to chemical problems. Core binding energy shifts have been correlated with Hammett σ constants. We present here results on substituted benzenes, which show that a correlation exists between the Hammett σ constants of para substituents and the C 1s shifts of the ring carbons as well as shifts of the group attached to the para carbon atom. Thus core binding energy shifts can be used to establish linear free energy relationships.

In Fig. 3.15, C 1s binding energies in the parasubstituted series of fluorobenzenes, benzonitriles, and aminobenzenes as well as the monosubstituted

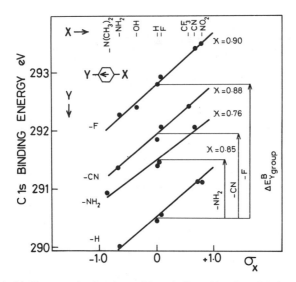

Fig. 3.15. C 1s binding energies for the positions indicated by the arrow in parasubstituted benzene derivatives versus Hammett σ substituent constants. The vertical arrows represent the group shifts of Y [274].

series are plotted against Hammett σ constants of the substituents. A good correlation exists; similarity of the slopes (κ) of the regression lines indicate that the shifts are additive.

If the group shifts of the Y substituents are truly additive, then the following relationship should hold:

$$E^B_{Y-\hexagon-X} = E^B_{\hexagon-X} + \Delta E^B_{Y\,\text{group}} \qquad (3.16)$$

Experimental results show:

$$E^B_{\hexagon-X} = \kappa_0 \sigma_X + E^B_{\hexagon} \qquad (3.17)$$

where κ_0, in the monosubstituted series, has a common value for the meta and para positions. Since the $\Delta E^B_{Y\,\text{group}}$ and $E^B_{\hexagon}$ are constants, one can express their sum as C and write

$$E^B_{Y-\hexagon-X} = \kappa_0 \sigma_X + C \qquad (3.18)$$

Thus, with additive group shifts, equal κ-values for all Y substituents may be expected; experimental results indicate such additivity of group shifts.

The second example involves binding energy shift measurements of vanadium complexes and correlation with their activity in homogeneous catalysis. Experimental V $2p_{3/2}$ binding energies and chemical shifts in some vanadium complexes are presented in Table 3.11. The charge on vanadium is estimated as follows. In vanadium carbide VC, the shift of C $1s$ binding energy with respect to carbon

Table 3.11. Binding Energies and Shifts of V $2p_{3/2}$ [275]

Compound	Binding Energy (eV)	Shift (eV)	q_V
V(V)			
V_2O_5	516.6	4.2	1.01
VO(OH) (oxinate)$_2$	516.7	4.3	1.03
V(IV)			
VOCl$_2$	516.4	4.0	0.96
VO(pc)[a]	516.2	3.8	0.91
VOSO$_4$	515.9	3.5	0.84
VO(oxinate)$_2$	515.5	3.1	0.79
VO(acac)$_2$[b]	515.1	2.7	0.65
V(III)			
V(acac)$_3$	514.2	1.8	0.43
V	512.4	–	0

[a] pc = phthalocyanine. [b] acac = acetylacetonate.

APPLICATIONS OF CHEMICAL SHIFTS 77

in hydrocarbons is $-2.6\,\text{eV}$. From these data and the results of CNDO calculations a charge of $-0.43\,\text{a.u.}$ on the carbon can be derived. This implies a charge of $+0.43\,\text{a.u.}$ on V in VC. The observed binding energy shift of V $2p_{3/2}$ in vanadium carbide, on the other hand, is $+1.8\,\text{eV}$. Assuming a proportionality between the binding energy shift ΔE and the charge q, the ratio $q/\Delta E = +0.43/1.8 = 0.24\,\text{a.u./eV}$ can be used to derive the charge on vanadium in vanadium compounds if the shift in the binding energy of vanadium is known. Charges on the vanadium atom derived in this manner on the basis of V $2p_{3/2}$ shifts are also presented in Table 3.11.

That the electron donating or accepting properties of surrounding ligands do have significant effect on the charge of the central atom in a complex is seen in the data of Table 3.11. One notes a large range in the binding energy shift in V(IV) complexes; the range, in fact, is larger than the difference between the V(IV) complex, which has the highest charge, and the V(V) compounds. The charges on vanadium atoms, however, are much less than their formal oxidation numbers. V(V) charges are about 20% of the formal oxidation state, and in the V(III) complex they are about 13%. Vanadylacetylacetonate V(IV) has the lowest q_V. In autooxidation reactions, if activation of the oxygen molecule is considered, then a low charge on the metal atom, which is equivalent to high electron population, would favor an electron donation from M to O_2, such as $M + O_2 \rightarrow M^{\delta+} + O_2^{\delta-}$. This can be taken as a possible explanation for the capacity of vanadylacetylacetonate(IV) as a catalyst in the autooxidation of primary alcohols to corresponding aldehydes.

Chemical shifts have been effectively employed in the elucidation of the structure of complex compounds; we present here one example on biguanide $H_2NC(=NH)NHC(=NH)NH_2$ complexes. There has been a controversy on certain aspects of the structure of biguanide complexes, the ligand forming two types of complexes, (1) in which the uncharged ligand LH forms complexes of the type $[M(LH)_m]^{m+}$, $m = 2$ or 3, and (2) in which the deprotonated ligand L^- forms complexes $[M(L)_m]$. Older views speculated about the presence of a quarternary nitrogen atom, but recent NMR data indicate its absence in bis (biguanide) nickel II chloride. Ultraviolet absorption studies support the formulation of the following structures for the neutral and charged species, involving π electron delocalization.

$$\begin{array}{cc}
\text{H}_2\text{N}-\text{C}\overset{\text{N}}{\diagup\diagdown}\text{C}-\text{NH}_2 & \text{H}_2\text{N}-\text{C}\overset{\overset{\text{H}}{\text{N}}}{\diagup\diagdown}\text{C}-\text{NH}_2 \\
\text{HN}\diagdown\diagup\text{NH} & \text{HN}\diagdown\hspace{-0.3em}+\hspace{-0.3em}\diagup\text{NH} \\
\text{M/2} & \text{M/2} \\
\text{I} & \text{II}
\end{array}$$

Some of the N $1s$ binding energies of metal biguanides are shown in Table 3.12. The respective linewidths (FWHM) are also shown; they happen to be consider-

Table 3.12. N 1s Binding Energies for Metal Biguanide Complexes and Related Nitrogen Compounds [276]

Compound	Nitrogen 1s Binding Energy(eV)	FWHM(eV)
Biguanide Complexes		
$[Cr(C_2N_5H_7)_3]Cl_3$	399.4	2.8
$[Cr(C_2N_5H_6)_3]H_2O$	399.2	3.0
$[Ag(III)(C_2N_5H_7)_2]_2(SO_4)_3$	400.0	2.4
$Cu(C_2N_5H_6\text{-}p\text{-}C_6H_4SO_3)_2$	400.0	2.7
$Ni(C_2N_5H_6\text{-}p\text{-}C_6H_4SO_3)_2$	400.2	2.7
$[Ni(C_2N_5H_6)_2]H_2O$	398.9	2.7
Other Nitrogen Compounds		
$C_4H_9NH_2$	398.1	—
$(n\text{-}C_4H_9)_3N$	398.1	—
C_6H_4N	398.0	—
C_6H_5CN	398.4	—
$[(CH_3)_4N]^+Cl^-$	401.5	—
$(NH_4)_2SO_4$	401.5	2.0
NH_4Cl	401.0	1.8
NH_4NO_3		
NO_3^-	405.7	1.8
NH_4^+	402.0	1.8

ably larger than N 1s widths found in compounds containing a single type of nitrogen atom, which are about 1.9 eV. The rather large linewidths in the complexes may be taken to indicate a composite spectral structure due to chemically shifted peaks of different types of nitrogen atoms. For chromium trisbiguanide, a best fit through a Gaussian model is obtained if two 1.9 eV peaks are taken separated by 1.5 eV with relative intensities of 2:3. Table 3.12 also shows that the N 1s binding energy in amines, nitriles, and pyridine is in the 398–398.5 eV range and is considerably less than those of simple ammonium salts, where the values are above 401 eV. In fact, it is possible to calculate the expected linewidth assuming that a quarternary atom is present. Using a conservative low value of 1.6 eV for the chemical shift between $-NH_3^+$ and $-NH_2$, a separation of 1.5 eV as observed in the spectrum of chromium trisbiguanide, and expected relative intensities of three types of nitrogen atoms as 1:2:2, a computer simulation predicts an N 1s FWHM linewidth of 3.8 eV when a quarternary nitrogen atom is present. This is much higher than any one of the widths observed in experimental spectra, and this result may be used to rule out the presence of any quarternary nitrogen atom. Further, if π delocalization does not extend to the N atoms outside the chelate region, differences in N 1s binding energy may be expected for the two types of amino nitrogen atoms. That delocalization actually extends to all the nitrogen atoms is indicated by N 1s binding energies of biguanidine sulfate and guanidinium chloride. Salts of biguanide complexes are thus quite well represented by the structure II.

The fourth example of core spectral study we present in this section involves the examination of differences in the core line peak shapes. The study casts light on the nature of backbonding of the axial and equatorial CO groups in the $Fe(CO)_5$ complex.

The valence-bond picture of backbonding between a transition metal and coordinated carbon monoxide is

$$^-M-C\equiv O^+ \longleftrightarrow M=C=O \tag{3.19}$$

This indicates a relationship between the C–O bond strength and the charge on the oxygen atom. Further, in these compounds a linear correlation between the O 1s binding energies and both multiplicity-weighted C–O stretching frequencies and weighted average stretching force constants may be expected. O 1s binding energy differences between structurally different CO groups can be calculated by using the relations between the binding energies and the weighted average force constants together with the force constants of structurally different CO groups. From such data, O 1s binding energy differences between E_B (axial) $-E_B$ (equatorial) have been calculated. For $Fe(CO)_5$, $Mn_2(CO)_{10}$, $Cl_3SiMn(CO)_5$, $HMn(CO)_5$, and $CH_3Mn(CO)_5$, these binding energy differences are, respectively,

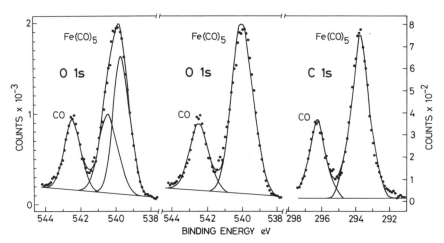

Fig. 3.16. O 1s and C 1s spectra of a mixture of $Fe(CO)_5$ and CO gas. O 1s spectrum at left is composed of two peaks having area ratio 2:3; O 1s spectrum in center is composed of a single peak. C 1s spectrum is at right. Adapted from S. C. Avanzino et al. [277] (copyright 1978, American Chemical Society).

+ 0.4, −0.2, −0.3, −0.4, and −0.6 eV. From the charge potential model, on the other hand, the binding energy difference is given by E_B(axial) $-E_B$(equatorial) $= k\Delta Q + \Delta V$. Taking axial and equatorial oxygen atom charges respectively as −0.360 and −0.396 as obtained from *ab initio* calculations, one obtains an O 1s chemical shift of +0.5 eV for CO groups in Fe(CO)$_5$. Corresponding calculations for carbon atom charges, however, do not show any significant shift of C 1s as also would be expected from the valence bond picture shown above.

Figure 3.16 shows the C 1s spectrum of a mixture of free CO, taken as reference compound, and the complex Fe(CO)$_5$. The C 1s peak of Fe(CO)$_5$ fits quite well (χ^2-test: $\chi^2 = 255$) using a single curve for the Fe(CO)$_5$ band. A slightly better fit ($\chi^2 = 244$) is obtained if two curves having an area ratio 2:3 are used. But as shown by an F-test, this improved fit has a low confidence level of only 92% and is therefore unacceptable. The O 1s peak, on the other hand, if fitted with one curve ($\chi^2 = 553$), shows clear asymmetry. Use of two curves having an area ratio of 2:3 shows an improved fit ($\chi^2 = 210$), and an F-test shows that it is statistically significant and has an acceptable confidence level of 99.999%. The spectral parameters are given in Table 3.13.

The conclusion about the O 1s binding energy difference at axial and equatorial CO positions is consistent with calculations that show significantly different charges for the equatorial and axial oxygen atoms but essentially identical charges for corresponding carbon atoms. That the deconvoluted O 1s peak of higher intensity is of lower binding energy is in agreement with the prediction that the three equatorial oxygen atoms are more negatively charged than the two axial oxygen atoms, or, in other words, the backbonding to the equatorial CO groups is greater than to the axial CO groups. This conclusion is consistent with the rule that in trigonal bipyramid d^8 complexes, strong acceptor ligands preferentially coordinate at equatorial positions.

Table 3.13. Carbon and Oxygen Core Binding Energies for Fe(CO)$_5$ [277]

	E_B (eV)	FWHM (eV)
C 1s		
Fe(CO)$_5$	293.72	1.17
CO reference	296.24	1.05
O 1s		
Fe(CO)$_5$ (treated as 1 peak)	540.02	1.42
Fe(CO)$_5$ (treated as 2 peaks	540.50	1.27
with area ratio 2:3)	539.71	1.09
CO reference	542.57	1.10

SHAKE-UP, SHAKE-OFF PROCESSES

A somewhat different aspect of core electron spectra deals with the occurrence of satellite lines that are observable in high resolution near the main core electron line, as shown in Fig. 3.17. A few high energy lines that appear in the neon spectrum shown in the figure are due to energy inhomogeneity of the incident X-rays having components of higher energy photons. Some of the lower energy satellite lines are pressure dependent. These arise because of collisions between ejected photoelectrons and neutral atoms and the consequent excitations of the latter resulting in an equivalent energy loss of the photoelectrons. The lines that are pressure *independent* have their origin in the ionization process itself. These satellite lines are due to valence electron excitation, which occur concurrently with the ejection of photoelectrons; the lines are thus called the *shake-up* lines. Depending on the extent to which the valence electrons are excited, the photoelectrons are ejected with that much less energy; hence, the kinetic energy spectrum of the photoelectrons have peaks at correspondingly lower energies from the main photoelectron line. An extreme situation of this shake-up process is the complete loss of a valence electron by ionization, and this is called the *shake-off* process. Since this loss of the valence electron generates a doubly charged ion in the ionization continuum, the photoelectron spectrum shows a broad continuous band due to the shake-off process.

In the Ne 1s high resolution spectrum shown in Fig. 3.17, the line 0 is the

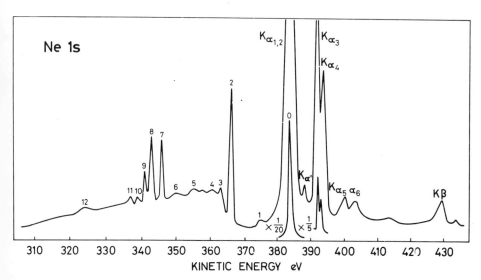

Fig. 3.17. Satellite lines in Ne 1s high-resolution core electron spectrum [4, 35, 36].

Table 3.14. Satellite Lines on the Low Kinetic Energy Side of Ne 1s [4]

Line	Relative Energy (eV) with Respect to the Main Peak	Process	Interpretation
0	0	Single-hole ionization	Ne 1s (Mg $K_{\alpha 1,2}$)
1	− 7.8	$K_{\alpha 3,4}$ satellite	to 2
2	− 16.8	Energy loss	$2p \to 3s$
3	− 20.0	Energy loss	$2p \to 3d, 4s$
4	− 22.0	Energy loss	$2p \to ns, nd$
5	− 28.0	$K_{\alpha 3,4}$ satellite	to 7
6	− 33.0	$K_{\alpha 3,4}$ satellite	to 8 + 9
7	− 37.3	Shake-up	$2p \to 3p$ ^{2}S lower
8	− 40.7	Shake-up	$2p \to 3p$ ^{2}S upper
9	− 42.3	Shake-up	$2p \to 4p$ ^{2}S lower
10	− 44.2	Shake-up	$2p \to 5p$ ^{2}S lower
11	− 46.4	Shake-up	$2p \to 4p$ ^{2}S upper
12	− 60.0	Shake-up	$2p \to 3s$ ^{2}S lower

NOTE: Besides the lines reported above, in the above energy range, a few more lines have been observed [36] with relative energies (with respect to the main peak) of − 33.35 ($2p \to 3s$ ^{2}P lower), − 45.10 ($2p \to 6p$ ^{2}S lower), and − 48.47 eV ($2p \to 5p$ ^{2}S upper).

main photoelectron line; 1, 5, and 6 are satellite lines due to other X-ray lines than MgK_α; and lines 2, 3, and 4 are pressure dependent lines that arise because of collisions with neutral atoms. Line 1 is a satellite of line 2, and lines 5 and 6 are satellites to line 7 and lines 8 + 9 respectively. Lines 7, 8, 9, 10, 11, and 12 are due to the shake-up process with transitions shown in Table 3.14 and also schematically in Fig. 3.18.

Electron shake-up lines are always present, in high resolution core electron spectra, for atoms as well as molecules. Correlation of the shake-up lines with the actual processes and the transitions involved requires accurate knowledge of the energy levels of the ion. Since the shake-up and shake-off are relaxation processes, a study of shake-up and shake-off spectra can lead to detailed infor-

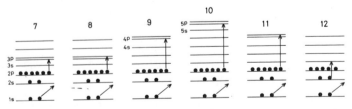

Fig. 3.18. Schematic description of Ne 1s shake-up processes [4, 36].

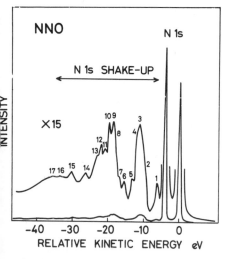 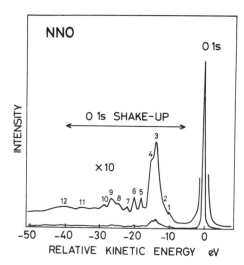

Fig. 3.19. N 1s and O 1s shake-up lines in the core electron spectra of nitrous oxide [35, 36].

mation about relaxation phenomena in atoms and molecules. Several elaborate analyses of shake-up states of atoms have been made. Similar studies have been undertaken for molecules; Fig. 3.19 shows the shake-up lines associated with N 1s and O 1s in the photoelectron spectra of N_2O.

VIBRATIONAL BROADENING OF CORE LINES

The widths of core electron spectral lines provide additional information on processes that take place during photoionization. The lifetimes of excited states are determined by probabilities of several types of electronic deexcitation transitions, such as transitions involving X-ray fluorescence, Auger transitions, and *Coster–Kronig transitions*, the last arising when the primary vacancy created by photoionization is filled by an electron from the same shell and the excess energy thus available is given to an electron in an outer shell. Since it is likely that chemical effects in molecules would influence transition rates through the valence electron population, the photoelectron lines from core electron levels, particularly those next to valence shells, may be expected to show *lifetime broadening*. For example, 1s levels should be affected most for the first row elements, and 2s, 2p for the second row elements. In such lifetime broadening, the peak shapes are symmetric.

There is another kind of broadening effect that has similarities with the electron shake-up. If the photoionization process of a molecule leads to vibrational excitation of the nascent molecular ion formed, one may expect the appearance of a line, to the low kinetic energy side of the main line, similar to the shake-up lines due to electronic excitation. But unlike the electronic shake-up lines, this line is expected to appear very close to the main photoelectron line because the vibrational energy differences are much less than the energy differences of electronic levels. If several such vibrational peaks appear close to the main line, their intensities would be dictated by the relevant Franck–Condon factors between the molecular and the molecular ionic potential energy curves. Further, one may expect that multiple vibrational shake-up peaks appearing close to the main peak will progressively be of lower amplitude, on the low kinetic energy side. If the resolution of the energy analyzer is very high and the ionizing photons are sufficiently monochromatic, all such lines should be separately identifiable. If not, a composite peak would be observed with an asymmetry towards the low kinetic energy side, that is, to the high binding energy side, of the main peak.

Experimental photoelectron spectra of C 1s from CH_4 and Ne 1s are shown in Fig. 3.20. Neon gives a symmetric line typical of Lorentzian lifetime broadening, whereas there is marked asymmetry in CH_4 C 1s. The relaxation following the C 1s ionization involves both a contraction of the orbitals around the core hole atom and a net flow of electron density to the latter. Such redistribution of valence electrons usually results in decreased bond length and a narrower

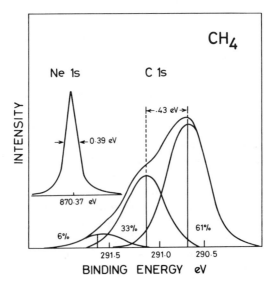

Fig. 3.20. Vibrational broadening of the C 1s line in the core electron spectrum of methane [36].

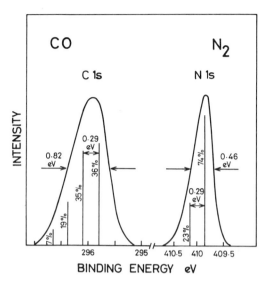

Fig. 3.21. Vibrational broadening of the C 1s line from carbon monoxide and N 1s from molecular nitrogen [36].

potential curve for the ionic ground state. Configuration interaction calculations for the C^*H_4 ion, the asterisk representing a C 1s hole, yields an equilibrium C–H internuclear distance which is 0.005 nm shorter than the C–H distance in the molecule indicating a shift of the potential energy surface of the molecular ion. This can explain the origin of two additional vibrational shake-up bands. Since the C 1s hole is symmetrically located, the C–H stretching mode is excited. The fundamental energy for the symmetric stretching is 0.07 eV larger in the ion than in the neutral molecule, which means a deeper potential energy trough for the molecular ionic state. Using numerical values of the Franck-Condon factors, the asymmetric C 1s peak can be quantitatively explained with respectively 61, 33, and 6% contributions due to population of the first three vibrational states of the molecular ion.

The C 1s line from CO and the N 1s line from N_2 are shown in Fig. 3.21. Note the striking difference in the peak width. In both C^*O and NN^* the electron distribution may be represented by NO^+ on the basis of equivalent core approximation. In the ground state, the CO bond distance is 0.0066 nm greater and the NN^+ bond distance 0.0036 nm greater than that of NO^+. But this small difference is quite adequate for exciting many more molecular ion vibrations in CO, hence a much wider peak. This occurs because the *vibrational broadening* is approximately quadratically dependent on the relative displacement of the potential curves. For the polyatomic molecules SF_6, CF_4, and CCl_4, whether the equilibrium internuclear distance changes or not depends on whether or not bonding electrons are involved in molecular flow relaxation processes. In all these molecules, subsequent to the core electron ejection, molecular flow relaxation originates from the electrons in numerous nonbonding ligand orbitals. With large coordination numbers, the effect per orbital is negligible and bonding electrons are not involved, hence the potential energy surface does not change substantially. As a result, a minimum of vibrational shake-up occurs. In the case of CH_4, molecular relaxation directly involves the bonding electrons, the potential energy surface shifts, and the vibrational excitation becomes important. By the same count, a terminal atom in a molecule is more likely to undergo change in bond length as only one bond is involved. Thus in NNO, due to vibrational broadening the core electron line width for the terminal nitrogen is substantially wider than that of the central nitrogen atom. Similar considerations apply to ammonium nitrate; the N $1s$ line from the ammonium radical is wider than that from the nitrate group.

COLLECTIVE RESONANCES

Core electron spectra are often influenced by *collective resonances*. Figure 3.22 shows the $4p$ energy region of xenon. Normally one would expect two photoelectron lines to appear as a result of $4p_{1/2}$ and $4p_{3/2}$, but actually only one line

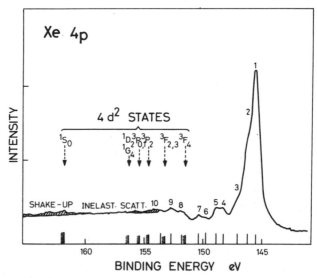

Fig. 3.22. Collective resonance effects in the core electron spectrum of xenon [36].

is observed with several components. The lines that are to the left of the main peak and contain broad maxima are not normal shake-up lines and are not due to inelastic scattering. This part of the spectrum contains a large number of weak discrete narrow lines due to $^2P_{3/2}$ states and a wide range of intense continuum. To locate the cause of such unusual structure we need to look into the other regions of the xenon spectrum.

The prominent peaks in the photoelectron spectrum of xenon up to about 170 eV of binding energy consists of the $5p_{1/2,3/2}$ doublet at 12 eV, the $5s_{1/2}$ peak at 23 eV, the $4d_{3/2,5/2}$ doublet at about 67 eV, the shake-up and inelastic scattering lines of the $4d_{3/2,5/2}$ peaks with an onset at about 75 eV, and the $4p$ structure at 145 eV. In order to explore the role of the normal shake-up, shake-off processes and also that of inelastic scattering of photoelectrons associated with the 145 eV peak, we first examine the satellite structure associated with the $4d_{3/2,5/2}$ peaks of the spectrum since the inelastic scattering and normal shake-up processes in the $4d$ and $4p$ ionizations may be expected to be similar.

The inelastic scattering peaks can be interpreted on the basis of the energies of optical transistions in Xe I. The lowest energy dipole transition $5p_{3/2} \rightarrow 6s$ $^2P_{1/2}$ has an energy of 8.45 eV; several other transitions with higher energies follow. At the given pressure the relative intensity of the inelastic scattering lines in the spectrum with respect to that of the main line is less than 1%. The electron shake-up region, which may be expected to have similarities with the Cs II optical spectrum, starts at 16 eV higher binding energy than the $4d_{5/2}$ main line. The first shake-up line, which involves the transition $5p_{3/2} \rightarrow 6p_{3/2}$ $^2D_{5/2}$, is the strongest one and has a relative intensity of 2%. A shoulder which appears at 16.5 eV is most likely due to the $5p_{3/2} \rightarrow 6p_{3/2}$ $^2D_{5/2}$ upper transition. More shake-up lines on the higher energy side in the range 16–19 eV follow and the shake-up lines of the $4d_{3/2}$ and $4d_{5/2}$ peaks overlap. In the Xe $3d$ ionization peaks, however, the two shake-up spectra do not overlap because of the larger (12.67 eV) spin-orbit splitting of the main lines ($3d_{5/2}$ and $3d_{3/2}$ appear at about 677 eV and 689 eV respectively). Here the first two shake-up lines involving the transitions $5p_{3/2} \rightarrow 6p_{3/2}$ and $5p_{1/2} \rightarrow 6p_{1/2}$ appear at 16.47 eV and 17.72 eV with relative intensities of 3.8% and 2.6% respectively; the corresponding Cs II energies are 16.50 eV and 17.91 eV. The next two lines are at 19.44 eV and 21 eV which correspond to the Cs II transitions at 19.55 eV and 20.8 eV. From the above considerations we note that in the $4p$ satellite the smallest excitation energy that may be expected is 8.45 eV and an inelastic scattering spectrum of low intensity; the first and the strongest shake-up transition should be at an excitation energy of about 16 eV. Further, on the basis of transitions from Cs II levels, the shake-up continua should start at 25–27 eV from the main line with a combined maximum intensity much smaller than the strongest shake-up line. The actual satellite structure immediately following the $4p$ peak at 145 eV, however, does not conform to any of the above expectations based on normal shake-up and inelastic scattering processes. Occurrence of an alternative physical process, along the lines presented below, is indicated.

We have earlier described the nature of ordinary Coster–Kronig transitions. When the available energy is so large that it is possible not only to excite, but also to eject an electron from the same shell, it constitutes a *super Coster–Kronig transition*. The super Coster–Kronig electron and the primary electron leave the ion practically simultaneously; this gives rise to a collective resonance interaction involving the two outgoing electrons and the hole state. As a result, the two ejected electrons share the available kinetic energy and the photoelectron spectrum shows a continuum instead of sharp lines, as happens in the case of xenon.

At an energy slightly higher than 145.51 eV, that is, at the binding energy of the main peak, super Coster–Kronig transitions involving two $4d$ electrons just become energetically feasible. This can be seen from the following table of the binding energies of the final doubly ionized $4d^2$ states — states obtained by loss of two $4d$ electrons, one filling the core hole and the other getting ejected. The energies required to form the $4d^2$ states are derived from $M_{4,5}N_{4,5}N_{4,5}$ Auger energies and the corresponding $M_{4,5}$ binding energies, and they are presented along with the respective spectroscopic term values of the doubly ionized states.

$4d^2$ term	$^3F_4$	$^3F_{2,3}$	$^3P_2$	$^3P_{0,1}$	$^1D_2 + {}^1G_4$	$^1S_0$
Binding energy, eV	151.5	153.2	154.6	155.4	156.3	161.8

The first super Coster–Kronig transition to form ionized $4d^2(^3F_4)$ when initiated by the loss of a Xe $3d$ electron requires about 6 eV higher excitation energy than is available from the singly ionized state that corresponds to the main peak in the $4p$ spectrum. This lowest energy limit of the $4d^2$ super Coster–Kronig transitions falls at the point where the spectrum shows the onset to the continuum. Since the spin–orbit splitting between the $4p_{1/2}$ and $4p_{3/2}$ levels is about 12.6 eV, the singly ionized higher binding energy $4p_{1/2}$ state should fall above most of the doubly ionized $4d^2$ states given in the table above. The origin of the observed phenomenon in the Xe $4p$ spectrum is thus due to a dynamic process. Calculations based on many-body theory explain the anomalous shape of the absorption spectrum above the $4d$ threshold in xenon and barium, and they also show evidence of strong collective resonances. Explanation of the discrete part of the spectrum requires better knowledge on the energy levels, and this can come from relativistic configuration interaction calculations.

CHAPTER

4

VALENCE ELECTRON SPECTRA

If core electron spectra are valuable for characterizing the chemical species present in a sample and the electronic surroundings of an atom, the valence electron spectra that involve electrons of lower binding energy are particularly useful for exploring the nature of bonding in molecules. They comprise direct sources of information on outer orbital energies and provide a powerful means of testing theoretical calculations of molecular orbital energetics. Besides the elements, the valence electron spectra of numerous compounds have been studied. Discussion of every individual spectrum and the information derived therefrom is beyond the scope of this chapter. Yet the subject has not reached a stage where the principles of interpretation of all ultraviolet photoelectron spectra can be summarized with just a few generalizations. The assignment of spectral bands to correct orbitals is often an arduous task requiring simultaneous consideration of experimental and theoretical data. Here spectra of a few sample molecules are presented to indicate some of the characteristics of valence electron spectra and the nature of information and correlations that can be derived. We start with spectral assignments of some simple diatomic molecules like N_2 and O_2 and then proceed to polyatomic molecules.

SPECTRA OF SMALL MOLECULES

Photoelectron spectra of N_2 obtained by using HeI radiation as well as those obtained by using MgK_α radiation are shown in Fig. 4.1. On the basis of N_2 molecular orbitals, the assignment of peaks in the XPS and UPS spectra are indicated in the figure. Note that the cross sections of photoionization of various levels are markedly different. The source of this difference is discussed in Chapter 6. Here we can only state that the cross section for the $2s$ subshell is about an order of magnitude larger than that of $2p$; this is known from studies of Ne. If this knowledge is applied to the N_2 MgK_α spectrum, one is led to the conclusion that the σ_u $2s$ orbital has the largest $2s$ character, followed

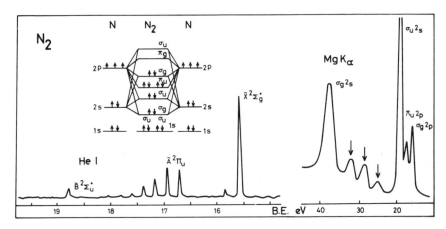

Fig. 4.1. Valence electron spectra of nitrogen using HeI and Mg K_α sources [4, 5].

by $\sigma_g\,2s$ and $\sigma_g\,2p$. For symmetry reasons, the $\pi_u\,2p$ orbitals do not have any s character. There is significant broadening in the $\sigma_g\,2s$ peak, probably due to some vibrational structure and Coster–Kronig processes, which partly explains the disparities in the apparent cross sections of this state with respect to that of the $\sigma_u\,2s$ state. The three small peaks intermediate between the $\sigma_g\,2s$ and $\sigma_u\,2s$, marked with arrows, may be interpreted as follows: The peak at 25 eV results from shake-up of the $\sigma_u\,2s$ state, the peak at 29 eV is due to Mg $K_{\alpha 3,4}$ and the one at 32 eV is due to a characteristic energy loss peak of the $\sigma_u\,2s$ line. It was through photoelectron spectroscopy that the $\sigma_g\,2s$ state was first experimentally observed. The detailed vibrational structure obtained in the low energy HeI photoelectron spectrum shows a small decrease in the vibrational spacing compared with the ground molecular state for the $\sigma_g\,2p$ band ($\tilde{X}^2\Sigma_g^+$), thereby indicating that it is almost nonbonding. The $\pi_u\,2p$ band ($\tilde{A}^2\Pi_u$) shows a large decrease in vibrational spacing; its long series length also indicates that it is a bonding orbital. For $\sigma_u\,2s$ ($\tilde{B}^2\Sigma_u^+$), a slight increase in the vibrational spacing indicates that it is weakly antibonding or an almost nonbonding orbital.

Valence shell photoelectron spectra of molecular oxygen obtained using MgK_α and HeI are shown in Fig. 4.2. When an electron is removed from the partly occupied $\pi_g\,2p$ orbital, a doublet term is obtained; this is observed experimentally in the MgK_α spectrum. It seems that the doublet state of the $\pi_u\,2p$ orbital overlaps with the quartet state of the $\sigma_g\,2p$ orbital. The lines A, B, and C are due to characteristic energy loss and shake-up processes. The intensities in the spectrum indicate that the $\sigma_u\,2s$ orbital has mostly an O $2s$ character and the $\sigma_g\,2p$ has the least O $2s$ character.

The oxygen photoelectron spectrum shows at least four bands when HeI

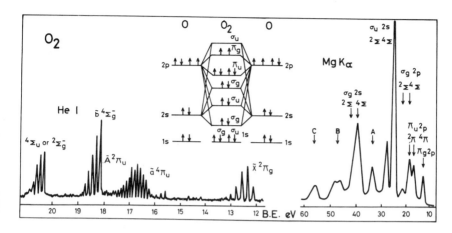

Fig. 4.2. Valence electron spectra of oxygen using HeI and Mg K_α sources [4, 5].

radiation is used. One expects the following O_2^+ states: $^2\Pi_g$ when a $\pi_g\,2p$ electron is removed, $^4\Pi_u$, $^4\Sigma_g$, and $^4\Sigma_g$ when an electron having its spin antiparallel to those in the $\pi_g\,2p$ orbital is removed in turn from the remaining valence shell orbitals, and $^2\Pi_u$, $^2\Sigma_g$, $^2\Sigma_u$, $^2\Sigma_g$ states formed by the removal of an electron of parallel spin. The band corresponding to the $^2\Pi_g$ state shows an increase in vibrational frequency compared with the ground molecular state and hence indicates some antibonding character of the $\pi_g\,2p$ orbital. The band involving removal of $\pi_u\,2p$ electron has strong bonding character and leaves the ion in the $^4\Pi_u$ state. The $^2\Pi_u$ state also falls in this region, with a vertical ionization potential of 17.7 eV. The band having adiabatic ionization potential 18.17 eV is due to removal of a $\sigma_g\,2p$ electron leaving the ion in its $\tilde{b}\,^4\Sigma_g^-$ state; a decrease in vibrational frequency indicates that it has some bonding character. The band with an adiabatic ionization potential of 20.29 eV corresponds to a Rydberg series limit at 20.308 eV and the dissociation limit of the $\tilde{b}\,^4\Sigma_g^-$ state, and this probably is the cause of the observed broadening.

The molecular orbitals of H_2O can be represented by $(O1s)^2(1a_1)^2(1b_2)^2(2a_1)^2(1b_1)^2$. The $1s$ orbitals of the two hydrogen atoms combine to give one even and one odd level which, with oxygen $2s$ and $2p$ orbitals, combine to give the molecular orbitals of H_2O. The HeI and MgK_α photoelectron spectra of water vapor are shown in Fig. 4.3; adiabatic ionization potentials, vibrational frequencies are presented in Table 4.1. The $1a_1$ orbital, due to strong O $2s$–H $1s$ bonding, possesses about 75% O $2s$ character according to CNDO calculations, and this probably explains its high intensity. The $1b_1$ orbital (oxygen $2p\pi$ lone-pair) peak, the narrowest, does not show any vibrational broadening.

The first band is due to removal of an essentially nonbonding electron $1b_1$;

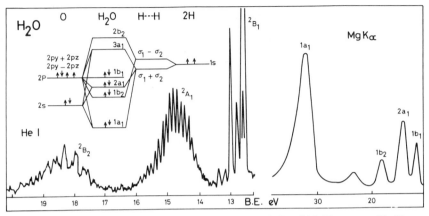

Fig. 4.3. Valence electron spectra of water using HeI and Mg K_α sources [4, 5].

this is associated with negligible change in molecular dimensions. Both the length of the series and the decrease in vibrational frequency of the band at 13.7 eV indicate that it involves removal of a strongly bonding orbital, leaving the ion in the $^2A_1$ state. A comparison with the molecular state shows that the vibrational frequency associated with the band is of the bending mode. A comparison with D_2O UPS indicates that the first few vibrational levels are smeared out and points towards some nonlinearity in the structure of the $^2A_1$ state that resembles NH_2. The 17.22 eV band shows broadened peaks, the first few sharper than the rest, indicating a short lifetime for the ionic state. Compared with the ground state frequency, the symmetric stretch frequency ν_1 decreases, whereas the bending mode ν_2 increases. This is consistent with an electron removal from an orbital having O–H bonding character and H–H antibonding character.

Table 4.1. Adiabatic Ionization Potentials and Vibrational Frequencies of Water [5]

		Ionization Potential to Form H_2O^+	ν_1 (Symmetric Stretch) cm^{-1}	ν_2 (Bending) cm^{-1}
$^1A_1$	H_2O		3652	1595
$^1B_1$	H_2O^+	12.61 eV	3200 ± 50	1380 ± 50
$^2A_1$	H_2O^+	13.7		975 ± 50
$^2B_2$	H_2O^+	17.22	2990 ± 100	1610 ± 100

SPECTRAL ASSIGNMENT OF LARGE MOLECULES

These spectral assignments for small molecules, and the derivation of bond properties using a molecular orbital energy level diagram, might appear quite straightforward. Surely of more interest would be an inquiry about general rules that could be applied, step by step, to completely assign ultraviolet photoelectron spectra of large molecules, both organic and inorganic. Unfortunately, as we have stated, no such simple procedure exists, and a series of related studies needs to be carried out before a valence electron spectrum can be completely assigned. In our sojourn through this chapter we shall encounter applications of molecular orbitals at almost every step. Depending on symmetry properties, molecular orbitals of one group of molecules may differ drastically from another and the simple extension to other groups of the results of investigating one group may not always be possible. Molecular orbital methods constitute the basic foundation of valence spectral interpretation; if one wishes to use such spectra for deriving information about molecular electronic structure, knowledge of the molecular orbital methods is indispensable. Here, by means of a fairly large collection of case studies, we attempt to provide an overview of the range as well as the methodology of using photoelectron spectral data to derive chemical information. In the sections that follow, we focus our attention particularly on the spectra of large organic molecules and their applications. At the conclusion of the chapter we discuss the valence electron spectra of some inorganic molecules.

The valence electron spectrum of even a moderate-sized organic molecule appears distinctly different from the valence electron spectrum of the small molecules shown in the preceding section. There are striking differences. Although there are far more bands in an organic molecule spectrum, they show a near total absence of vibrational structure. This paucity of structure is very rarely a reflection of the width of the exciting line, since the width of the exciting radiation is usually far smaller than the width between the vibrational levels. Lack of vibrational structure, however, may be expected if one looks into the complex processes that are involved in the photoionization of an organic molecule.

First, most organic molecules have low symmetry, C_1, C_2, or C_s. As a result, many electronic configurations that lie in a given energy range will mix. Second, the process of photoionization is associated with excitation of one or more of the $3n-6$ normal modes of vibration that the molecular ion possesses. All the spectra that involve excitation of various normal modes differ from one another only marginally, and the result is obliteration of vibrational fine structure. Further, owing to the large number of valence electrons present in an organic molecule, a large number of radical cation states are formed, each yielding a set of photoelectron bands, and overlapping of different sets of bands takes place. Conformers and valence isomers also lead to slightly different photoelectron spectra, and the overall experimental spectra may contain overlaps of several such individual spectra. Additional complications may arise with any fragmentation or isomerization of the sample molecules occurring in the ionization process, and, as a consequence, the photoelectron spectrum of a sample may consist of contributions from extraneous molecules. As a result, some of the details of electronic configuration and structural information are lost.

In the analysis of photoelectron spectra of large molecules, spectra that are devoid of vibrational structures and are convoluted with overlapping bands cause considerable difficulties. Interestingly, with such spectra it is possible to perturb the molecule slightly by substituents, study the consequence of the perturbations on orbital energetics, and evolve correlation diagrams. This is one of the most useful aspects of photoelectron spectroscopic studies of organic molecules; much useful chemical information can be obtained by studying the spectra of closely related compounds. Basic to such studies are the molecular orbital models used to estimate orbital energetics.

Besides the results of theoretical studies of molecular orbitals, experimental studies of photoelectron band intensities and angular distribution measurements of photoelectrons also help in the interpretation of valence electron spectra. Wherever vibrational structures are observed, the nature of changes in vibrational frequencies in ionization can indicate the type of orbital from which an electron is ejected. Study of the effects of certain types of substituents on spectra also assist in band assignments. In addition, the study of charge-exchange mass spectra can aid in the identification of photoelectron spectral bands. Bond rupture and the production of fragment ions take place at energies where the electrons lost are strongly bonding. Knowledge of the energy needed to generate the fragment ions can be used to identify bands in photoelectron spectra that correspond to certain bonding electrons. Similarly, Rydberg spectra can be helpful in the assignment of photoelectron spectral bands. But before we consider various experimental approaches to the interpretation of valence electron spectra, we should briefly discuss the role of the molecular orbital models.

ROLE OF MOLECULAR ORBITAL MODELS

The first point to comprehend here is the relationship between orbital energies and experimental ionization potentials. One can write a single determinantal wave function representation for the neutral molecule consisting of ψ_j's, which are one-electron SCF spin orbitals obtained by minimizing the total energy of the neutral. This energy minimization is achieved by solving the Hartree–Fock equations $\mathscr{F}\psi_j = \Sigma_i \lambda_{ij} \psi_i$. If this set of ψ_j's is subjected to a unitary transformation, another set of ψ_j's can be obtained, $[\{\psi_j'\} = \mathbf{U}\{\psi_j\}]$, but in such a transformation the total wave function remains unchanged. If, however, the unitary transformation employed is such that the λ_{ij}'s obtained by operating $\mathscr{F}$ on the ψ_j's derived through the unitary transformation become a diagonal matrix, then these special ψ_j's — let us call them ϕ_j's — have the property that $\mathscr{F}\phi_j = \lambda_{jj}\phi_j = \epsilon_j\phi_j$. These ϵ_j's are referred to as orbital energies, and the ϕ_j's are called the canonical orbitals. If one makes the further simplifying assumption that ionic orbitals denoted by $\tilde{\psi}_j$ and neutral orbitals ϕ_j are the same, $\tilde{\psi}_j \equiv \phi_j$, ionization is then equivalent to the loss of just one electron from the neutral, say from jth orbital; the rest remain unaffected and constitute the ion. Then one derives the result that the vertical ionization potential I_{vj} is equal to the negative of the jth orbital energy of the neutral, $I_{vj} = -\epsilon_j$, which, of course, is Koopmans's theorem. These canonical orbitals ϕ_j's could therefore be looked upon as a convenient basis for understanding ionization energies, if not directly then somehow as constituents of specially constructed orbitals whose energies correspond to the experimental ionization energies, such as those obtained from photoelectron spectra.

For a *true* assignment of a photoelectron spectrum, much basic information is required. One would at least need knowledge of the symmetry groups of the neutral and the molecular ion, vertical and adiabatic ionization potentials for the molecular ion formation, the internal coordinates of the radical cation after relaxation, and transition probabilities for the basic ionization process — including Franck–Condon factors for the different normal modes excited in M^+ and the normal modes of the relaxed molecular ion. Even a small part of this information is rarely available. One is therefore constrained to accept partially satisfactory alternatives, the use of *semiempirical models*. Examples of such models are CNDO, INDO, MINDO, SPINDO, and HAM. The procedure of each is designed to fit a particular property of the molecule. SPINDO and HAM, for example, are designed for the photoelectron band positions. A band assignment derived from one such model thus may be drastically different from that provided by another. Neither may be rejected as untrue — the assignment, in a way, is an effect of the model itself — but one accepts that model which has more desirable characteristics, such as a satisfactory explanation of band intensities, band shapes, spin–orbit splittings, vibrational fine structure, the prediction of the spectra of other molecules, or the prediction of physicochemical properties of the sample molecules from spectral assignments.

One way of comparing different semiempirical models is as follows. One chooses a matrix $\mathbf{U} \equiv \mathbf{L}$ such that the canonical orbitals ϕ_j are transformed into localized orbitals λ_j, that is, $\{\lambda_j\} = \mathbf{L}\{\phi_j\}$, using a criterion such as

$$\sum_j \left\langle \lambda_j(1)\lambda_j(2) \left| \frac{e^2}{r_{12}} \right| \lambda_j(1)\lambda_j(2) \right\rangle = \text{maximum} \qquad (4.1)$$

which seeks the maximum in the coulombic repulsion integral above between the electrons 1 and 2 in the same orbital λ_j; under that condition maximum localization occurs. Using this **L**, one can then form a transformed Hartree–Fock matrix $\mathbf{F}_\lambda$ as follows. $\mathbf{F}_\lambda = (F_{\lambda,ij}) = \mathbf{L}\mathbf{F}_\phi \mathbf{L}^\dagger$, $\mathbf{L}^\dagger$ being the transpose of **L**, and $\mathbf{F}_\phi$ the Hartree–Fock matrix. The matrix elements $F_{\lambda,ij}$, although differing in absolute values from one model to another, show a high degree of transferability from one compound to another in a given semiempirical model. The localized orbitals λ_i's thus derived can now be used as a basis for describing another molecule in an essentially Hückel type of approximation:

$$\chi = \sum_j c_j \lambda_j \qquad \langle \lambda_i | \mathscr{H} | \lambda_i \rangle = A_i \qquad \langle \lambda_i | \mathscr{H} | \lambda_j \rangle = B_{ij} \qquad (4.2)$$

The first equation represents a linear combination of localized orbitals, the second equation represents self-energies of localized orbitals, and the third defines the localized orbital interaction terms. Such models are extensively used in the interpretation of photoelectron spectroscopic data.

With these observations on molecular orbital models, we may now proceed to a consideration of some practical aspects of valence electron spectra, including semiempirical rules usable in the assignment of photoelectron spectral bands. Following that, we consider a case study on the detailed interpretation of the benzene spectrum before taking up the construction of correlation diagrams from photoelectron spectral data.

VIBRATIONAL STRUCTURE, LONE-PAIR PEAKS, AND PERFLUORO EFFECT

We have noted that changes in vibrational frequency of the molecular ion as well as the extent of vibrational progressions provide useful indications about the nature of the orbital from which the electron is lost in photoionization. Removal of a nonbonding electron does not usually cause a change in the internuclear distance, and in such a transition hardly any vibrational progression is observed. The vibrational progressions, on the other hand, are quite long when bonding or antibonding electrons are removed, with an associated change in ionic vibrational frequency. This property of ionic vibration can be effectively used in photoelectron band assignment.

Although devoid of much vibrational structure, nonbonded orbital peaks — for example, the lone-pair peaks — are quite sharp, and this characteristic can sometimes be used to distinguish them from other orbitals. If the energies and symmetries of these lone-pair orbitals are comparable to those of adjacent π orbitals, interaction may take place, resulting in the formation of new molecular orbitals with somewhat higher as well as lower energies compared with that of the initial orbitals. On ionization, then, splitting of the original level takes place and the nonbonding orbital electron peak may show splitting due to this interaction. Photoelectron peaks may also show spin–orbit interaction, which is dependent on the initial state orbital. The extent of this interaction depends on the nuclear charge of the atom. Furthermore, as we saw in Chapter 1, Jahn–Teller splitting also takes place when symmetry of the initial state is destroyed by photoionization.

Certain effects of substitution in molecules often indicate the nature of the orbital from which the photoelectron is lost. For example, in molecules containing planar π systems, if all hydrogen atoms are replaced by fluorine atoms, as happens in the conversion from ethylene to perfluoroethylene or benzene to hexafluorobenzene, the ionization potentials of σ orbitals increase drastically (2–3 eV) whereas those of the π orbitals hardly change. This is because a fluorine atom draws electrons towards it strongly, and a strong inductive effect is caused throughout the molecule. Since there is strong mixing of fluorine σ atomic orbitals and σ molecular orbitals, the result of the inductive effect is to cause stabilization of the σ orbitals and hence substantially more of an increase of binding energy than in the π orbitals. This is called the *perfluoro effect*. (Often the change in ionization potential is a combination of inductive and mesomeric effects; the actual shift in ionization energy then depends on which of the effects dominates.)

SUM RULE

Besides the methods mentioned above, there is also a sum rule that can be expressed as

$$\sum(-I_i) = \sum E_i \qquad (4.3)$$

where I_i's are experimentally observed vertical ionization potentials of the molecular orbitals, and E_i's are characteristic energies of the orbitals from which the electron is lost. E_i's can be obtained from empirical fitting with data of simple molecules. Some E_i's are listed in Table 4.2. In the table, n stands for nonbonding orbital.

One might take the example of methylene halide CH_2X_2 series of compounds to demonstrate the validity of the sum rule. The orbital assignment of the experimentally observed ionization potentials I_1–I_8 of the eight experimentally observed photoelectron bands in CH_2Cl_2, which is supported by CNDO/2 calculations, is given below Table 4.2.

Partial Sum

	b_1			a_1	
	Calculation $(n+\pi)$	Experiment (I_1+I_7)	Calculation $(n+\sigma)$	Experiment (I_2+I_6)	
CH_2Cl_2	27.10	28.17	27.23	27.34	
CH_2Br_2	26.15	26.86	25.35	25.57	
CH_2I_2	25.03	24.92	23.21	23.43	

Partial Sum / Total Sum

	a_2		b_2		Total Sum	
	Calculation (n)	Experiment (I_4)	Calculation $(n+\sigma)$	Experiment (I_3+I_5)	Calculation $(4n+2\sigma+\pi)$	Experiment
CH_2Cl_2	12.78	12.22	27.23	27.52	94.34	95.25
CH_2Br_2	11.83	11.28	25.35	25.40	88.68	89.11
CH_2I_2	10.71	10.50	23.21	22.96	82.16	81.87

OTHER ADDITIVE PROPERTIES: s-TYPE AND p-TYPE BANDS

Table 4.2. E_i Values (eV) [313, 314]

$n(F)$	16.05	$\sigma(C-N)$	14.42
$n(Cl)$	12.78	$\sigma(C-C)$	11.75
$n(Br)$	11.83	$\sigma(N-N)$	14.90
$n(I)$	10.71	$\sigma(O-O)$	16.10
$n(N)$	10.80	$\sigma(O-H)$	16.40
$n(O)$	12.61	$\pi(CH_3)$	14.32
$\sigma(C-F)$	17.14	$\pi(CH_2)$	14.32
$\sigma(C-Cl)$	14.45	$\pi(C=C)$	10.51
$\sigma(C-Br)$	13.52	$\pi(C\equiv C)$	11.40
$\sigma(C-I)$	12.50		

CNDO/2 Results (eV)	Orbital	Experimental Bands (eV)
11.05 (b_1)	$n(Cl)$	11.40 (I_1, I_2)
11.42 (a_1)	$n(Cl)$	
11.89 (b_2)	$n(Cl)$	12.22 (I_3, I_4)
13.15 (a_2)	$n(Cl)$	
15.22 (b_2)	$\sigma(C-Cl)$	15.30 (I_5)
16.74 (a_1)	$\sigma(C-Cl)$	15.94 (I_6)
18.37 (b_1)	$\pi(CH_2)$	16.77 (I_7)
20.83 (a_1)	$3s(Cl)$	20.30 (I_8)

Similar assignments can be made to the photoelectron spectral bands of CH_2Br_2 and CH_2I_2. Partial and total sums, obtained in p-type orbitals of CH_2Cl_2, CH_2Br_2, and CH_2I_2 using the E_i values from Table 4.2, are given on p. 98 and a comparison with the corresponding sums obtained from experimental ionization potentials shows that the sum rule holds for partial as well as total sums.

This sum rule has been found to hold for many types of compounds. Whereas, in general, no unambiguous assignment of orbitals is possible by using this sum rule alone, often the rule is helpful in the overall assignment process, especially for overlapping bands.

OTHER ADDITIVE PROPERTIES: s-TYPE AND p-TYPE BANDS

There have been other applications of photoelectron spectroscopy in exploring the additive behavior of atomic contributions to the binding energies of molecular orbitals of simple hydrocarbons. When binding energies of all the valence orbitals are known, it is possible to check whether the sum of the assigned binding energies of the atomic orbitals associated with a molecule equals the sum of the experimentally determined binding energies of the molecular orbitals.

If such a relationship is established, the atomic orbital contributions can be used effectively to predict total binding energies. Such a method has been tried with a wide range of hydrocarbons. It involves first recognizing two kinds of orbitals, on the basis of which atomic orbital contributions are estimated.

A knowledge of ionization energies of all the valence orbital electrons is helpful for the assignment of orbitals of any molecule, and therefore also of simple hydrocarbons. This permits one to start from the lowest valence shell of the hydrocarbon molecule, where the molecular orbital is an in-phase combination $(s + s)$ of C $2s$ atomic orbitals. Then one can proceed to the $2s$-derived outer orbitals, which contain increasing numbers of nodal planes (such as $s-s$). Along with this, the assignment of orbitals built from combinations of C $2p$ and H $1s$ can be made, which considers, besides in-phase and out-of-phase combinations of the atomic orbitals, the possibilities of other types of overlaps, for example, broadside-on, in-plane, and out-of-plane combinations. These two classes of orbitals, which have been called *s-type* and *p-type* orbitals, have considerable differences in ionization energies. For example, in CH_4, these s-type and p-type bands appear at 23 eV and 13 eV respectively. In polyatomic molecules they spread, by in-phase and out-of-phase combinations of the atomic orbitals, but the two groups of bands remain distinctly separated at least up to hydrocarbons containing C_5 as seen in 30.4 nm HeII spectra of the methane series presented in Fig. 4.4. Such s- and p-type bands are also observed in photoelectron spectra of long chain cycloalkanes.

In the methyl methanes, a group of bands are present around 22 eV. Since bands at about this energy are present whenever there is a carbon atom in the molecule, it is reasonable to associate these bands with combinations of C $2s$ atomic orbitals. With a nitrogen atom in the molecule, bands at around 28 eV

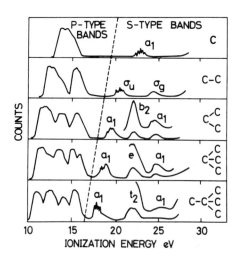

Fig. 4.4. He II spectra of methyl methanes [311].

OTHER ADDITIVE PROPERTIES: s-TYPE AND p-TYPE BANDS 101

are observed that may be assigned to nitrogen $2s$ atomic orbital. Presence of an oxygen, which results in a band at 32 eV, may be associated with the oxygen $2s$ atomic orbital. This may be taken to indicate some atomic properties characteristic of core electron spectra transmitted to valence electron bands. The s-type bands are seen to be broadened by the bonding interactions that are characteristic of valence shell orbitals. In the case of molecules having high skeletal symmetry, assignment of s-type bands are made on the basis of the bonding order expected from the number of nodal surfaces in the molecular orbitals. The s- and p-types of bands can also be identified for molecules containing atoms like N, the halogens, and so on.

According to the Hückel theory, for a complete set of orbitals where overlap between the $2s$ orbital of the carbon atom and the $1s$ of the hydrogen atom, and also overlap of the C $2s$ with the $2p$ of other carbon atoms, is neglected such that the resonance integrals are zero, the summed orbital energies should be equal to the summed coulomb integrals. This implies neglect of the off-diagonal terms in the secular determinant. Separation of the s and p regions would therefore be justifiable if a reasonably straightforward relationship can be shown to exist between ΣE_s (or ΣE_p), the summed experimentally obtained vertical ionization potentials (E_s or E_p) of s-type (or p-type) bands of a molecule, and the sums $\sum_x \alpha_x^s$ (or $\sum_x \alpha_x^p$), where α_x^s (or α_x^p) is the vertical ionization potential of the s-type (or p-type) band of the simple hydride $X H_m$ of the element X. The α_x^s and α_x^p derived from experimental ionization potentials for various elements are shown in Table 4.3. Plots of ΣE_s versus $\sum_x \alpha_x^s$ and ΣE_p versus $\sum_x \alpha_x^p$ are given in Fig. 4.5. It shows good agreement with the simple theory and therefore justifies separation of the bands into two regions. Deviations arise because of neglect of overlap, interaction between s and p orbitals,

Table 4.3. Values of α_x in eV for Some Elements of Groups III, IV, V, and VI [311]

	α_x^s	α_x^p		α_x^s	α_x^p
H	—	(7.99)	P	19.0	13.93
F	39.01	44.01	As	19.0	12.41
Cl	25.7	34.01	Sb	17.3	10.32
Br	24.4	31.26	C in CH_3	22.30	11.09
I	21.7	27.75	C in CH_2	21.53	11.09
O	32.2	29.89	C in CH	20.25	11.09
S	22.2	23.29	C	19.44	11.09
Se	21.0	21.45	Si	18.17	6.50
Te	18.6	18.41	Ge	18.41	5.42
N	27.0	19.69	Sn	16.88	3.20
B	—	5.38			

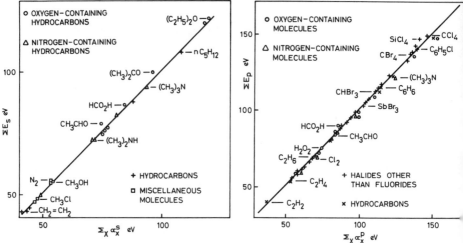

Fig. 4.5. Plots of ΣE_s as a function of $\sum_x \alpha_x^s$, and ΣE_p as a function of $\sum_x \alpha_x^p$ [311].

and any variations of α_x^s (or α_x^p) that occur with chemical environment.

The p-type bands generally fall below 21 eV. As with the s-type orbitals, α_x^p contributions are deduced from the known ionization potentials of simple hydrides of the elements. For example, $\alpha_H^p = 15.98/2$ eV, where 15.98 is the vertical ionization potential of the $(1\sigma_g)^{-1}$ band of H_2^+. Division by 2 appears because the band ionization potential should be equal to the sum of coulomb integrals which are two in number for H_2. The hydrogen 1s atomic orbital is treated as a p-type orbital. The α_{Cl}^p is evaluated as follows. Ionization in HCl starts with the strong double peak at 12.74 and 12.82 ± 0.01 eV. It corresponds with the ionization from $p\pi$ orbital, $(p\pi)^{-1}$, leading to $^2\Pi_{3/2, 1/2}$ states of the HCl$^+$ ion. 16.44 eV is the vertical ionization potential for the $(p\sigma)^{-1}$ band of HCl$^+$. Therefore, $\alpha_{Cl}^p + \alpha_H^p = 12.74 + 12.82 + 16.44$ eV; that is, $\alpha_{Cl}^p = 2 \times 12.78 + 16.44 - \alpha_H^p = 34.01$ eV. The $\Sigma E_p - \Sigma \alpha_x^p$ agreement is best in molecules without fluorine and those without strong multiple bonding. In general, the agreement is found to be good for unsaturated hydrocarbons, including some benzene derivatives.

ASSIGNMENT OF BENZENE PHOTOELECTRON SPECTRUM

In order to have a somewhat better idea about the way one proceeds in interpreting molecular valence electron bands, we now follow the band assignment procedure of a reasonably simple molecule in some detail. To this objective, we consider here certain aspects of the UV photoelectron spectrum of benzene.

Although, as will be seen, the process of assignment is not exactly simple and straightforward, and draws substantially on results of experiments other than photoelectron spectroscopy, it is expected to show several facets of the assignment procedure.

The symmetry orbitals of benzene can be derived on the basis of the irreducible representations of the point group D_{6h}. The orbitals can be classified in terms of the following types: π, t (tangential), r (radial) or s. These are built up from various combinations of C $2s$, $2p_z$, $2p_x$, $2p_y$, and hydrogen orbitals, represented respectively as s, z, r, t, and h. x and y stand for radial and tangential directions respectively, in relation to the center of the benzene molecule. A representation of the orbitals within the range of HeI radiation show, on the basis of their constituent orbitals, the following properties:

Orbital (Approximate Form) [327]	Type	Nature of Bonding
$1e_{1g}$ $(z_1 + 2z_2 + z_3) - (z_4 + 2z_5 + z_6)$ $1e'_{1g}$ $(z_3 + z_4) - (z_6 + z_1)$	π	Weakly C–C bonding
$3e_{2g}$ $(t_1 - t_2) + (t_4 - t_5) + rh_3 + rh_6$ $3e'_{2g}$ $(rh_1 - rh_2 + rh_4 - rh_5 - t_3 - t_6)$	t (with some r contribution)	Weakly C–C bonding, weakly C–H bonding
$1a_{2u}$ Σz	π	Strongly C–C bonding
$3e_{1u}$ $(rh_1 + 2rh_2 + rh_3) - (rh_4 + 2rh_5 + rh_6)$ $3e'_{1u}$ $(rh_3 + rh_4) - (rh_6 + rh_1)$	r	Strongly C–H bonding and C–C nonbonding
$1b_{2u}$ $(t_1 - t_2 + t_3 - t_4 + t_5 - t_6)$	t	Strongly C–C bonding and C–H nonbonding
$2b_{1u}$ $\Sigma (s + rh)(-1)^n$	s	Strongly C–H bonding and C–C antibonding

Orbital (Approximate Form) [327]	Type	Nature of Bonding
$3a_{1g}$ Σrh	r	Strongly C–H bonding and weakly C–C bonding
$2e_{2g}$ $(s_1 + s_2 - 2s_3 + s_4 + s_5 - 2s_6)$ $2e'_{2g}$ $(s_1 - s_2 + s_4 - s_5)$	s	Weakly C–H bonding and weakly C–C antibonding

The nature of bonding depends on the disposition of the orbitals on the carbon and hydrogen atoms. For example, distribution of $1e_{1g}$ constituent orbitals are all z's oriented perpendicular to the benzene ring plane, therefore it will form π-type bonding, and that also is weak C–C bonding because of the nodes present. $1a_{2u}$, however, where there are no nodes, may be expected to form stronger C–C bonding although also of the π type. The orientation of the tangential t orbitals in $1b_{2u}$ indicates strong C–C bonding, but since it does not spread out at all radially, it leads to a CH nonbonding property. Properties of the other orbitals can be similarly derived.

Schematic diagrams of the orbitals are shown below. A He I photoelectron spectrum of benzene with details of vibrational band structure is shown in Fig. 4.6.

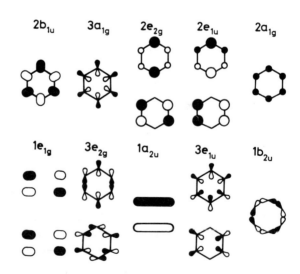

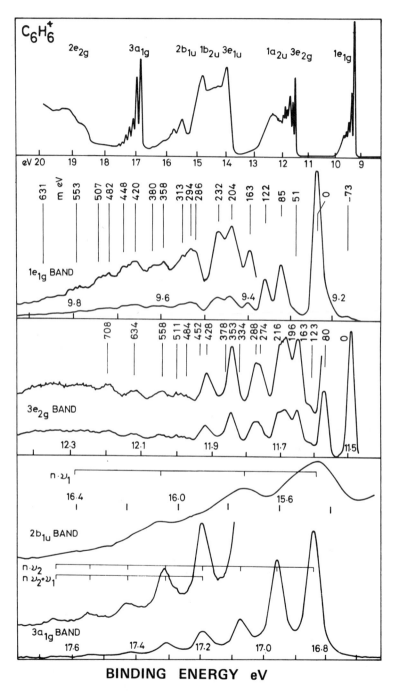

Fig. 4.6. He I photoelectron spectrum of benzene [331].

If the orbitals formed from $1s$ electrons of carbon atoms are neglected, then the orbital structure of benzene and the presently accepted band assignments are as follows:

[327] $2a_{1g}^2$ $2e_{1u}^4$ $2e_{2g}^4$ $3a_{1g}^2$ $2b_{1u}^2$ $1b_{2u}^2$ $3e_{1u}^4$ $1a_{2u}^2$ $3e_{2g}^4$ $1e_{1g}^4$

Orbital type   s   s   s   r   s   t   r π t π

Band assignment   (25.9)   (22.58)   19.2   16.9   15.4   14.7   13.8   12.1   11.4   9.3 eV

The band ionization potentials shown within the parentheses were obtained from a 40.8 eV He II spectrum. We now can describe the assignment procedure.

Theoretical energy estimates of the orbitals constructed from C $2s$ orbitals $2a_{1g}$, $2e_{1u}$, $2e_{2g}$ and $2b_{1u}$ give respectively the following order of magnitude values: 30 eV, 26 eV, 22 eV, and 18 eV. Due to mixing with H orbitals, the energy of the last two orbitals may be expected to somewhat decrease, to about 19 eV and 16 eV respectively. On this basis, the experimental photoelectron spectral bands at 19.2 and 15.4 eV may be assigned to $2e_{2g}$ and $2b_{1u}$ respectively. For molecular orbitals constructed from C $2p$ orbitals, a high ionization potential, of the order of 19 eV, is quite unlikely.

The presence of nodal planes in the orbital can be used to make inferences about relative energies of various orbitals. For example, it is reasonable to assert that the $1e_{1g}(\pi)$ orbital, which has a nodal plane, must be having a higher energy hence a lower ionization potential than that of $1a_{2u}(\pi)$, and similarly the ionization potential of $3e_{1u}(r)$ must be lower than that of $3a_{1g}(r)$. Additionally, one may assume the ionization potential of $3e_{2g}(t)$ to be lower than that of $1b_{2u}(t)$, as has been obtained in some theoretical calculations. Since by all calculations the $1e_{1g}$ orbital is known to have the lowest ionization potential, the 9.3 eV band is therefore assigned to this orbital.

Mass spectrometric fragmentation patterns of benzene can be quite useful in assigning the photoelectron bands. At 13.8 eV energy, the mass spectrometric breakdown pattern shows a rapid increase of $C_6H_5^+$. This scission of the CH bond implies that the photoelectron peak at this energy must be due to an orbital that is strongly CH bonding, that is, an r type, and therefore one concludes that it is either $3e_{1u}$ or $3a_{1g}$. At 14.7 eV, rapid increases of $C_4H_4^+$ and $C_3H_3^+$ are observed in the breakdown curves, therefore, the orbital responsible for the 14.7 eV band must be strongly C–C bonding, either $1a_{2u}(\pi)$, or $1b_{2u}(t)$.

The 9.3 eV, 11.4 eV, and 16.9 eV peaks show vibrational structure corresponding to the ν_2 breathing vibration. The selection rule requirement that in allowed transitions only totally symmetric vibrations are excited implies that the three orbitals responsible for these bands are somewhat C–C bonding or C–C antibonding, which means that they are from $1e_{1g}(\pi)$, $3e_{2g}(t)$, $1a_{2u}(\pi)$, $1b_{2u}(t)$, $2b_{1u}(s)$ or $3a_{1g}(r)$. The 9.3 eV band has already been assigned to $1e_{1g}(\pi)$. The 11.4 eV band cannot be due to $1b_{2u}(t)$ or $3a_{1g}(r)$, as in that case neither the $3e_{2g}(t)$ nor the $3e_{1u}(r)$ orbital can be assigned a lower binding energy band, as concluded from nodal plane considerations. Since as an s-orbital, $2b_{1u}(s)$ is expected to have much higher ionization energy, the 11.4 eV orbital is probably due to either $3e_{2g}(t)$ or $1a_{2u}(\pi)$. The observation that the 13.8 eV band does not have much vibrational structure is an additional reason to believe that it is an essentially C–C nonbonding orbital, $3e_{1u}(r)$ or $3a_{1g}(r)$.

That the 11.4 eV band is most likely due to $3e_{2g}(t)$ rather than $1a_{2u}(\pi)$ receives support from spectral measurements of electron impact and energy loss. In these measurements, monoenergetic electrons introduced in the system cause transitions between energy states in the sample molecules and consequently suffer energy losses themselves. The extent of these losses provides a measure of transition energies in the sample system. Higher excited states of benzene are mostly Rydberg states. Results of theoretical calculations on benzene Rydberg state energies and on allowed transitions from a given molecular orbital permit prediction about where the electron energy loss peaks should be observed. Among the allowed transition of Rydberg states in benzene, there are transitions from $3e_{2g}(t)$ to Rydberg orbital npe_{1u} at 9.3 and 10.4 eV corresponding to $n = 3$ and $n = 4$. These transitions are detected in an electron-impact energy-loss spectrum, and this observation would not fit an assignment of 11.4 eV band to $1a_{2u}$.

The 13.8 eV band could be due to either $3e_{1u}(r)$ or $3a_{1g}(r)$. If an assignment to $3e_{1u}(r)$ is made, then one expects Rydberg transitions to $3sa_{1g}$, $3da_{1g}$, $3de_{1g}$, and $3de_{2g}$ at 11.0 eV, 11.6 eV, 12.3 eV, and 12.2 eV respectively, all of which are actually observed in the electron-impact energy-loss spectrum. Assignment of $3a_{1g}(r)$ to the 13.8 eV band somewhat explains the experimentally observed 11.6 eV and 12.1 eV energy loss peaks due to transitions to Rydberg states $3pe_{1u}$ and $3pa_{2u}$, but it provides no explanation for the 11 eV energy loss peak. This means that the 13.8 eV band cannot be due to $3a_{1g}(r)$ and therefore must be due to the alternative $3e_{1u}(r)$. Similarly, the 16.9 eV peak is interpreted to be due to $3a_{1g}(r)$, and this explains the Rydberg transitions expected at 14.7 and 15.2 eV for the excitations $3a_{1g}(r) \to 3pe_{1u}$ and $3a_{1g}(r) \to 3pe_{2u}$ respectively.

The mass spectrometric fragmentation pattern indicates that the 14.7 eV band is due to strongly C–C bonding electrons, it cannot be assigned to $2b_{1u}(s)$. Since this orbital is formed from C $2s$ orbitals, the 12.1 eV band is of low energy to be assigned to it. A more appropriate band is that with 15.4 eV ionization potential. In that case, the 14.7 eV band must be due to either $1b_{2u}(t)$ or $1a_{2u}(\pi)$. The assignment of the 14.7 eV band to $1b_{2u}(t)$ means that a Rydberg transition to nde_{2g} at 13.1 eV with $n = 3$ may be expected, and this is observed in electron energy loss spectrum. The Rydberg transitions expected for $1a_{2u}(\pi)$ with 14.7 eV are not observed. In that case, the 12.1 eV band must be due to $1a_{2u}(\pi)$. Rydberg transitions at 9.3 eV and 10.5 eV corresponding to this assignment are observed in the electron energy loss spectrum.

Until recently, there was lack of general agreement on the interpretation of some UPS bands of benzene, notably about the 11.4 eV band assignment to $3e_{2g}$. This orbital is of sufficiently σ bonding character that one should expect Jahn–Teller distortion of the ionic potential surfaces if an electron is removed from this orbital; transition to the resulting nondegenerate ionic states then should lead to splitting in the band. The apparent absence of Jahn–Teller splitting in the 11.4 eV peak was a reason for alternative assignments. From vacuum ultraviolet data on a band corresponding to that in 11.4 eV UPS band, it is now clear that Jahn–Teller splitting actually takes place, although the magnitude of splitting is rather small. Thus agreement has been reached on a $3e_{2g}$ assignment of the 11.4 eV band.

Turning our attention to the detailed vibrational structure in the photoelectron spectrum of benzene, we see in Fig. 4.6 that the vibrational structure of the first two bands, due to $1e_{1g}$ and $3e_{2g}$, are very similar in terms of energies and intensities. The pattern shows a strong adiabatic transition followed by short and weak progressions, which is characteristic of little change in molecular geometry consequent to ionization. On this basis, it is reasonable to expect vibrational energies in these ionic states to be similar to those of the neutral molecule, and by matching the observed frequencies of the ion with that of the neutral molecule, appropriate normal modes can be assigned. In the detailed vibrational structure of the $1e_{1g}$ and $3e_{2g}$ bands shown in Fig. 4.6, the energy difference of various peaks with respect to the mean peak are indicated in meV. The diffuse bands above 11.916 eV can be accounted for by adding 350 meV to the bands that appear at lower energies. Since 350 meV happens to be the frequency of ν_1 vibration, it is possible that the bands above 11.916 eV are continuations of a structure observed at lower energies with an additional quantum of ν_1 normal mode. The next band, due to $2b_{1u}$, is characterized by very broad peaks. The spacings are slightly less than the ν_1 frequencies, which indicates the antibonding nature of the orbital. In this band, the increase of interpeak spacing with increasing energy indicates a negative anharmonicity in the molecular potential of the electronic state; this may be expected from the antibonding properties of the $2b_{1u}$ orbital. In the band at 16.841 eV, due to the $3a_{1g}$ orbital, eight vibrational bands have been detected. The uniform spacing of 117 meV suggests that a ν_2 progression is excited. The intensity ratios between the bands, however, deviate from what is expected from a single progression. This is due to the excitation of a single ν_1 vibration having an energy of ~ 350 meV in this electronic state.

CORRELATION DIAGRAM OF ACENES

As an illustration of the role of photoelectron spectroscopy in establishing correlation diagrams, we now take the example of acenes. In acenes the total number of π orbitals is given by $4N + 2$, where N stands for the number of benzene rings. The bonding molecular orbitals, which number $(4N + 2)/2$, have the following designations:

$$\psi^0 \quad N \text{ odd } \tfrac{1}{2}(N+1)b_{3g}$$
$$N \text{ even } \tfrac{1}{2}(N+2)b_{1u} \qquad (4.4)$$

ψ_j^+ j odd $\tfrac{1}{2}(j+1)b_{1u}$ $\qquad$ ψ_j^- j odd $\tfrac{1}{2}(j+1)b_{2g}$

 j even $\tfrac{1}{2}j\,b_{3g}$ $\qquad\qquad$ j even $\tfrac{1}{2}j\,a_u$

where j can assume the values $1, 2, \ldots, N$. According to this description, the bonding orbitals of benzene, naphthalene, anthracene, tetracene, and pentacene are as given below. One notes that only the highest and lowest j orbitals are different; intermediate ones, wherever they exist, do not change from one molecule to another.

Benzene $N = 1, j = 1$
$\psi^0 = 1b_{3g}$ $\quad \psi_1^+ = 1b_{1u}$
$\phantom{\psi^0 = 1b_{3g} \quad} \psi_1^- = 1b_{2g}$

Naphthalene $N = 2, j = 1, 2$
$\psi^0 = 2b_{1u}$ $\quad \psi_2^+ = 1b_{3g}$
$\phantom{\psi^0 = 2b_{1u} \quad} \psi_2^- = 1a_u$

Anthracene $N = 3, j = 1, 2, 3$
$\psi^0 = 2b_{3g}$ $\quad \psi_3^+ = 2b_{1u}$
$\phantom{\psi^0 = 2b_{3g} \quad} \psi_3^- = 2b_{2g}$

Tetracene $N = 4, j = 1, 2, 3, 4$
$\psi^0 = 3b_{1u}$ $\quad \psi_4^+ = 2b_{3g}$
$\phantom{\psi^0 = 3b_{1u} \quad} \psi_4^- = 2a_u$

Pentacene $N = 5, j = 1, 2, 3, 4, 5$
$\psi^0 = 3b_{3g}$ $\quad \psi_5^+ = 3b_{1u}$
$\phantom{\psi^0 = 3b_{3g} \quad} \psi_5^- = 3b_{2g}$

The next step in obtaining a theoretical correlation diagram is to use appropriate theoretical orbital energies. The bonding orbital energies are given by Hückel molecular orbital (HMO) theory:

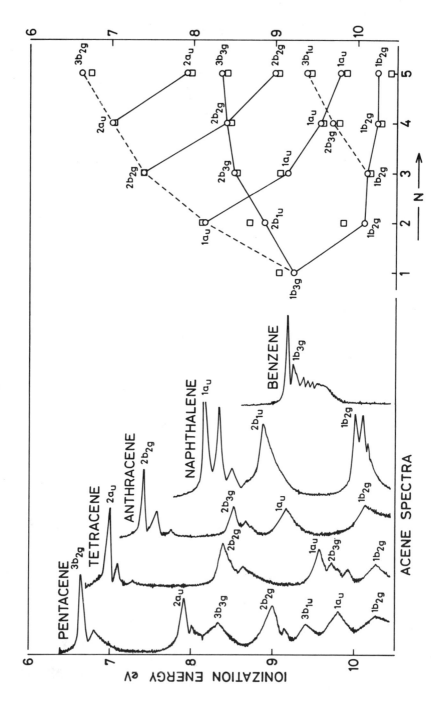

Fig. 4.7. He I photoelectron spectra and correlation diagram of π ionization energies of acenes. ○ experimental values; □ theoretically calculated values [332].

$$\epsilon_j = \alpha + x_j \beta \qquad (4.5)$$

where α and β are Hückel parameters, and

$$x^0 = 1$$
$$x_j^+ = \tfrac{1}{2}(r_j + 1) \qquad (4.6)$$
$$x_j^- = \tfrac{1}{2}(r_j - 1)$$

with r_j given by

$$r_j = \left[9 + 8\cos\left(\frac{\pi j}{N+1}\right)\right]^{1/2} \qquad (4.7)$$

Using these equations, energies corresponding to the ψ_j's for each of the acenes can be calculated. The ψ^0 and ψ_j^+ orbital energies calculated in this manner are presented in Fig. 4.7. The correlations thus obtained are entirely theoretical and there is no certainty that the actual energies follow theoretical calculations; and it is here where photoelectron spectroscopic data become valuable. HeI photoelectron spectra of the first five acenes are also shown in Fig. 4.7. Since the π molecular orbitals are expected to be those with the lowest binding energies, it is a relatively simple task to assign the low binding energy photoelectron spectral peaks. One can then introduce the relevant ionization potentials in Fig. 4.7 and obtain the experimental correlation diagram.

CORRELATIONS IN BENZENOID HYDROCARBONS

Another application of valence electron spectra has been in an experimental verification of π orbital energies of fused benzenoid hydrocarbons obtained from HMO calculations and then expressed in a parameterized form, using first-order bond localization through a perturbation method.

In a π-electron system, if a complete $\sigma-\pi$ separation is assumed, and for a molecule M if $E_\sigma(M)$ represents energy of the σ core and $E_\pi(M)$ is the energy of the π-electron system proper, then the total energy $E_T(M)$ can be written as $E_T(M) = E_\sigma(M) + E_\pi(M)$. If one further assumes that the σ energy can be written

$$E_\sigma(M) = \sum_{\mu\nu} \frac{k_0}{2}(r_{\mu\nu} - r_0)^2 \qquad (4.8)$$

where k_0 is the force constant, $r_{\mu\nu}$ the interatomic distance between the atoms

μ and ν, and r_0 the equilibrium bond length of a pure sp^2-sp^2 σ bond, and the π-energy can be written as

$$E_\pi(M) = n\alpha + 2\sum_{\mu\nu} P_{\mu\nu}\beta_{\mu\nu} \qquad (4.9)$$

where n is the number of π centers, α is the coulomb integral, $P_{\mu\nu}$ the bond order between the atoms μ and ν, and β the resonance integral, then $E_T(M)$ becomes the sum of independent contributions $E_{\mu\nu}(M)$ from each bond, that is, $E_T(M) = \sum_{\mu\nu} E_{\mu\nu}(M)$. This is valid as long as the first-order bond fixation is taken into account. If the orbital ψ_j is described by $\psi_j = c_\mu^j \phi_\mu^j + c_\nu^j \phi_\nu^j$, then bond order $p_{\mu\nu}^j$ for an orbital j between atoms μ and ν is given by $p_{\mu\nu}^j = c_\mu^j c_\nu^j$. For all the orbitals together, we have $P_{\mu\nu} = \sum_j p_{\mu\nu}^j$. The minimum of $E_T(M)$ is attained for a set of interatomic distance $R_{\mu\nu}$ [$r_{\mu\nu}$ at minimum of $E_T(M)$] under the condition

$$\frac{\partial E_T(M)}{\partial r_{\mu\nu}} = \frac{\partial \sum_{\mu\nu} E_{\mu\nu}(M)}{\partial r_{\mu\nu}} = 0 \qquad (4.10)$$

Differentiating $E_T(M)$, that is, the sum of $E_\sigma(M)$ and $E_\pi(M)$ from Eq. (4.8) and Eq. (4.9) respectively and equating to zero, we obtain

$$R_{\mu\nu} = r_0 - \left(\frac{2}{k_0}\frac{\partial \beta}{\partial r_{\mu\nu}}\right)P_{\mu\nu} = r_0 - \left(\frac{2}{k_0}\beta'_{\mu\nu}\right)P_{\mu\nu} \qquad (4.11)$$

where we denote $(\partial \beta/\partial r_{\mu\nu}) = \beta'_{\mu\nu}$. Similarly, for the radical cation obtained by the removal of an electron from orbital ψ_J, we have

$$R^+_{\mu\nu,J} = r_0 - \left(\frac{2}{k_0}\beta^*_{\mu\nu}\right)P^+_{\mu\nu,J} \qquad (4.12)$$

where $P^+_{\mu\nu,J}$ is the bond order of the bond μ,ν in the radical cation, and $\beta^*_{\mu\nu} = (\partial \beta/\partial r_{\mu\nu})$ at $R^+_{\mu\nu,J}$.

In HMO theory it is assumed that all $\beta_{\mu\nu} = \beta$, which implies that a fixed value R_0 ($R_0 = 0.140$ nm, if for the reference compound benzene with bond order $P_{\mu\nu} = P_0 = 2/3$ is used) is assigned to both $R_{\mu\nu}$ and $R^+_{\mu\nu}$. The true vertical ionization potential, however, would be different from the ionization potential expected from a vertical transition from the interatomic distance R_0. The correction that needs to be incorporated in the HMO ionization potential will have two components. First, a term $\delta E_{\mu\nu}$ because at the starting point R_0 the neutral molecule would be located higher on the potential energy surface than when the starting point is $R_{\mu\nu}$, and second, a term $\delta E^+_{\mu\nu,J}$ because the terminal point on the ionic potential energy surface in a vertical transition from R_0 will reach a different energy level than what it would be in the true vertical transition from $R_{\mu\nu}$. This total correction turns out to be

$$\delta E_{\mu\nu} + \delta E^+_{\mu\nu,J} = k_\pi \left(\frac{2\beta'}{k_0}\right)^2 (P^+_{\mu\nu,J} - P_{\mu\nu})(P_0 - P_{\mu\nu}) \quad (4.13)$$

where k_π is the force constant of a π bond (for example, a C⋯C bond in benzene). A summation over all bonds, using Eq. (4.13), gives the second term on the right side of the following equation, the total correction factor to HMO energy:

$$I'_{v,J} = I^0_{v,J} + k_\pi \left(\frac{2\beta'}{k_0}\right)^2 \sum_{\mu\nu} (P^+_{\mu\nu,J} - P_{\mu\nu})(P_0 - P_{\mu\nu}) \quad (4.14)$$

where $I^0_{v,J}$ is the standard HMO energy, which can be written as $-(\alpha + \beta X_J)$. It is this $I'_{v,J}$ that may be compared with the experimental vertical ionization potentials.

The $I'_{v,J}$'s calculated from the above equation are shown in Table 4.4 against $I_{v,J}$, the experimentally obtained vertical ionization potentials for the following isomeric molecules having molecular formula $C_{18}H_{12}$:

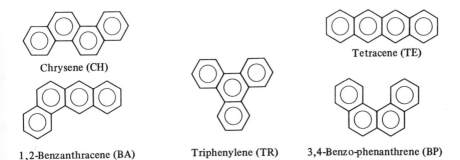

Chrysene (CH)

1,2-Benzanthracene (BA)

Triphenylene (TR)

Tetracene (TE)

3,4-Benzo-phenanthrene (BP)

The results show very good agreement, better than that obtained with unperturbed HMO treatment. However, the relative improvement of the correlation decreases with increasing size of the π systems, owing to the perturbation term in the above equation. This is so because the relative influence of π electron removal decreases with the π-system size. The regression also supports the assumption that the first six bands in spectra of $C_{18}H_{12}$ (four bands in the case of triphenylene) are due to π orbitals. This is corroborated by band shapes and relative band intensity studies.

Table 4.4. Experimental and Calculated Ionization Potentials (eV) of Isomeric Benzenoid Hydrocarbons $C_{18}H_{12}$ [333]

PE Band	CH ψ_J	CH $I_{v,J}$	CH $I'_{v,J}$	BA ψ_J	BA $I_{v,J}$	BA $I'_{v,J}$	TR ψ_J	TR $I_{v,J}$	TR $I'_{v,J}$	TE ψ_J	TE $I_{v,J}$	TE $I'_{v,J}$	BP ψ_J	BP $I_{v,J}$	BP $I'_{v,J}$
1	a_u	7.61	7.60	a''	7.42	7.40	e''	7.86	7.98	a_u	7.01	6.88	b_1	7.62	7.71
2	a_u	8.10	8.19	a''	8.03	8.14	a_1	8.63	8.68	b_{1g}	8.41	8.34	a_2	8.00	8.00
3	b_g	8.68	8.64	a''	8.82	8.78	e''	9.66	9.67	b_{1u}	8.6	8.59	a_2	8.96	8.93
4	b_g	9.44	9.41	a''	9.34	9.29	a_2''	10.05	9.93	a_u	9.56	9.53	b_1	(9.13)	9.05
5	a_u	9.73	9.69	a''	9.90	9.87				b_{2g}	(9.7)	9.78	b_1	9.95	10.05
6	b_g	10.52	10.58	a''	10.40	10.38				b_{1g}	10.25	10.31	a_2	10.26	10.25

NOTE: $-\alpha = 5.782\,\mathrm{eV}$, $-\beta = 3.199\,\mathrm{eV}$, $k_\pi(2\beta'/k_0) = 5.106\,\mathrm{eV}$. The orbital designation ψ_J refers to the symmetry CH, C_{2h}; BA, C_s; TR, D_{3h}; TE, D_{2h}; BP, C_{2v}.

CORRELATION IN SUBSTITUTED π SYSTEMS

Extensive studies have now been made of photoelectron spectra of π systems and substituted π systems. A model that systematically describes the nature and extent of influence of substitutents on the π-ionization potentials would therefore be of considerable interest. In this regard, a study that uses alkyl substituents R has the advantage that their basis orbitals $\phi(R)$ are located at much lower energies than most of the π orbitals, and therefore the π bands of substituted π systems do not overlap with the σ bands of the alkyl substituent. Consequently, the effect of a substituent on π bands of a parent π system can be conveniently followed.

A relationship between the change $\Delta I_{v,J}$ in the vertical ionization potential of a π compound caused by a substituent and the corresponding change in orbital energy can be expressed, within the frame of Koopmans's approximation, as

$$\Delta I_{v,J} = -\Delta \epsilon_j \qquad (4.15)$$

$\Delta \epsilon_j$ may be calculated through a perturbation treatment, the perturbation being caused by the alkyl substitution. A *linear free energy relationship* (LFER) of the type involving alkyl-substituted compounds XR can be written

$$I_{a,1}(XR) = I_{a,1}(XMe) + \chi_X \mu_R \qquad (4.16)$$

where X is the moiety from which the electron is ejected, R is the alkyl group, $I_{a,1}(XR)$ represents the first adiabatic ionization potential of XR, and χ_X and μ_R are parameters characteristic of X and R group respectively. μ_R is defined as

$$\mu_R = I_{a,1}(IR) - I_{a,1}(IMe) \qquad (4.17)$$

which is equivalent to adopting, in Eq. (4.16), $\chi_I = 1.00$ for alkyl iodides. Experimental data show that the series of molecules XR (X = HCO, R'CO, OH, I, and OR'; R' = alkyl) conform to the relationship Eq. (4.15).

A relationship such as Eq. (4.16) can also be constituted using the vertical ionization potentials. In Fig. 4.8 the first vertical ionization potentials $I_{v,1}(RX)$ for X = Br, OH, $CH_2 = CH$ (vinyl), $HC \equiv C$ (ethynyl), and C_6H_5 (phenyl) have been plotted against $[I_{v,1}(R)]_{av}$, the mean of $^2\Pi_{3/2}$ and $^2\Pi_{1/2}$ ionization energies of electron ejection from $5p$ orbitals of iodine in corresponding alkyl iodides:

$$[I_{v,1}(IR)]_{av} = \tfrac{1}{2}[I_{v,1}(IR, {}^2\Pi_{3/2}) + I_v(IR, {}^2\Pi_{1/2})] \qquad (4.18)$$

Existence of a linear relationship is evident from the figure. μ_R's obtained from a least square analysis of the data of Fig. 4.8 (denoted as μ_R'') are H 0.89, Me 0.00, Et -0.19, Pr -0.29, iPr -0.35, Bu -0.31, iBu -0.37, and tBu -0.47.

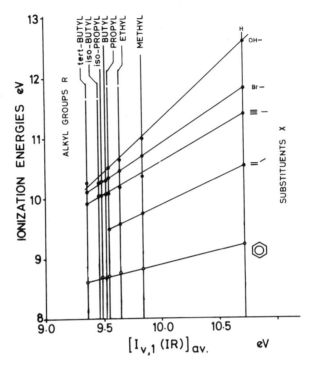

Fig. 4.8. Plot of first ionization energies $I_{v,1}(RX)$ versus $[I_{v,1}(IR)]_{av}$ values. Adapted from E. Heilbronner and J. P. Maier [61] (copyright 1977, Academic Press Ltd., London).

Where multiple substitutions in X are possible, Eq. (4.16) can be extended as

$$I_v(XR_1R_2\ldots) = I_v(XMe_n) + \chi_X \sum_{R_\rho} \mu''_{R_\rho} \qquad (4.19)$$

where XMe_n represents a fully methylated reference compound and R_ρ means a substitution by R at the ρth position of the moiety X. A more convenient reference compound, however, is the hydride. Then one may write

$$I_v(XR_1R_2\ldots) = I_v(XH_n) + \chi_X \sum_{R_\rho} (\mu''_{R_\rho} - \mu''_H) \qquad (4.20)$$

$$= I_v(XH_n) + \chi_X \sum_{R_\rho} \mu^H_{R_\rho} \qquad (4.21)$$

$\mu^H_{R_\rho} = \mu''_{R_\rho} - \mu''_H$ are similar to the classical Hammett constants (where $\mu^H_H = 0$ rather than $\mu''_{Me} = 0$). In alkyl substituted ethylenes, experimental results show that this additivity rule Eq. (4.21) is almost obeyed; there is only a slight departure from linearity.

In view of these experimental results, Eq. (4.15) may be taken as valid, in which case the LFER type of dependence of $I_{v,j}$ on alkyl substitution may be parameterized using an independent electron molecular orbital treatment.

If the normalized π orbitals of X are represented by $\psi_j = \sum_\mu c_{j\mu}\phi_\mu$ and those of alkyl groups denoted by $\phi(R)$, $\mathscr{H}$ is the HMO hamiltonian for the parent compound X for the separated alkyl groups R, R',... then the corresponding orbital energies are

$$\epsilon_j = \langle \psi_j | \mathscr{H} | \psi_j \rangle \tag{4.22}$$

$$A_R = \langle \phi(R) | \mathscr{H} | \phi(R) \rangle \tag{4.23}$$

Making substitutions in the compound X at positions ρ, ρ' by alkyl groups R, R' to form the molecule XRR' results in a new HMO hamiltonian $\mathscr{H}' = \mathscr{H} + \mathfrak{h}$, where $\mathfrak{h}$ is the perturbation operator into which we absorb all those effects that we wish to take into consideration, such as inductive effect and hyperconjugative effect. We then have the following:

$$\delta\alpha_\rho(R) = \langle \phi_\rho | \mathfrak{h} | \phi_\rho \rangle \quad \text{(inductive effect)} \tag{4.24}$$

$$\beta_{\rho R} = \langle \phi_\rho | \mathfrak{h} | \phi(R) \rangle \quad \text{(hyperconjugative effect)} \tag{4.25}$$

that is, a change in the atomic coulomb term α_ρ in the moiety X [$\alpha_\rho = \langle \phi_\rho | \mathscr{H} | \phi_\rho \rangle$] and a resonance integral due to substitution of X at position ρ by R. From such a treatment it can be shown that $|\Delta\epsilon_j|$, which is equal to the change in vertical ionization potential $|\Delta I_{v,j}|$, is given by

$$\Delta\epsilon_j = \sum_\rho c_{j\rho}^2 \, \delta\alpha'_\rho(R) \tag{4.26}$$

where

$$\delta\alpha'_\rho(R) = \delta\alpha_\rho(R) + \frac{\beta_{\rho R}^2}{\epsilon_j - A_R} \tag{4.27}$$

From a knowledge of the coefficients $c_{j\rho}^2$, and from $\Delta I_{v,j}$ ($\equiv \Delta\epsilon_j$) observed for a series of alkyl substituted derivatives XR, we can obtain $\delta\alpha'_\rho(R)$, which represents the sum of the inductive and the hyperconjugative contributions. However, as long as the assumption $|\epsilon_j - A_R| \gg |\Delta\epsilon_j|$ is valid, the precise extent of the two component contributions cannot be separately ascertained.

NORBORNADIENE

We now consider photoelectron spectral data as well as the results of theoretical calculations of the norbornadiene system to establish the π-orbital sequence in norbornadiene and related hydrocarbons. In norbornadiene, interaction between the two double bonds can be either the through-bond or through-space type. To discriminate between these types of bonds, alternative band positions for the photoelectron spectrum of 7-isopropylidene norbornadiene are derived and are compared with experimental results.

Here we examine, in steps, the various interactions that are possible between the three π bonds in 7-isopropylidene norbornadiene and explain the orbital energetics. We then compare the theoretical orbital energies with experimental energies obtained from photoelectron spectra, from which we may derive an understanding about the nature of the interactions that actually exist in the molecule.

The three π bonds in 7-isopropylidene norbornadiene are schematically shown in the following figure. As a result of the interaction of Π_a and Π_b, two

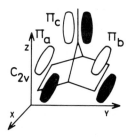

orbitals may be expected, one bonding Π_+ and the other antibonding Π_-, as follows:

$$\psi_+ \quad \Pi_+ \to \frac{(\Pi_a + \Pi_b)}{\sqrt{2}} \qquad (4.28)$$

$$\psi_- \quad \Pi_- \to \frac{(\Pi_a - \Pi_b)}{\sqrt{2}} \qquad (4.29)$$

If the Π_a, Π_b interaction is through-space, then the bonding orbital is of lower energy than the antibonding orbital; this is the normally observed order. If, however, the Π_a, Π_b interaction is through-bond, the order of the orbitals is reversed and the bonding orbital is of higher energy.

The antibonding orbitals Π_- whose charge density away from the cyclohexane ring is more than that of the bonding orbital Π_+, can interact with Π_c and may be expected to result in the following two orbitals $\psi_{-,1}$ and $\psi_{-,2}$:

$$\psi_+ = a_1 \Pi_+ \qquad (4.30)$$

$$\psi_{-,1} = b_2 (\Pi_c - \lambda \Pi_-) \qquad (4.31)$$

$$\psi_{-,2} = b_2 (\Pi_- + \lambda \Pi_c) \qquad (4.32)$$

λ is positive if Π_c lies above Π_- and is negative if Π_c lies below Π_-.

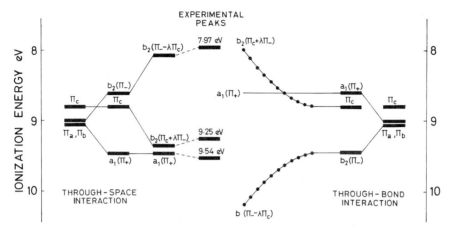

Fig. 4.9. Predicted and experimental π orbital energies in 7-isopropylidene norbornadiene [336].

The left side of Fig. 4.9 shows the location of Π_+ and Π_- orbitals in the through-space case. The bonding $a_1 \Pi_+$ is of lower energy, and the antibonding $b_2 \Pi_-$ is of higher energy. Interaction of Π_c with the antibonding orbital $b_2 \Pi_-$ results in the two orbitals $b_2(\Pi_- - \lambda \Pi_c)$ and $b_2(\Pi_c + \lambda \Pi_-)$, the interaction matrix element being $\langle \Pi_c | \mathscr{H} | b_2(\Pi_-) \rangle$. If, however, the Π_a, Π_b interaction is through-bond, then the bonding orbital $a_1(\Pi_+)$ will be of higher energy and the antibonding $b_2(\Pi_-)$ of lower energy, as shown on the right side of the figure. Interaction, then, between Π_c and $b_2 \Pi_-$ can be expected to lead to energy level separations shown by the dashed line in the figure, exact energy separation depending on the choice of the interaction parameters. The experimental binding energies 7.97 eV, 9.25 eV, and 9.54 eV are shown in the central part of the diagram. It is clear that only in the sequence where $b_2(\Pi_-)$ is above $a_1(\Pi_+)$, (through-space interaction between Π_a and Π_b) does one expect final energy levels that can result in the photoelectron spectrum observed in practice.

π-IONIZATION ENERGIES AND DIHEDRAL ANGLES IN BIPHENYLS

Another area where photoelectron spectroscopic methods have been used is in conformational studies. We consider in this section an example involving the biphenyls. The nonbonded through-space steric effects play an important part in the conformation of this class of compounds, and electronic perturbation by the constituents and the size effect occur together. By determining all the occupied energy levels of a complete set of related reference compounds, the substituent electronic effects may be estimated and thus the contribution due to steric factors ascertained.

Correlation of π-orbital energies of substituted biphenyls with the dihedral angles between the two phenyl moieties constitutes an interesting application of photoelectron spectral data. The π-molecular orbitals $1a_{2u}$ and $1e_{1g}$, the two degenerate orbitals of two benzene rings, interact to form the π orbitals of biphenyl. These orbitals are shown in Fig. 4.10; the nodal surfaces are indicated by dashed lines. Among these orbitals, π_4 and π_5 are degenerate. In the HMO treatment, orbital energies depend on the coulomb integral α and the resonance integral β, the energy expressed in the form $\alpha + k\beta$. The difference in energy between π_6 and π_3, $\Delta E(\pi_6 - \pi_3)$, would therefore be proportional to the resonance integral β, that is, $\langle \psi_i^R | \mathcal{H} | \psi_i^S \rangle$ or $\langle \psi_i^R | \mathcal{H} | \psi_j^S \rangle$ where ψ_i and ψ_j's are the molecular orbital (MO) wave functions of the benzene ring, and R and S represent the two benzene moieties in biphenyl. Negative of $\langle \psi_i | \mathcal{H} | \psi_i \rangle$ then would represent the experimental ionization energy of an electron ionized from the π-MO ψ_i of benzene, assuming Koopmans's theorem.

From the solution of the secular determinant arising in HMO treatment, one obtains for the difference in energy of π_6 and π_3 a quantity proportional to the resonance integral. This can be written

$$\Delta E(\pi_6 - \pi_3) = 2B \tag{4.33}$$

where B is a standard interaction parameter of the form $\langle \pi_i | \mathcal{H} | \pi_j \rangle$. But the resonance integral, or B, is proportional to the overlap integral S_{ij}, that is,

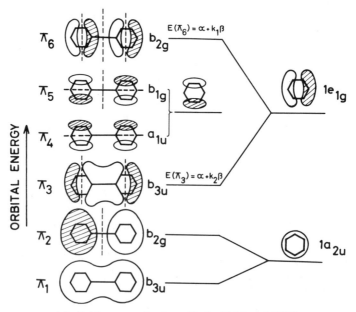

Fig. 4.10. π molecular orbitals of biphenyl [338].

$\langle\psi_i|\psi_j\rangle$. In the case of biphenyl, when the two rings are coplanar (dihedral angle $\theta = 0°$) the overlap between the π orbitals is expected to be maximum. On the other hand, when the two rings are perpendicular, the overlap should be minimum, that is, zero. The overlap in those cases where the dihedral angle is $0 < \theta < \pi/2$ may therefore be expected to follow $\cos\theta$ dependence. This means that in non-coplanar cases of substituted biphenyls the off-diagonal elements in the secular determinant may be written $B_\theta = B\cos\theta$. If π–σ mixing is ignored in these molecules, then one can write

$$\Delta E = 2B \cos\theta \qquad (4.34)$$

and one should expect a correlation between the energy difference $\Delta E(\pi_6 - \pi_3)$ and $\cos\theta$. In the photoelectron spectra of biphenyls, seven bands are observed with ionization potentials $I_1, I_2, \ldots, I_7$, of which I_1 and I_4 are assigned to π_6 and π_3 respectively. A plot between the experimental ionization energy difference of the π_3 and π_6 bands of substituted biphenyls and the dihedral angle θ is presented in Fig. 4.11. The θ's are taken from electron diffraction data, and the linearity of the plot indicates a clear relationship between $\Delta E(\pi_6 - \pi_3)$ and $\cos\theta$. One could use such linear regression to determine dihedral angles of any substituted biphenyl if its photoelectron spectrum is known.

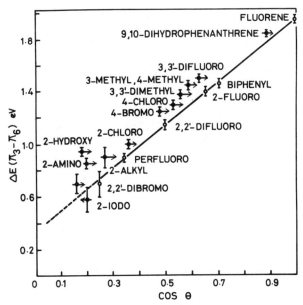

Fig. 4.11. Plot of $I_4 - I_1$ against the dihedral angles for biphenyls. $E(\pi_3 - \pi_6) \equiv \Delta E \equiv I_4 - I_1$. (2-IODO means 2-iodo biphenyl). ○ ΔE values from UPS and θ values from electron diffraction; ● ΔE values from UPS [338].

ARYLCYCLOPROPANES

We may take another example of a recent application of UPS to conformational studies, on arylcyclopropanes. There has been extensive work on conjugative interaction between the cyclopropyl group and adjacent p or π electron systems, the interaction being generally visualized as an overlap between the π system and the frontier Walsh-type orbitals of the cyclopropane system. The Walsh orbitals arise as follows.

The C–C–C angle strain in the cyclopropane molecule is large. This strain gets effectively reduced if the p-character in C–C bonds increases to about 75%, which means that the C–H bonds would then have a correspondingly larger degree of s-character. If these three sp^2 hybridized orbitals (along with a pure p orbital) are used on the carbon atoms, two of these can be used for forming two C–H bonds (the H–C–H plane would be perpendicular to the cyclopropane ring) and the third sp^2 orbital can be directed towards the center of the cyclopropane ring. The fourth carbon orbital, the p orbital, would be in the plane of the cyclopropane ring but directed perpendicular to the $-CH_2$ plane. The remaining sp^2 orbitals (one from each carbon atom, a total of three sp^2 orbitals) and the three p orbitals (one from each carbon atom, a total of three) combine to provide bonding, nonbonding, and antibonding Walsh orbitals of the cyclopropane system. These orbitals, in which the six electrons are distributed, and their relative energies are shown in Fig. 4.12. The larger the number of nodes, the higher is the energy.

In arylcyclopropanes the maximum interaction occurs between the aryl π system and the cyclopropyl e_a and e_a^* orbitals in the bisected conformation where the aromatic and cyclopropyl rings are mutually perpendicular. NMR studies of p-deutero phenylcyclopropane in CS_2 solution show a temperature-dependent shift of ortho protons and indicate a rapid equilibrium between the bisected conformation, where the H_0 position is shielded by the magnetic anisotropy effect of the cyclopropane ring, and the nonbisected conformation, in which the shielding effect is absent. Bisected conformation is favored at low temperature. The presence of an α-methyl group leads exclusively to the nonbisected conformation, which reduces the steric repulsion between H_0' and the α-alkyl group. Electron diffraction measurements on phenylcyclopropane in gas phase, however, indicate the exclusive presence of bisected molecules.

The data on vertical ionization potentials on arylcyclopropanes are given in Table 4.5. The first two bands I_1 and I_2 can be assigned to the highest π orbitals of the aromatic system, degeneracy of which is lifted as a result of overlap with the cyclopropane orbitals. I_3 and I_4 can be attributed to ionization from levels that are primarily upper Walsh type orbitals (e_a, e_s) in character. This implies that the difference between I_1 and I_2, that is, ΔE_π, as well as,

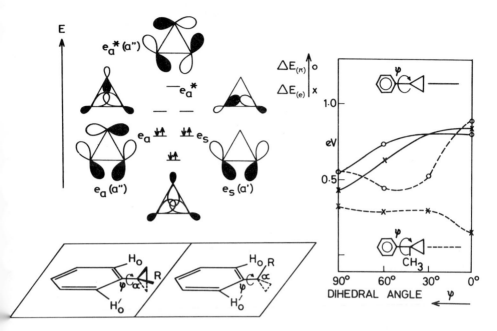

Fig. 4.12. *Top left*: The frontier Walsh-type orbitals of the cyclopropyl system. The symmetry labels refer to the C_s point group. *Bottom left*: The bisected ($\phi = 0°$) and non-bisected ($\phi = 90°$) conformations of the arylcyclopropane system; ϕ is the dihedral angle between the plane of the aromatic system and the plane through C_1, C_α, and R. *Right*: Calculated splitting of the upper π-type levels and the cyclopropane e-type levels as a function of ϕ in phenylcyclopropane and in α-methyl phenyl cyclopropane [340, 341, 342].

between I_3 and I_4, that is, ΔE_e, should be sensitive functions of the dihedral angle ϕ, the relative orientation of the benzene and cyclopropane rings.

The results of some theoretical calculations using a modified iterative extended Hückel method on ΔE_π and ΔE_e as a function of the dihedral angle are shown in Fig. 4.12. The rather large π splitting for nonbisected conformation is not compatible with the simple Walsh scheme of Fig. 4.12. An analysis shows, however, that the Walsh e_s type of orbital contains a rather large C $2p$ contribution in the cyclopropane ring plane at the carbon atom connecting it to the phenyl ring. This would explain the possible interaction between the cyclopropane Walsh orbitals and the π system, through the C $2p$ contribution, even in the nonbisected conformation.

The presence of alkyl groups destabilizes the cyclopropane Walsh levels, but the splitting ~ 0.5 eV remains essentially constant. The π level energy

Table 4.5. Vertical Ionization Potentials Determined from HeI Spectra [342]

Structure		$I_1{}^a$	$I_2{}^a$	$I_3{}^b$	$I_4{}^b$ (eV)
Benzene		9.23			
Toluene		(8.9)	9.13		
Cumene		8.98	9.20		
Cyclopropane		10.90			
Methylcyclopropane		10.10	10.90		
p-Methylanisole		8.18	9.11		
p-Chlorotoluene		8.90	9.57		
X = H	R = H	8.66	9.21	10.53	11.11
	Me	8.73	9.17	10.09	10.59
	Et	8.70	9.17	9.95	10.50
	i-Pr	8.63	9.12	9.74	10.38
	t-Bu	8.63	9.15	9.63	10.33
X = CH$_3$O	R = H	8.05	9.08	10.13	10.67
	Me	8.09	9.05	9.79	10.46
	Et	8.11	9.02	9.69	10.31
	i-Pr	8.10	9.00	9.68	10.25
	t-Bu	8.05	9.05	9.64	10.15
X = Cl	R = H	8.64	9.47	10.49	10.98
	Me	8.67	9.42	10.11	10.74
	Et	8.64	9.42	10.04	10.61
	i-Pr	8.64	9.39	9.89	10.52
	t-Bu	8.64	9.35	9.80	10.44

^{a} Highest π orbitals of the aromatic system, $\Delta E_\pi = I_2 - I_1$.
^{b} Upper Walsh-type orbitals, $\Delta E_e = I_4 - I_3$.

or ΔE_π is also unaffected. Para substitution, on the other hand, affects I_1, I_2, and ΔE_π, but practically none of the rest. All these indicate the absence of any significant conjugation between the π molecular orbitals of the aromatic ring and the cyclopropane Walsh orbitals. This weak conjugation may be taken to indicate nonbisected conformation. This is in agreement with the modified iterative extended Hückel calculations, where for $\phi = 90°$ (nonbisected), substitution of H by CH$_3$ does not significantly influence ΔE_e, that is, the I_3, I_4 splitting, while ΔE_π, the I_1, I_2 splitting, remains essentially constant at

~ 0.5 eV. For bulkier groups this effect is further augmented, and the non-bisected conformation is shown to be the clear alternative.

BIOMOLECULES AND ABSOLUTE DONOR CAPABILITIES

In the field of biomolecules, a recent application of UPS has been the study of a large range of heteroatomic biological molecules and the verification of Pullman k indexing. According to the Pullman index, a simple Hückel approach can be used to obtain Koopmans's equivalent of ionization energy $I(\epsilon)$, that is, $\epsilon(\pi)$. This is expressed in terms of the average coulomb integral α and exchange integral β in the form $\epsilon(\pi) = \alpha + k\beta$, where k, the Pullman index, contains information about the π-donor capabilities.

In Fig. 4.13 a plot is presented of experimental ionization potentials against the Pullman k coefficient of a number of biomolecules and a series of aromatic hydrocarbons. The biomolecule regression line vindicates the Pullman k scale. On the basis of this diagram, obtained directly from UPS data, absolute electron donor capabilities defined with respect to the ionization continuum can be derived, as shown in Fig. 4.14.

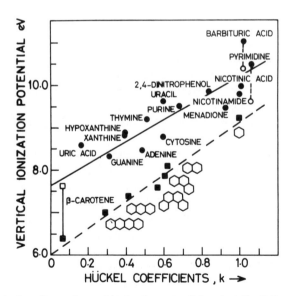

Fig. 4.13. A plot of experimental ionization potential against the Pullman k coefficient for a number of biomolecules and for a series of aromatic hydrocarbons; ● $I(\pi$, vertical), ○ $I(n$, vertical); for β-carotene ■ $I(\pi$, adiabatic), □ $I(\pi$, vertical) [67, 346].

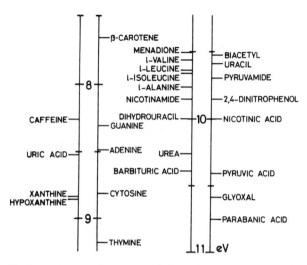

Fig. 4.14. Absolute electron donor capabilities defined with respect to the ionization continuum [67, 346].

STUDY OF TAUTOMERIC EQUILIBRIA

Photoelectron spectral data have also been utilized in equilibrium studies. We consider next an example of the determination of equilibrium constants in tautomeric equilibria, using temperature-dependent photoelectron spectroscopic studies. The lowest few ionization energies that lead to bands in the photoelectron spectra of the molecules acetylacetone, 3-methylacetylacetone, and 3,3-dimethylacetylacetone, along with their orbital assignments, are shown in Fig. 4.15.

The assignment is based on the following rationale. The two bands that are observed in the nonenolizable 3,3-dimethylacetylacetone must be due to keto group n_- and n_+ (antibonding and bonding respectively, due to the oxygen lone-pair contribution; MINDO/2 calculations place n_- above n_+). Correlating these two bands with the corresponding bands of 3-methylacetylacetone and acetylacetone, one obtains the keto orbital assignments of these two molecules. For the enol forms of acetylacetone and 3-methylacetylacetone, one band is expected from π MO and the other from the carbonyl oxygen lone-pair, nonbonding (n) MO. The correlation is based on MINDO/2 calculations, which predict a destabilization of the π orbital, a lower ionization energy, in 3-methylacetylacetone by 0.45 eV, which is precisely the difference observed between the 9.00 eV band of acetylacetone and the 8.55 eV band of 3-methylacetyl-

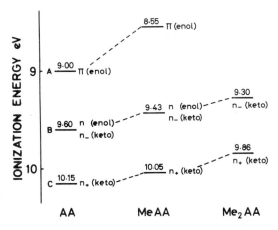

Fig. 4.15. The lowest few ionization energies of acetylacetone, 3-methyl acetylacetone, and 3,3-dimethyl acetylacetone [349].

acetone. This means that the n(enol) band and the n_-(keto) band appear unresolved.

The photoelectron spectrum of 3,3-dimethylacetylacetone, which does not enolize, is invariant with temperature. This is expected, as the keto form does not change into any other form with temperature change. In contrast, the spectra of acetylacetone and 3-methylacetylacetone strongly depend on temperature. Among the three bands in acetylacetone, the intensity I_A of the 9.00 eV band A is determined by the concentration of the enol species, the intensity I_B of the band B at 9.60 eV is due to both enol and keto forms and thus depends on the concentration of both species, and the intensity I_C of the 10.15 eV band C is due to the keto form. Experimental results show that both the relative band intensities $n_-:n_+$ in the nonenolizable β-diketones and $\pi:n$ in molecules containing simultaneously a π(olefinic) MO and an oxygen lone-pair MO — as are present in the enolizable species considered — are approximately 1:1. Since I_A and the enol contribution that occurs in I_B are the same, the total enol concentration may be taken as $2I_A$. Further, if I_A is subtracted from the intensities $I_B + I_C$, the intensity due to keto can be derived. Thus it is possible to write

$$K = \frac{[\text{enol}]}{[\text{keto}]} = \frac{2I_A}{I_B + I_C - I_A} = \frac{2I_A/(I_B + I_C)}{1 - I_A/(I_B + I_C)} \qquad (4.35)$$

Equilibrium constants have thus been determined as a function of temperature, and from such data ΔH^0 has been obtained. For acetylacetone, ΔH^0 is found to be equal to -1.9 ± 0.1 kcal/mole, and for 3-methylacetylacetone ΔH^0 is

−1.3 ± 0.1 kcal/mole. This compares well with the ΔH^0 value of −2.4 kcal/mole for acetylacetone obtained from infrared and with −2.0 kcal/mole obtained from mass spectrometric measurements. For 3-methylacetylacetone, the mass spectrometric value is −1.3 kcal/mole. The photoelectron spectroscopic values imply that at room temperature, 76% of acetylacetone and 50% of 3-methylacetylacetone exists in enol form.

An alternative method of determining the equilibrium constant depends on the measurement of peak areas under the A and C bands for a fixed intensity of the band B. This method has been found to yield a somewhat lower value of ΔH^0 (≈ -4 kcal/mole) for acetylacetone, which could, however, be due to the fact that in this work the experimental data that go to higher temperatures show a curvature in the $\log K - (1/T)$ plot, with a higher slope at higher temperatures. In this method the intensity I_A can be written as equal to k_A [enol] and the intensity I_C as equal to k_C [keto] where the proportionality constants k_A and k_C incorporate ionization cross section as well as effects due to asymmetry parameter β of photoejection. Since the total concentration [total] remains the same, it is possible to write

$$[\text{total}] = \frac{I_A}{k_A} + \frac{I_C}{k_C} \tag{4.36}$$

or,

$$I_C = -\frac{k_C}{k_A} I_A + k_C [\text{total}] \tag{4.37}$$

A plot of I_C against I_A as the temperature is varied yields k_C/k_A. Once the ratio k_C/k_A is known, K can be found out at any temperature if the intensities I_C and I_A are experimentally measured:

$$K = \frac{[\text{enol}]}{[\text{keto}]} = \frac{I_A}{I_C} \frac{k_C}{k_A} \tag{4.38}$$

OTHER CORRELATIONS

It should be evident by now that a dominant theme in the use of photoelectron spectral data is the correlation of experimental ionization energies with various physicochemical properties of the atoms and molecules involved. We have seen several examples of such correlations and examined some of them in considerable detail. To further explore such correlations in chemical applications, let us briefly review a wide range of such use of photoelectron spectral data.

The sprays of linear regressions presented in Fig. 4.16 refer to the following. Nonbonding n_0 orbital energies of parasubstituted benzamides obtained from

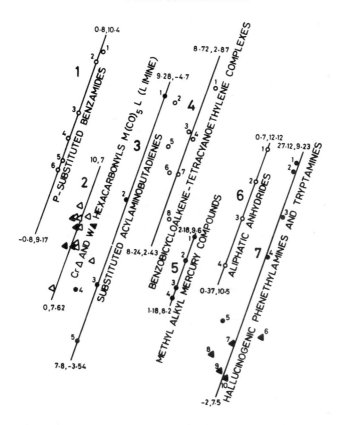

Fig. 4.16. Correlations: (1) Nonbonding n_0 orbital energies of p-substituted benzamides versus Hammet σ constants. (2) Ionization potentials of the metal $e(d)$ in $M(CO)_5 L$ (M = Cr, W; L = imine) versus pK_a of the ligand. (3) Reaction rates of acylaminobutadienes with methyl acrylate versus first ionization potentials of dienes. (4) Charge transfer transition energies versus ionization potentials of benzobicycloalkene–TCNE complexes. (5) Ionization potential of methyl alkyl mercury compounds versus oxidation potential of the corresponding Grignard reagent. (6) First ionization potentials of anhydrides versus FWHM of the first ionization potentials. (7) First and second ionization potentials versus minimum effective brain levels for interfering with conditioned avoidance response in rats [377, 409, 373, 375, 378, 376, 374].

photoelectron spectra are plotted against Hammett σ constants in regression 1. The coordinates shown at the extremities of the line represent values of the Hammett σ constant and ionization potential in eV, in that order. Separate linear relationships on ionization potentials are found for parasubstituted and metasubstituted benzamides. Similar correlations exist for the π_N orbital. Substituents influence both the n_0 orbital and π_N ionization energies. The changes in π_N ionization potentials from p-CH$_3$O benzamide to p-NO$_2$ benzamide is 0.67 eV, which is comparable to that observed for the n_0 orbital. For all orbitals, the introduction of donor groups causes an increase in orbital energy, that is, a decrease in ionization potentials, and acceptor groups lower the orbital energy levels, that is, increase the ionization potentials.

Photoelectron spectral data of a number of M(CO)$_5$L complexes with M = Cr, W, and L = nitrogen donor ligand are now available. Ionization potentials of the metal $e(d)$ are plotted in regression 2 against the pK_a of the ligand, here various imines. The coordinates are pK_a of the ligand and ionization potential $e(d)$ in eV. It is observed that the ionization potentials of the metal d orbitals in these complexes depend on the σ and π bonding properties of the ligands. If the extent of π interactions within a series of nitrogen donor ligands is assumed to be invariant, then the d orbital energy is expected to shift to higher energy with a larger σ donating ability of the ligand. This relationship holds within the series of amines and imines.

Rates of reaction of several acylaminobutadienes and pentadienoic acid with methyl acrylate are plotted in regression 3 against the first ionization potentials of the dienes. The coordinates are diene ionization potential in eV and log k(110°C). The correlation shows a relationship between the ionization potentials of the dienes and the theoretical Diels–Alder reactivities. It is seen that there is excellent correspondence between the rates of these cycloadditions and the electron donor abilities of these dienes.

The next correlation involves photoelectron and charge-transfer spectral data of a number of benzocycloalkenes. Charge transfer complexes of these compounds with tetracyanoethylene (TCNE) give broad overlapping absorption bands. A plot of the charge transfer transition energies versus the experimental ionization potentials, both in eV, is presented in regression 4. The coordinates are ionization potential and charge transfer energy, both in eV. The correlation $h\nu_{CT} = 0.89\, IP_{\text{donor}} - 4.90$ eV holds. The compound 2 deviates from the correlation. This is characterized by a significantly higher charge transfer transition energy than expected. For this type of compound and for others involving donors with relatively high ionization potentials, difficulties arise from the overlap of the lowest energy transition with one at shorter wavelengths.

The ionization potentials of a series of dialkyl mercury compounds (R'HgMe) obtained from HeI UPS have been plotted in regression 5 against the oxidation potentials of the corresponding Grignard reagent. The coordinates are the solution oxidation potential of RMgBr and the first ionization potential of RHgR' in eV. Similar plots of R'HgEt, R'Hgi-Pr, R'Hgt-Bu, highlight the sensitivity of a series of mercurials RHgR' to alkyl substitution. The slope decrease in order R'HgMe > R'HgEt > R'Hgi-Pr > R'Hgt-Bu. This indicates that partial delocalization of charge by the first group R in the molecular ion diminishes the effect of the second group R'.

Regression 6 shows a plot of the first ionization potential of a series of anhydrides against the FWHM of the first ionization band. The coordinates are the FWHM, and the first ionization potential, both in eV. This correlation illustrates the delocalization of the molecular orbital with increasingly electronegative substituents. $(HCO)_2O$ does not fit on this line since the baricenter of the carbonyl lone pairs is particularly stabilized. This is due to a conjugative interaction of the carbonyl groups with the central oxygen lone pair and is facilitated by the planar geometry of the molecule.

The final correlation involves an example of quantitative relationships between the structure and activity of hallucinogens. The coordinates in regression 7 are the minimum effective brain levels, in nmoles/g, to interfere with conditioned avoidance response in rats, and the average of first and second ionization potentials in eV. Extensive studies of the relationship between structure and activity have also been carried out for hallucinogenic amphetamines, and many good correlations between activities and physical properties have been found.

In the foregoing sections, we have looked at a wide range of examples from organic chemistry. We now take up two examples from inorganic molecules.

SPECTRA OF HEXAFLUOROACETYLACETONE COMPLEXES

Let us first consider a case study that illustrates how photoelectron spectra may be used in investigating the nature of bonding in inorganic complexes. In the example considered, photoelectron spectra data show how the metal d orbitals attain enhanced stability in a series of transition metal complexes involving hexafluoroacetylacetone as the ligand. The data also provide relative energies of the highest occupied ligand orbitals.

Hexafluoroacetylacetone exists in enol form in the gas phase as well as in the liquid phase, and it has C_{2v} symmetry. The π system of the six-membered ring is formed from the $p\pi$ orbitals of the carbon and oxygen atoms, as shown in Fig. 4.17. The figure shows that in the π-system, five orbitals are formed. Three belong to the set $a_2 + 2b_2$ and two to the set $a_2 + b_2$ in the C_{2v} point group. From Hückel theory, the orbital energies are $\pi_1 < \pi_2 < \pi_3 < \pi_4 < \pi_5$, with π_4, π_5 separated enough in energy from π_3 not to be important in bonding in metal chelates. From the orbital disposition of C and O atoms, it is seen that all CO orbitals in π_1, π_2, π_3 involve localized π bonding. From this, π_2 may be considered to be approximating an antisymmetric combination of two localized C–O π-bonding orbitals with a nodal plane vertically bisecting the ring plane, and π_3 is an antibonding orbital formed from the symmetric combination of

132 VALENCE ELECTRON SPECTRA

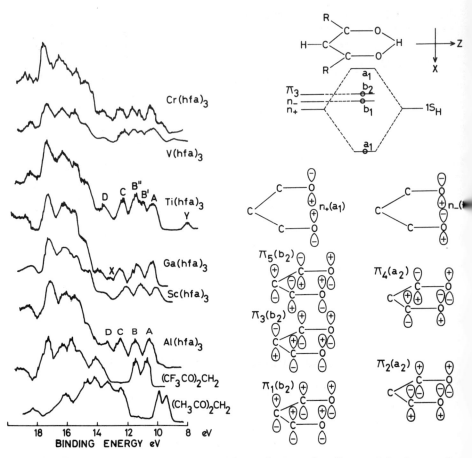

Fig. 4.17. He I photoelectron spectra of acetylacetone, hexafluoroacetylacetone, and hexafluoroacetylacetonato complexes of Al, Sc, Ga, Ti, V, and Cr. *Right, from the top:* The structure of the enol form of a β-diketone, a molecular orbital diagram for the enol form, and the π molecular orbitals and lone-pair combinations for the enol form [396].

π_{CO} localized orbitals and the p_π orbital of the central carbon atom. The oxygen lone-pair orbitals can be taken as p_x orbitals on the six-membered ring plane. Symmetric combination of two oxygen p_x orbitals yield $n_+(a_1)$ and the antisymmetric combination yields $n_-(a_1)$. The interaction of the hydrogen 1s orbital and $n_+(a_1)$ is primarily responsible for binding of the proton. The $n_-(b_1)$ participates only in the C-O π bonding.

The photoelectron spectra of the complexes indicate the energy level scheme shown in Fig. 4.17. The UPS of acetylacetone, also shown in Fig. 4.17, shows two bands at 9.2 eV and 9.7 eV, one of them due to $b_2(\pi_3)$ and the other due to oxygen lone-pair electrons $b_1(n_-)$. The 9.2 eV peak is most probably not due to $a_1(n_+)$, because then 0.56 eV splitting between n_+ and n_- would be too small. Hexafluoroacetylacetone exhibits a similar spectrum with a general shift to high ionization energies due to the perfluoro effect. The 14.0 eV band is most probably due to $a_1(n_+)$.

The photoelectron spectra of Al(hfa)$_3$, Ga(hfa)$_3$ and Sc(hfa)$_3$ are very similar, the structure beyond 14 eV being common to all simple hexafluoroacetylacetonate complexes. The peak around 17 eV observed in the complexes corresponds to ionization from some fluorine lone-pair combination, which appears at 17.44 eV in hexafluoroacetylacetone. Molecular orbital energy diagrams constructed by considering interactions of the tris-ligand orbitals with metal s, p, and d valence electrons are shown in Fig. 4.18. Assuming that the closed shell tris-hexafluorylacetylacetonate complexes possess trigonal symmetry in the D_3 point group with octahedral configuration of oxygen atoms around the metal atom, the ligand orbitals π_3, n_-, n_+ correspond respectively to $e + a_2$, $e + a_2$, and $e + a_1$ ligand symmetry orbitals. $e(n_+) - e(np)$ and $a_1(n_+) - a_1(ns)$ ligand–metal interactions are expected to be the major source of metal–ligand binding; ns, np, nd refer to metal valence electrons. The np atomic orbital overlaps with $e(\pi_3)$ and $a_2(\pi_3)$ are expected to be similar, hence only a small splitting of π_3 ionization is anticipated and qualitatively indicated in Fig. 4.18. The interaction $e(n_-) - e(np)$ being smaller than $a_2(n_-) - a_2(np)$, the e and a_2 molecular orbitals arising from n_- are expected to have a larger splitting. The largest separation is expected between the $e(n_+)$ and $a_1(n_+)$.

On the basis of the above, the lowest ionization potential band A in Fig. 4.17

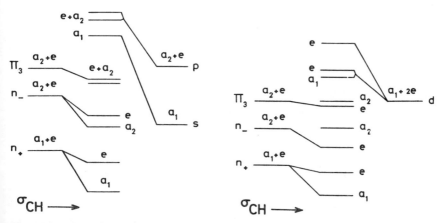

Fig. 4.18. A molecular orbital energy diagram showing the interaction of the *tris*-ligand system with metal s, p, and d valence orbitals [396].

is due to high orbital energy unresolved $e(\pi_3)$ and $a_2(\pi_3)$ ionizations, and the band B involves ionization from $e(n_-)$ and $a_2(n_-)$. Either the C and D bands originate from ionization of the main bonding electrons [primarily localized as ligand n_+, that is, $e(n_+)$ for the C band and $a_1(n_+)$ for the D band] or the C band is due to both $e(n_+)$ and $a_1(n_+)$, and the D band due to some ligand σ or π_2 ionization. The former assignment helps in understanding the Ga(hfa)$_3$ spectrum. The splitting of the band B is probably due to an increased interaction between the n_- ligand of the symmetry orbitals and metal $4p$. The X band in Ga(hfa)$_3$ is due to ionization of $a_1(n_+)$ electrons. Since no such band between the C and D bands is observed in Al(hfa)$_3$, this indicates increased covalence in Ga(hfa)$_3$ relative to Al(hfa)$_3$ because $a_1(n_+)$ is a deeper lying molecular orbital. The B band of the Sc(hfa)$_3$ spectrum shows a slight splitting. This can be understood if the $3d$ orbitals play a significant part in metal–ligand bonding, and from Fig. 4.18, it can be seen that $e(n_-)$ and $a_2(n_-)$ have a larger splitting when d electrons are involved.

The transition metal complexes Ti(hfa)$_3$, V(hfa)$_3$, Cr(hfa)$_3$ are in a series of t_{2g}^n configuration. Figure 4.17 shows that the peculiarity of the spectra of these compounds with respect to that of Sc(hfa)$_3$ is a band at 7.94 eV for Ti, which progressively shifts in V and Cr. It is reasonable to assign this band Y to metal $3d$ electrons. The assignment is supported by a comparison of intensity increase of Y with increasing number of $3d$ electrons. In the D_3 point group, $3d$ transforms as $a_1 + 2e$, with one e set e_a correlating with $t_{2g}(t_{2g} \rightarrow a_1 + e_a)$ and the other e_b directly with the e_g orbitals ($e_g \rightarrow e_b$). The interaction between $e_b(d)$ and $e(n_+)$, $e(n_-)$ controls the bonding since the $e_a(d)$, $a_1(d)$ (both belonging to the t_{2g} set) interaction with the ligand symmetry orbitals are expected to be of minor importance. One expects only a small interaction between the nonbonding $a_2(\pi_3)$ and $a_2(n_-)$ orbitals and $e(\pi_3)$ and metal $e(d)$, leading to the possible overlapping of $e(\pi_3)$ and $a_2(\pi_3)$ bands.

The B band in Sc(hfa)$_3$ appears in Fig. 4.17 as two separate B'' and B' peaks, indicating increased stabilization of $e(n_-)$ with respect to $a_2(n_-)$, as shown in Fig. 4.18b. The relative intensity of the bands B'' and B' supports the B assignment mentioned earlier. The increasing trend in this separation correlates well with the Y band shift towards higher ionization energy, that is, lower d-energy, as one goes from Ti(hfa)$_3$ to Cr(hfa)$_3$. One may expect a corresponding increase in $e(n_-) - e(d)$ interaction with increasing stability of $3d$ orbitals. In Al(hfa)$_3$, where d interaction is absent, the magnitude of the $a_1(n_+)$ and $e(n_+)$ separation is larger than when such interaction is present. If the $3d$'s have a more important role in bonding than the $4s$ orbitals, $a_1(n_+)$ and $e(n_+)$ should lie closer than is seen in Al(hfa)$_3$; in such a situation, the band C is more properly assigned to both these ionizations. This assignment is justified by a large separation of C and D in transition metal complexes. One can detect an increase in covalency from the higher energy shift of C, B', B'' bands as

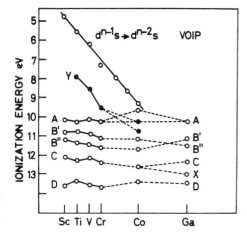

Fig. 4.19. A plot of ionization energies for the *tris*-hexafluoroacetylacetonato complexes of the transition metals [396].

one goes from Sc(hfa)$_3$ to Cr(hfa)$_3$, coupled with a size decrease of the central M(III) cation. The behavior of the A peak, however, is complex.

Figure 4.19 summarizes the behavior of photoelectron bands in hexafluoroacetylacetone complexes and shows stabilization of d orbitals. However, the figure tends to exaggerate the way d orbitals are stabilized. For example, for Cr(hfa)$_3$, the transition involved in ionization is $^4A_{2g}(t_{2g}^3) \rightarrow {}^3T_{1g}(t_{2g}^2)$ connecting the ground states of neutral t_{2g}^3 and t_{2g}^2. But since $^4A_{2g}$ is more stable relative to the baricenter of t_{2g}^3 than is $^3T_{1g}$ to the t_{2g}^2 baricenter, the fully spin-and-space-randomized transition process will involve less energy. In comparison with the observed ground term ionization energies, the average of configuration ionization energies will be lower before the half-filled t_{2g} shell. This difference in ionization energies increases up to t_{2g}^3 configuration, then it changes sign and the difference progressively decreases to zero at the closed shell t_{2g}^6 configuration; thus the $t_{2g}(3d)$ metal orbitals do not stabilize very drastically.

SPECTRA OF Se$_2$ AND Te$_2$

This final section concerns the photoelectron spectra of Se$_2$ and Te$_2$, which exist as dimers in high temperature vapor of the elements. We consider here briefly the salient aspects of Se$_2$ and Te$_2$ spectra and discuss how energy level structure is obtained and correlated with the O$_2$, S$_2$ system. In this system, spin–orbit and multiplet splitting play important roles.

The Se$_2$ valence molecular orbital configuration is $(\sigma_g 4s)^2 (\sigma_u 4s)^2 (\sigma_g 4p)^2 (\pi_u 4p)^4 (\pi_g 4p)^2$. Possible molecular terms for a partly filled π_g^2 configuration are $^3\Sigma_g^-$, $^1\Sigma_g^+$, $^1\Delta_g$. For Se_2 (as in S_2, Te_2) the ground state is $^3\Sigma_g^-$, and the $^1\Sigma_g^+$, the next higher state, lies about 0.3 eV higher. If λ splitting is sufficiently large, as can happen in heavy atoms, the coupling can go from Hund's case (a) and case (b) to case (c). In Se_2, the ground state $X^3\Sigma_g^-$ is split into $X1g$ and XO_g^+, the former being 45.5 meV higher than the latter. For O_2 and S_2 the triplet splitting is not significant, O_2 ($2\lambda = 3.9696 \text{ cm}^{-1}$) and S_2 ($2\lambda = 23.68 \text{ cm}^{-1}$). But for Te_2, $2\lambda \leqslant 2230 \text{ cm}^{-1}$, and for Se_2, it is of the order of kT at about 700°K. As a result, the λ split components are equally populated in O_2 and S_2, and the $X1g$ component in Te_2 is hardly populated, but in Se_2, according to Boltzmann distribution, the $X1g$ population is about 0.47 times the population of XO_g^+ and hence both contribute to the spectrum — although ordinarily the finer lines are not resolved.

A part of the low energy photoelectron spectra of Se_2 and Te_2 vapor is shown in Fig. 4.20. In Se_2, the first band (which is not shown in the figure) is due to the loss of one of the $\pi_g 4p$ electrons, which leaves the ion in the $X^2\Pi_g$ state; the spin–orbit resolved lines are at 8.89 eV and 9.13 eV assignable to $^2\Pi_{g,1/2}$ and $^2\Pi_{g,3/2}$ as the final states. The probable substates involved are $O_u^+(Se_2^+) \stackrel{np\sigma}{\leftarrow} XO_g^+(Se_2)$ and $O_u^+(Se_2^+) \stackrel{np\pi}{\leftarrow} XO_g^+(Se_2)$ respectively. Owing to the larger magnitude of spin–orbit splitting compared with the splitting of Ω substates, the transitions are similar to those from a conventional ground state to two spin–orbit split ion states. Comparison of spectral intensities show that the intensity ratio of $^2\Pi_{g,1/2}$ to $^2\Pi_{g,3/2}$ for Se_2 is between those of S_2 and Te_2, but with 2λ approaching the spin–orbit splitting, as happens with increasing

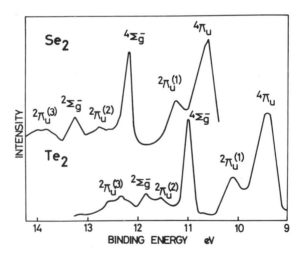

Fig. 4.20.   He I photoelectron spectra of Se_2 and Te_2 [397].

atomic weight of the chalcogen, the coupling tends to Hund's case (c) and the $^2\Pi_{g,1/2}$ state is strongly favored.

Loss of an electron from the second highest orbital leaves the ion in the configuration $(\sigma_g 4s)^2(\sigma_u 4s)^2(\sigma_g 4p)^2(\pi_u 4p)^3(\pi_g 4p)^2$ with five possible states: $^4\Pi_u$, $^2\Phi_u$, and three $^2\Pi_u$ states. For the $^2\Phi_u$ state to be reached requires an excited neutral state, and this has been observed in the spectrum of the excited oxygen molecule. The 10.68 eV peak is due to the $^4\Pi_u$ state and the 11.27 eV due to the $A\,^2\Pi_u$ state, and on the basis of calculated ionization energy of $^2\Pi_u^{(2)}$ in O_2, SO, and S_2, relative to the $^2\Pi_u^{(1)}$ and $^2\Pi_u^{(3)}$ bands, the ionization energy of $^2\Pi_u^{(2)}$ (Se_2^+) can be predicted to be about 12.9 eV.

Ionization to the configuration $(\sigma_g 4s)^2(\sigma_u 4s)^2(\sigma_g 4p)^1(\pi_u 4p)^4(\pi_g 4p)^2$ results in the $^4\Sigma_g^-$ and $^2\Sigma_g^-$ states. By analogy with S_2^+, the 12.27 eV peak can be assigned to $b^4\Sigma_g^-$ and the 13.31 eV peak to $^2\Sigma_g^-$. Two diffuse maxima occur at 12.59 eV and 12.81 eV, and in pursuance of the aforementioned prediction the latter can be assigned to $^2\Pi_u^{(2)}$. The tellurium spectrum is very similar to that of selenium. In Te_2 the weak band at 11.87 eV is assigned to $^2\Sigma_g^-$ and that at 12.42 eV to $^2\Pi_u^{(3)}$.

Ionization from ns orbitals results in an ion configuration $(\sigma_g ns)^2(\sigma_u ns)^1 (\sigma_g np)^2(\pi_u np)^2(\pi_u np)^4(\pi_g np)^2$, which corresponds to the states $^4\Sigma_u^-$ and $^2\Sigma_u^-$. The configuration $(\sigma_g ns)^1(\sigma_u ns)^2(\sigma_g np)^2(\pi_u np)^4(\pi_g np)^2$ correspond to the states $^4\Sigma_g^-$ and $^2\Sigma_g^-$. Some of these have been identified in the case of oxygen and shown earlier in this chapter. The 16.20 eV region, not shown in the figure, has a large number of peaks. They are relatively more distinct in

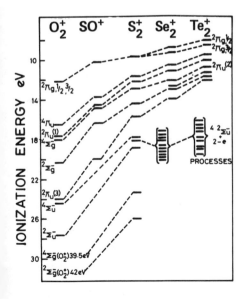

Fig. 4.21. Correlation diagram of O_2^+, SO^+, S_2^+, Se_2^+, and Te_2^+ levels [397]

tellurium spectra, contains $^4\Sigma_u^-$ and $^2\Sigma_u^-$ states as well as peaks from shake-up processes. The $^4\Sigma_g^-$ and $^2\Sigma_g^-$ are not seen in HeI spectra and are expected to appear beyond 21 eV. The analysis of shake-up bands, however, is very much limited by the lack of information on excited ionic states. A correlation of orbital energies for O_2^+, S_2^+, SO^+, Se_2^+, Te_2^+ on the basis of photoelectron spectra is given in Fig. 4.21.

From the preceding examples we find that applications of valence electron spectra, although varying in detail, are primarily helpful in locating and identifying the nature of molecular orbitals of sample molecules. From the case studies we have considered here, it would seem that the assignment of photoelectron spectral bands to various molecular orbitals is not always a simple matter. In the interpretation of the benzene spectrum, for example, we saw that complementary techniques such as mass spectrometry, energy loss spectrometry, and vacuum ultraviolet spectroscopy are of significant help. Although the identification of bands from experimental data alone would be most desirable, it turns out that one is constrained to take assistance from the results of theoretical calculations on orbital energetics, and the role of this aspect has been seen in several cases discussed here. The orbital energy correlations that can be established using photoelectron spectra are seen in the examples of the acenes, the biphenyls, Se_2, and Te_2, and also in the example of hexafluoroacetylacetonate complexes. In summary, we note that whereas each individual case imposes somewhat peculiar constraints on the exact procedure that leads to complete information on molecular energy levels and structure, for direct determination of the valence orbital energies and the bonding characteristics of orbitals, the photoelectron spectroscopic technique plays a particularly powerful role.

CHAPTER

5

SPECTRA FROM SOLIDS AND SURFACES

From the time the photoelectron spectroscopic method was introduced, studies on solids have received intense attention. This consists, first, in determining the binding energies of electrons in various shells of elements, most of the elements being solids at ordinary temperature. There has been extensive activity in this area because of the superiority of the photoelectron spectroscopic method over other methods that use, for example, X-ray absorption. The photoelectron spectroscopic method provides direct information on band structures of solids, hence it also presents a powerful way to test the results of band structure calculations. Not only have the band structures of metals and alloys been studied, but band structures of semiconductors and insulators have also been investigated, and there have been studies to ascertain the angular momentum character of conduction bands. The photoelectron spectroscopic method has been applied equally effectively to organic and inorganic solids.

Whereas photoelectron spectroscopy is extensively used for the determination of bulk properties of solids, it turns out that the method is particularly potent for probing the region close to the surface. This property arises out of the constraints of the electron mean free path inside the solids studied, which can restrict emergence of the photoelectrons that are formed deep inside the bulk of the sample. For the same reason, the potential application of photoelectron spectroscopy to surface studies is enormous; the sensitivity to surface constituents can be so high that one is able to examine monolayers of adsorbed material on surfaces. The photoelectron spectroscopic method has found extensive use as a tool in investigations of surface properties, including minutely detailed studies of catalyst surfaces and adsorption phenomena. In this chapter, we first discuss the nature of the information on bulk electronic structure that can be derived from photoelectron spectroscopic studies, and then we consider applications of the method to studies of surfaces. Angular distribution of photoelectrons from solids is considered in Chapter 7.

VALENCE BAND PHOTOEMISSION

Preliminary to an understanding of the photoelectron spectra of solids and surfaces is a consideration of the band structure of solids and the process of photoionization itself as it applies to a solid. A schematic of energy levels and bands that one encounters in a non-metallic solid is shown in Fig. 5.1a. Here E_{vac} represents the *vacuum level*, E_F the *Fermi level*, E_C the bottom edge of the *conduction band*, E_V the top edge of the *valence band*, and E_1 and E_2 a set of core levels. In order that an electron may escape from the solid, it is essential that the energy of the incident photon is at least equal to the difference between

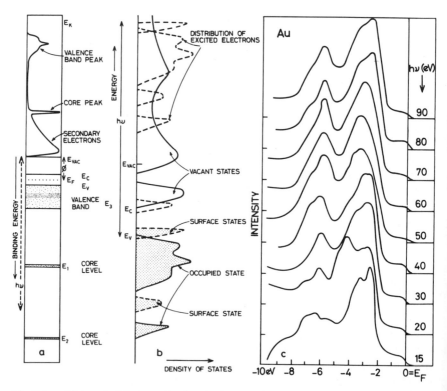

Fig. 5.1. (a) Spectral features in photoionization of a solid. (b) Bulk and surface states; photoexcitation of electrons. (c) Spectra of polycrystalline Au for a range of incident photon energies. Adapted from R. H. Williams [65] (copyright 1978, Taylor & Francis), and from J. Feeouf et al. [427] (copyright 1973, Pergamon Press Ltd).

the energy of the state the electron is in and E_{vac}. In a metal, the minimum energy necessary for ionization of the most loosely bound electron is the difference between E_{vac} and the energy of the Fermi level. This is the *work function* ϕ.

In photoelectron spectroscopic studies, the incident photoionizing radiation is monochromatic. If a radiation having energy equal to the length of the dashed line in Fig. 5.1a is used, all those levels below E_{vac} that fall within the length of the dashed line will be amenable to photoionization. In this schematic, it means that photoionization up to the level E_1 will be possible but the binding energy of E_2 will be too high for any ionization. As a result of such photoionization in a metallic sample, the electrons at the Fermi level E_F, which have the least ionization energy, will result in the highest kinetic energy electrons, and those from the core level E_1, which have the highest binding energy, will result in the lowest kinetic energy electrons. If the binding energy E of the electron lost is measured from the Fermi level, its kinetic energy E_K may be expressed approximately by the relationship $E_K = h\nu - (E + \phi)$.

Apart from the electron mean free path constraints, the intensity of the photoelectron spectral peaks depends primarily on two factors: photoionization cross section of the initial state at the frequency of the radiation used, and population density of the initial state. In Fig. 5.1a, the expected electron kinetic energy spectrum, that is, the *energy distribution curve* (EDC), is shown schematically above the vacuum level E_{vac}. Two major peaks are seen; the high-energy peak is due to the valence band electrons, and the low-energy peak is due to the core electrons in the level E_1. The core electron peak is located on a background of low-energy electrons that peaks near zero kinetic energy. This background is due to inelastically scattered photoelectrons, which have their origin inside the solid. Such scattering may be due to a number of processes.

The photoelectrons that are elastically scattered do not lose any energy in collision; only their directions undergo change. The electrons that are inelastically scattered by lattice vibrations or phonons cause only meV changes in energy, and therefore such inelastic energy losses are negligible. The electron–electron scattering processes include the process of excitation of a target electron to generate in a bound–bound transition an electron–hole pair, the electron in the state to which it is excited and the hole where the electron is lost, but both still in bound states. In a nonmetal this energy would be equal to the band gap. A highly energetic ("hot") free electron present in the bulk of the solid may lose in this manner arbitrary amounts of energy over and above that needed for the creation of the electron–hole pair.

Electron energy may also inelastically scatter because of the excitation of *plasmons*, which are collective oscillations of valence electrons. The frequency of such collective oscillations is given by $(4\pi n e^2/m)^{1/2}$ where n is the density of free electrons, m is the electronic mass, and e is the electronic charge. There may be surface plasmons as well as bulk plasmons, and for an ideal surface the frequency of surface plasmons is $1/\sqrt{2}$ times their frequency in the bulk. When the electron energy is less than the plasmon energy or the energy needed to create an electron–hole pair, the loss of energy due to collisions is less, and therefore the mean free path between inelastic collisions is large. This happens to the low-energy electrons in semiconductors and insulators. For example, solid Kr has an energy gap of 11.5 eV and a mean free path of 100 nm for electrons generated by 12.7 eV photons. In contrast, for electrons having energy greater than that needed for plasmon excitation, the mean free path is small, < 1 nm, which is comparable to the thickness of a few atomic layers. This is the case for electrons in the energy range 20–200 eV. This means that of the electrons in this energy range that escape from the solid, most must originate from about 1 nm of the surface and hence may be expected to carry information predominantly from this region of the solid. Similarly, electrons having large mean free paths may be expected to carry more information from the bulk. Later in this chapter we discuss in detail the question of this *escape depth* in connection with photoelectron peak intensities expected from the surface and bulk regions of a solid. Here we merely note that electrons that are scattered inside the solid, repeatedly losing energy, may finally emerge through the surface as low-energy electrons, and it is these secondary electrons that constitute the background seen in the photoelectron spectrum of Fig. 5.1a.

Whether the photoelectron peak intensities truly reflect the density of initial states depends in an important way on the availability of the final states. This may be understood by a consideration of Fig. 5.1b, which schematically shows the bands near the vacuum level. There are basically two types of states in a solid: the *bulk states* and the *surface states*. The former is due to electronic structure of the bulk of the solid and the latter due to surface electronic structure, which may be quite different from the bulk electronic structure. Both these types of states can be either vacant or occupied by electrons. A photoionization transition from a particular initial state can take place only if there exists a vacant state above the occupied one having an energy higher by an amount equal to the energy of the incident photon. Thus, if a band gap exists at the final state energy, no photoionization can take place and photoelectrons are not ejected. In such a situation the photoelectron spectrum does not reflect the density $N(E)$ of initial states. The nature of the final states available may thus drastically alter the pattern of the photoelectron spectrum observed. This effect is particularly severe near the vacuum level, where band gaps are numerous. But with increasing final-state energy a continuum of final states becomes available — through level broadening due to the uncertainty principle — and photoionization for all initial states becomes possible so long as the photon energy is sufficiently high. With low-photon energies the photoelectron spectral data may only display a pseudo density of states; with high-photon energies the spectrum is more likely to be a true reflection of the density of initial states. An illustration of this is seen in Fig. 5.1c, which shows photoelectron spectra of Au with incident photon energies in the range 15–90 eV. It is clearly seen how with increasing photon energy the spectra converge into an invariant pattern representing the true *density of states* (DOS). With this introduction, we now look at a few examples of valence band spectra.

VALENCE BAND SPECTRA

The XPS valence band spectrum of an Na sample prepared by evaporation on Al substrate is shown in Fig. 5.2. As the photoemission cross section of valence electrons in Na is low, the spectrum was recorded over a period of several hours and for this reason shows high fluctuations. A parabola drawn through the experimental points is consistent with free electron representation of the valence electrons of Na and yields a valence bandwidth of 3.2 ± 0.1 eV. If a correction is applied because of the linear background of secondary electrons, one obtains a lower value of the bandwidth equal to 2.8 ± 0.1 eV. An average of the two values is 3.0 ± 0.2 eV. Calculated bandwidth with a free electron model or using a pseudopotential method gives 3.2 eV. Temperature broadening is found to be negligible.

The AlK_α XPS valence band spectra of Pd, Cd, and Te single crystals are shown in Fig. 5.3. Spectra clearly show the doublet structure of $4d$ levels. When normalized with respect to Sb, the experimentally observed separation between the $4d$ doublet components seen in Ag, Cd, In, Sn, Sb, and Te is in agreement

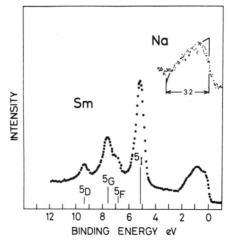

Fig. 5.2. Valence band spectra (XPS) of Na and Sm [428, 431].

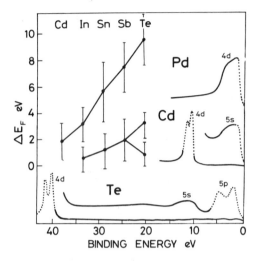

Fig. 5.3. Valence band spectra (XPS) of Pd, Cd, and Te. The inset shows the positions of $5s$ and $5p$ bands relative to the Fermi level [430].

with theory. Cd shows two and Te shows three peaks. In the high-sensitivity spectrum of Te, the band that is closest to the Fermi level is a $5p$ band and the peak at higher binding energy is due to $5s$ states. The inset curves show positions of $5s$, $5p$ peaks of Cd, In, Sn, Sb, and Te relative to the Fermi level. Analysis of spectra in Fig. 5.3 and also spectra of In, Sn, and Sb shows how atomization of $5s$ electrons, from band structure to isolated atom structure, occurs as one proceeds from Cd to Te. Measurements show that the ratio of the areas under the characteristic $5s$, $5p$ lines is the same as the ratio of occupation numbers for free atoms with a configuration of $5s^n 5p^m$.

We also consider here the valence band spectrum of Sm, shown in Fig. 5.2. In rare earths the principal features of XPS spectra are due to $4f$ electrons, the latter having a localized character due to high orbital quantum number l. In these metals the valence bands consist of hybridized $6s$ and $5d$ states, and near the Fermi level the partial density due to d states is significantly higher than the corresponding density of s states. Samarium, with an electronic configuration of $4s^2 4p^6 4d^{10} 4f^5 5s^2 5p^6 5d^1 6s^2$, is considered to be trivalent, having five $4f$ electrons with parallel spins. With d electrons constituting the band this implies, considering the five $4f$ atomic electrons, an initial ground state of $^6H_{5/2}$, and with one electron less the final ionic states of ^5I, ^5F, ^5G, and ^5D. The valence band region of the photoelectron spectrum of Sm in Fig. 5.2 shows that the valence band and the $4f$ structures are well separated, and lines due to the ^5I, ^5F, ^5G, and ^5D final states are clearly observed.

VALENCE BAND SPECTRA

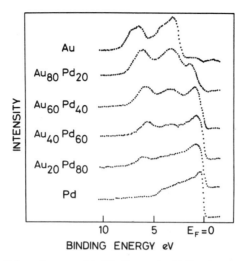

Fig. 5.4. Spectra of AuPd alloys using HeII radiation [432].

Figure 5.4 shows UPS valence band spectra of a series of Au-Pd alloys using He II 40.8 eV photons. The Au–Pd system forms a solid solution, an fcc alloy, for the entire range of compositions. Both Pd and Au possess broad d bands, extended over energy ranges that show considerable overlap. The change in bandwidth and the effective spin–orbit splitting of the Au $5d$ band depend only on its concentration in the alloy, implying that the Au $5d$ electrons interact only weakly with the Pd electrons. With increased Pd concentration, however, the Au d bands decrease in both width and splitting. This may be taken to indicate the importance of like-neighbor interactions. States associated with Pd d electrons change from a virtual bound state — a state at low concentration where a large concentration of Au perturbs the potential at Pd sites — to form a d band at 40% Pd. At higher concentrations of Pd, this band shifts closer to the Fermi level.

A very close approximation to an undistorted electron density of states function can be obtained from high resolution XPS data. We now examine a few more valence band spectra of single crystals with a bit more detail on band structures. Spectra of AgCl, AgBr and AgF taken at 100°K are shown in Fig. 5.5 with the Fermi level taken as zero of binding energy. The XPS (AlK$_\alpha$) spectra of AgCl and AgBr are very similar; the difference is mainly in a minor shift in the absolute energies and a broadening of the chloride peak in comparison with that of the bromide. Results of recent band structure calculations are also shown along with the spectra.

We may add here that AgCl, AgBr, and AgF are known to exist in NaCl-type structures, that is, in two face-centered lattices displaced with respect to each

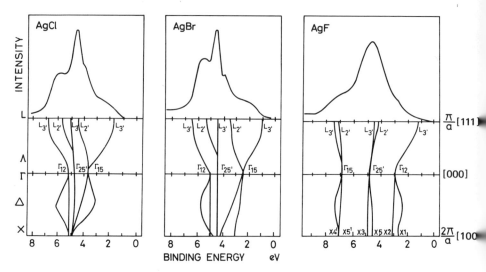

Fig. 5.5. Valence band spectra (XPS) and energy band structure of AgCl, AgBr, and AgF [433].

other by half the body diagonal. The Brillouin zone for fcc lattice with origin Γ, axes k_x, k_y, k_z in the k space, and the critical points are shown in Fig. 5.6. The various high symmetry points arise from the irreducible representations of the crystal space group. In this figure X is the point on the surface of the Brillouin zone in the direction Δ, that is, [001] ($k_x = 0, k_y = 0, k_z = 1$), and equivalent directions such as [100] and [010]. Hence there are three nonequivalent X-points. The point L lies along the direction Λ([111] and equivalent directions). There are four nonequivalent L points. K is a point along the direction

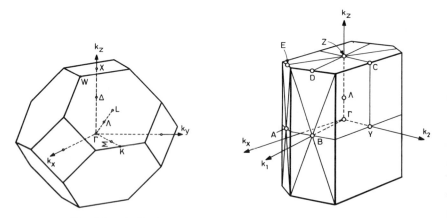

Fig. 5.6. Brillouin zones for fcc and simple monoclinic crystals.

Σ([110] and equivalent directions). W is along [102] and equivalent directions; there are six nonequivalent W points in groups of four. The coordinates of the symmetry points X, L, W, and K are $(0, 0, 2\pi/a)$, $(\pi/a, \pi/a, \pi/a)$, $(\pi/a, 0, 2\pi/a)$, and $(3\pi/2a, 3\pi/2a, 0)$ respectively where a is the lattice constant, that is, the side of the unit cube of the lattice.

In AgCl and AgBr, the lowest binding energy feature is due to a strongly hybridized halogen $2p$ level, which is broadened by mixing with the lowest Ag $4d$ state. A second component of halogen p level results in the shoulder at 3.8 eV. The intense maximum that follows is due to an Ag $4d$ state, which remains atomiclike throughout the Brillouin zone. The second component of the crystal field split Ag $4d$ state, broadened by mixing with the lowest binding halogen p state, results in the highest binding energy peak. A comparison of the results from UPS and XPS given in Table 5.1 indicates good agreement.

In AgF, the broad band that is observed near the threshold region of AgCl and AgBr spectra is absent. Instead, one appears on the high binding energy side. This is interpreted as an indication that in contrast to the situation in AgCl and AgBr, the Ag $4d$ levels form the topmost portion of the valence band in AgF and the broad structure at higher energy represents the F $2p$ derived states. This assignment is consistent with band structure calculations as well as optical absorption spectra. The maxima in the density of states are manifest in a peak at

Table 5.1. XPS and UPS Data (eV) on Valence Bands of AgF, AgCl, AgBr, and AgI [433]

	XPS	UPS	Assignment
AgF	3.0 ± 0.4		Ag $4d$
	4.6 ± 0.1		Ag $4d$
	6.7 ± 0.4		F $2p$
AgCl	2.7 ± 0.4	2.3 ± 0.1	Cl $3p$
	3.8 ± 0.3	4.1 ± 0.1	Cl $3p$
	4.5 ± 0.1	4.8 ± 0.1	Ag $4d$
	5.8 ± 0.2		Ag $4d$
AgBr	3.0 ± 0.4	2.3 ± 0.1	Br $4p$
	3.86 ± 0.1	4.3 ± 0.1	Br $4p$
	4.57 ± 0.1	5.1 ± 0.1	Ag $4d$
	5.6 ± 0.3	6.4 ± 0.1	Ag $4d$
AgI	2.1 ± 0.2	2.0 ± 0.2	I $5p$
	4.88 ± 0.1		Ag $4d$
	5.57 ± 0.1	5.9 ± 0.1	Ag $4d$
	2.5 ± 0.5		I $5p$
	5.4 ± 0.3		Ag $4d$

4.6 eV with a shoulder at 3 eV, due to a crystal field splitting of the Ag $4d$ states. This splitting of 1.6 eV is identical to the exciton splitting that has been observed, and these are most likely due to direct–forbidden transitions whose origins are Γ_{12} and Γ_{25}' points of the valence band and termination near the conduction band minimum at Γ_1.

Unlike the other halides, AgI does not crystallize in simple NaCl structure. At room temperature it exists either in β phase, a hexagonal Wurtzite structure, or in γ phase, a cubic zinc blende structure. Above 419°K, it exists in the α phase, a complicated structure wherein the Ag ions are statistically distributed among the vacant sites of a body-centered cubic iodide lattice. In the photoelectron spectrum of AgI at 100°K (not shown in the figure) a broad weak peak appears near the threshold and is assigned to the I $5p$ levels; the Ag $4d$ states cause a strong doublet at higher binding energy. The broadening of the Ag $4d$ doublet into a single peak occurs well below 300°K but is not connected with

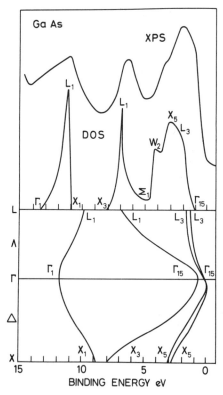

Fig. 5.7. Valence band spectrum (XPS), density of states, and energy band structure of GaAs [27, 435, 436, 437].

a phase transition. The broadening of the I $5p$ levels and the shift of the threshold to lower binding energy take place sharply at the transition temperature. This is taken to indicate a decreased band gap of the high temperature α phase. The ratio of integrated intensity of Ag $4d$ to I $5p$ bands is 5:1 in contrast to the 5:3 observed for similar ratios in AgBr and AgCl. This is due to a relatively low photoionization cross section for I $5p$ levels and is supported by results of other experiments.

In this interpretation, spin—orbit splitting has been ignored. This is justified by some recent theoretical calculations which show that the largest splitting is expected between the components of the highest Ag $4d$ band, which turns out to be only 0.1 eV. These results also indicate that spin—orbit splittings fall off rapidly away from $k = 0$ and that throughout most of the Brillouin zone, these splittings are most likely smaller than the instrumental resolution.

GaAs has a zinc blende structure (fcc), with Ga at $(0,0,0)$ and As $(\frac{1}{4},\frac{1}{4},\frac{1}{4})$. The Brillouin zone of fcc lattice, along with the symmetry axes and high symmetry points, has already been presented in connection with silver halides. That UPS and XPS methods provide direct information about the density of states of valence bands can be seen from Fig. 5.7, which shows the XPS photoemission from crystalline GaAs, the results of a density-of-states calculation, and a GaAs energy band diagram. It will be seen from Table 5.2 that the energies of the critical points obtained from experiment and those obtained from calculations using the pseudopotential method are in fair agreement.

Finally, we see in Fig. 5.8 the XPS spectrum of the unusual polymeric monoclinic crystal consisting of $(SN)_x$ chains, which shows metallic properties and is superconducting at low temperatures. The weak shoulder at A in the spectrum is due to the π conduction band and the peak B is from overlapping σ and π bands. The peaks C, D, E are due to σ electrons. The uppermost occupied level is a broad π band with 2 eV width. Theoretical density of states calculated using S_2N_2 and then extended to $(SN)_x$ are also shown in Fig. 5.8. Band structure calculations have been performed on the single $(SN)_x$ chain and also for the two-

Table 5.2. Energies (eV) for the High Symmetry Points of GaAs Valence Band [435, 436, 437]

	UPS	XPS	Pseudo-potential		UPS	XPS	Pseudo-potential
L_3	0.8	1.4 ± 0.3	0.9	W_1	–	6.1 ± 0.1	6.6
X_5	–	2.5 ± 0.3	2.5	$X_3(L_1)$	6.9	7.1 ± 0.2	6.8
W_2	–	4.0 ± 0.2	3.5	X_1	10.0	10.7 ± 0.3	11.4
Σ_1^{min}	4.1	4.4 ± 0.2	3.9	L_1	–	12.0 ± 0.5	–
				Γ_1	12.9	13.8 ± 0.4	12.8

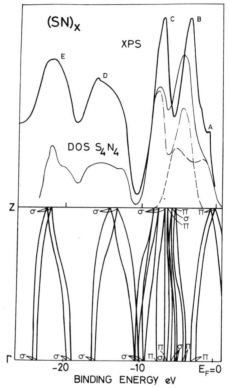

Fig. 5.8. Valence band spectrum (XPS), density of states, and energy band structure of $(SN)_x$ [438, 441].

dimensional crystal. For the latter, two chain geometries are considered which correspond to interactions in the $(\bar{1}02)$ and (100) planes of the crystal. The one-dimensional calculation yields a symmetry-induced degeneracy at the zone edge for all bands. The Fermi level lies at the top of the second highest band. In the two-dimensional $(\bar{1}02)$ plane and (100) plane calculations, each band of the one-dimensional case becomes split into bands due to interchain interactions. To a good approximation, the band structure for the three-dimensional crystal can be constructed by superimposing the σ bands from the calculations on the $(\bar{1}02)$ plane and π bands from the calculations of the (100) plane. Results thus obtained for the ΓZ direction, which is the direction of the chain axis, are shown in Fig. 5.8. Calculations also show strong dispersions of the energy bands along ΓZ and directions parallel to ΓZ, such as YC. The Fermi surface cuts two bands, in the ΓZ and YC directions, and this explains the origin of metallic behavior.

SURFACE STATES

A knowledge of surface states is important; they have a major role in determining the properties of solids, for example, in metal—semiconductor junctions. The presence of surface states can be detected by photoelectron spectroscopy. A good example of the effectiveness of the photoelectron spectroscopic method applied to investigations of surface states involves studies of the Si (111) surface. Low energy electron diffraction (LEED) studies show that the surface Si atoms constitute a 2 × 1 pattern, that is, the surface unit cell dimensions are 2 × 1. The photoelectron spectrum of the Si (111) surface is shown in Fig. 5.9, with the binding energies presented in relation to the valence band edge E_V. The dashed line shows the spectrum that is expected from the bulk. The lowest binding energy peak cannot originate from the bulk structure, but can be explained in terms of a surface state. A possible mode of surface rearrangement has been shown in the inset of Fig. 5.9 where alternate rows of atoms are at a lower level. The photoelectron emission observed from the surface state could be due to the dangling bonds, remnants of broken bonds caused by cleavage of the crystal. That the first peak S_0 is due to a surface state receives support from the fact that the surface state emission undergoes significant change if a small amount of chlorine is admitted into the system; the next peak shifts to higher binding

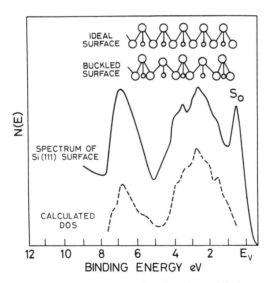

Fig. 5.9. He I spectrum of silicon (111) surface (continuous line) compared with calculated local density of states (dashed line). The peak close to E_V in the experimental spectrum is due to the dangling bond surface states [65].

energy indicating that some adsorption takes place. Theoretical band positions on the basis of a rumpled surface model explain the experimentally observed photoemission results. Bonding electrons between the atoms of the surface layer and the layer underlying it are expected to be of different distribution than those between two bulk layers. LEED studies after annealing of the samples show a change in the surface structure; a 7 × 7 surface unit mesh is observed. This leads to a change in surface states but not bulk states.

Other electron spectroscopic methods also provide information about vacant surface states. A beam of monochromatic electrons incident on a Si (111) surface may excite electrons from the occupied to vacant states and this shows up in the electron energy loss spectrum. In Si, such a peak is observed at a loss energy of 0.52 eV. Since the band gap (the gap between the lowest unoccupied and the highest occupied levels) of bulk Si is 1.1 eV, the 0.52 eV loss peak is interpreted as due to a surface state transition.

We mentioned earlier that if at the final energy there are no vacant states available, then such a transition cannot take place. Since secondary electrons occupy vacant energy states (they are free electrons and therefore occupy levels above Fermi level), absence of such states in a certain energy range for a certain crystallographic direction means that for that crystallographic direction no secondary electrons will be emitted. This is true, for example, for the (001) surface of W, and angularly resolved secondary electron spectroscopy can detect such surface states.

Photoelectron spectroscopy using variable photon energy can also detect unoccupied surface states. In this *photoemission partial yield spectroscopy*, the number of secondary electrons is monitored as a function of increasing photon energy. For sufficiently high radiation energy, core holes are created and the conduction band is populated. This core hole may be filled by Auger process. The emanating Auger electron can cause additional secondary electrons (partial yield), and since it depends on the strength of the initial transition from the bound to unoccupied states in the surface region, this could be a way to study surface states.

CORE LEVEL SPECTRA

Although core level photoelectron peaks from solids have binding energies characteristic of the element involved and also show chemical shifts as in free atoms, they do not show simple lifetime-broadened Lorentz profiles. Metal photoelectron peaks, for example, show complex shapes as the conduction electrons provide a collective response following the creation of a core hole. Further, as the charge on an atom of the solid suddenly increases by one owing to the ejection of a core electron, and the potential energy surface of the hole state moves toward smaller internuclear distances, an electronic transition from the initial neutral state to the steep part of the final hole state potential curve causes core peak broadening; this is somewhat similar to the vibrational broadening of the core lines of free atoms. Any change in internuclear distance due to vibrations in the ground state shows up as a broadening of the core peak by excitations of different numbers of phonons in the final state; the latter may be termed as a Franck–Condon excitation of the lattice. Besides these features of core peaks in solids, there are shake-up lines, configuration interaction satellites, and satellites due to multiplet splitting.

The origin of shake-up satellites is similar to that described in Chapter 3, but in solids they involve filled and empty states near Fermi energy, the transitions involving definite initial and final states. Relaxation of the atom and the lattice around the core hole takes place, and the magnitude of the relaxation energy determines the extent to which the excited states are populated. The relaxation energy decreases as the photohole gets closer to the valence shell, and this causes reduced shake-up intensities in such transitions. Shake-up satellites are absent in metals as the outer shell excitations are prevented by rapid screening of the core hole by conduction electrons. In the shake-up processes in solids, electron transfer from ligands to the ionized atom may also make substantial contributions, and an isolated atom description is not adequate.

Lines have been observed in core spectra that are associated with a single valence shell core level, and they are therefore not due to a shake-up process such as in alkali metal ions in K $3s$ and Rb $3s$ lines. Such lines are due to configuration interaction involving two-hole final states. In the case of K $3s$ line, the states $3s^1 3p^6$ ^2S and $3s^2 3p^4 3d$ ^2S, both two-hole states having ^2S symmetry, are involved. An essential condition for such interaction is that the symmetries of the two states be identical. For appreciable intensity of the satellite line the energies of the one-hole parent and the two-hole excited state must be close.

Multiplet splitting is observed when interaction takes place between the unpaired electron generated in the solid following core ionization and electrons in incomplete outer d or f shell, as in transition metals and rare earth atoms. The interaction is between the spin and angular momentum of the unpaired core electron and the total spin and orbital angular momenta of the electrons in the outer shells of the atom or the ion. In general, the number of resulting final states is relatively small when a core s electron ($l = 0$) is ionized, and for the outer shells the total angular momentum is $L = 0$. But the total number of final states is very large; this can be determined by the usual rules of angular momentum coupling, when one has $l \neq 0$ for the core electron and also $S \neq 0$, $L \neq 0$ for the outer electrons. Such split lines are a common feature in the spectra of transition metals and rare earth compounds. The agreement between theory and experiment on the extent of splitting and intensity ratios to peaks is not satisfactory in all cases; inadequate incorporation of electron correlation effects, which play an important part, could be one reason for such discrepancies. Configuration interaction effects may also be considered as electron correlation effects caused by electrostatic interaction of the electrons in the solid.

PLASMON PEAKS

Plasmon peaks have been seen in many XPS spectra of solids. The spectrum of Al + Al$_2$O$_3$ showing plasmon peaks is shown in Fig. 5.10a. In this spectrum we see a series of four volume plasmon peaks following the 2s peak of the metal, and two plasmon peaks for the 2p peak. From the completely resolved 2s and 2p peaks of the metal and the oxide, the relevant chemical shift can be easily determined.

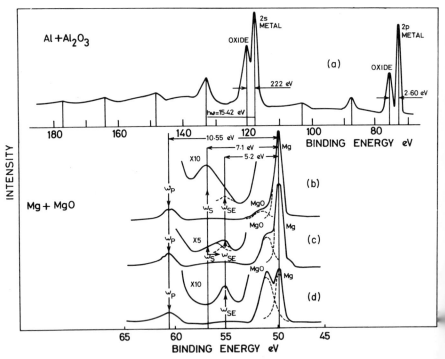

Fig. 5.10. A series of volume plasmon peaks in the Al + Al$_2$O$_3$ spectrum, one volume plasmon peak (ω_p) and surface plasmon peaks (ω_s) in the Mg + MgO spectrum at different stages of oxidation of Mg. For a thick oxide layer the relaxed position is at $\omega_p/\sqrt{1 + \epsilon}$, where ϵ is the dielectric constant of the layer [35].

The bottom portion of Fig. 5.10 shows Mg + MgO XPS spectra at various stages of oxidation. As in the aluminum spectrum, several peaks are due to volume plasmons. The first one (ω_p), 10.55 eV removed from the Mg peak, can be seen in Fig. 5.10b. The spectrum also shows a surface plasmon peak at about 7.1 eV from the Mg peak, which corresponds closely to the theoretical value of $\omega_p/\sqrt{2} \sim 7.4$ eV for a metal–vacuum interface. A low-intensity plasmon distribution extending toward the fully relaxed surface plasmon $\omega_{s\epsilon}$ is discernible at $\omega_p/\sqrt{1+\epsilon} \sim 5.2$ eV, which corresponds to a dielectric constant of $\epsilon = 3.1$ for the MgO layer. With an increase in the oxide layer thickness, the surface plasmon distribution is strongly influenced; the original peak at ω_s is attenuated and moves towards $\omega_{s\epsilon}$, as seen in Fig. 5.10c. Figure 5.10d shows that at 1–2 nm oxide layer thickness, the surface plasmon peak merges with $\omega_{s\epsilon}$. Such observations can be used to monitor surface oxidation phenomena by following the course of plasmons.

PHOTOELECTRON SPECTROSCOPIC DETERMINATION OF FERMI LEVEL

Recently the photoelectron spectroscopic technique has been applied for a determination of position of the Fermi level E_F in amorphous hydrogenated silicon, a-Si:H. Since there is no true band gap in a-Si:H, it is difficult to ascertain the relationship between a shift in E_F and the activation energy ΔE that can be obtained from conductivity measurements as a function of temperature (using the expression $\sigma = \sigma_0 \exp -\Delta E/kT$ where σ is the conductivity). This is particularly so if the charge carriers occupy states in an extended energy range in the pseudogap. In this regard, the determination of E_F from photoemission data is fairly straightforward.

Amorphous hydrogenated silicon can be substitutionally doped with either boron or phosphorus, much like crystalline silicon. The conductivity of the doped Si changes by as much as four orders of magnitude. In a recent experiment on measurement of Fermi level using photoemission data, the a-Si:H samples were grown as thin films on Mo substrate by glow discharge in SiH_4 and Ar. When the a-Si:H was to be doped, the discharge mixture contained respectively PH_3 or B_2H_6, depending on whether n-type or p-type material was desired. Since Mo forms ohmic contact with both a-Si:H and the electron energy analyzer material, the binding energies are obtained with respect to the Fermi level of a-Si:H. The valence band edge is obtained from the intercept of the steepest descent of the leading edge of the valence band spectrum, as shown in the inset of Fig. 5.11. Peak I originates mainly from Si $3p$, and peaks C and D are due to Si–H bonding peaks. If in the photoelectron kinetic energy spectrum the onset of the valence band is at E_V^{kin} eV then the location of the valence band edge with respect to vacuum level is $h\nu - E_V^{\text{kin}} = E_{\text{vac}} - E_V$ where E_V is the energy of the valence band edge referred with respect to the vacuum level. If Mo work function is denoted by ϕ_{Mo}, then the location of Mo Fermi level is given by $E_{vac} - \phi_{Mo}$. Since in this case Mo and a-Si:H Fermi levels are at the same energy, the location of the Fermi level of a-Si:H with respect to its valence band edge is given by $(h\nu - E_V^{kin}) - (E_{vac} - E_{F,Mo}) \equiv E_F - E_V$. $E_F - E_V$ thus obtained for a range of doped a-Si:H are shown in Fig. 5.11, where the shifts in the E_F can be clearly seen. Additional evidence that such indeed is the nature of E_F variation is provided by the binding energy values of the core Si $2p$ level obtained from the same series of samples. Here also the binding energies are obtained with respect to the Fermi level. Since it is reasonable to expect that the core level binding energies remain invariant in the process of doping, any change $E_F - E_B$ therefore may be taken to reflect changes in the position of the Fermi level. These $E_F - E_B$ values are also plotted in Fig. 5.11 by shifting the binding energy scale, by 99.6 eV, to match the $E_F - E_B$ and $E_F - E_V$ of the undoped material. The shifts of the Fermi level determined from photoemission measurements in the two different ways are within ± 0.06 eV, which is the accuracy of the measurements.

Using these data along with the data on activation energy ΔE for the various

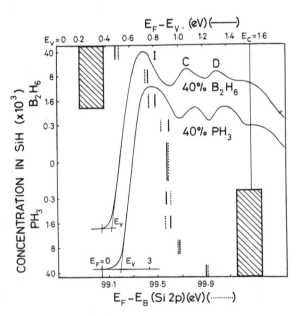

Fig. 5.11. Valence band spectra of boron- and phosphorus-doped samples of amorphous hydrogenated silicon, and position of the Fermi level E_F in the band gap. Adapted from B. von Roedern et al. [464] (copyright 1979, Pergamon Press Ltd).

doped a-Si:H, it is possible to determine the location of the charge carriers. It is known that for the n-type samples ΔE varies from 0.65 to 0.35 as one goes from the least doped to the most doped a-Si:H; the corresponding variation is from 0.65 to 0.2 eV for the p-type samples. If one takes for the n-type activation energy $\Delta E = E_C - E_F$, then for the highest doping case from data in Fig. 5.11 $E_C = \Delta E + E_F = 0.35 + 1.3 = 1.65$ eV. For p-types $\Delta E = E_F - E_V$, $E_V = E_F - \Delta E = 0.5 - 0.2 = 0.3$ eV where E_F's are measured from E_V. If the band gap is taken as 1.6 eV, then it shows that whereas the n-type carriers are located at the conduction band edge, p-type carriers are about 0.3 eV inside the gap, an observation that is supported by field effect measurements.

TEMPERATURE-DEPENDENT PHOTOEMISSION

The presence of magnetic moment in d-band ferromagnetic metals is explained by the existence of two sets of separate valence bands for the two opposite electron spins. The energy difference between these two valence band maxima is called the *exchange splitting*. At ordinary temperatures the Fermi level is located such that one of the spin bands is less filled than the other; as a result, there is a net magnetic moment. We consider here the results of a photoemission experiment in which the critical behavior of d-band photoelectrons near the Curie temperature of polycrystalline nickel has been investigated and the extent of exchange splitting determined.

UPS measurements in the temperature range 100–400°C show an anomalous temperature dependence of the d-band EDC near the Curie temperature $T_c = 358$°C. The temperature dependence of the Ni and Pd d-band peak energies (shown with respect to the respective Fermi levels) are presented in Fig. 5.12. The specific heat is also known to show an anomalous behavior near this temperature. The strongest effects are observed when the incident photon energy is in the 4–5 eV range. According to the band structure model of a ferromagnetic material, a strong temperature dependence of the electronic spectra is predicted between room temperature and T_c because at the latter temperature the exchange splitting δE_{ex} reduces to zero. Magnetic-order fluctuations, which are responsible for specific heat and other anomalies, also require, near the Fermi energy, a similar behavior of the electronic states.

To explain the Ni d-band temperature dependence, besides spin–exchange interaction, electron–phonon interactions and volume expansion need to be considered. Whereas the d-band EDC peak should increase as the exchange splitting tends to zero, electron–phonon interactions and volume expansions decrease the bandwidths and band gaps and thus tend to decrease the d-band energy, as in Pd. An estimate of the exchange splitting of Ni is derived in the following manner.

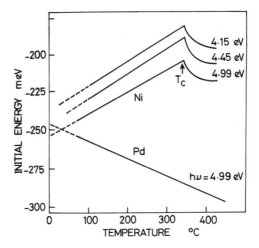

Fig. 5.12. Temperature dependence of Ni and Pd d-band peak energy [466].

Variation of the d-band peak energy is shown in Fig. 5.12. The net d-band peak shift in the case of Ni reflects the resulting effect of both magnetic and nonmagnetic temperature dependence. Assuming that Ni and Pd both have the same nonmagnetic temperature dependence, the magnetic part of the Ni d-band temperature dependence – which would amount to a shift upwards $\Delta_\uparrow$ toward the Fermi level – can be obtained by subtracting the Pd d-band peak energy from that of Ni, that is, $\Delta_\uparrow = \delta E_{tot}(\text{Ni}) - \delta E_{tot}(\text{Pd}) = \delta E_{tot}(\text{Ni}) + |\delta E_{tot}(\text{Pd})|$ where $\delta E_{tot}(\text{Ni})$ and $\delta E_{tot}(\text{Pd})$ represent the magnitude of the d-band peak energy with respect to the Fermi level. This upward shift $\Delta_\uparrow$ in Ni, the magnetic part of the Ni d-band temperature dependence, has two components, one due to the upward shift $\delta E_{d\uparrow}$ of the filled spin band, and the other due to the downward shift $\delta E_{d\downarrow}$ of the unoccupied spin band. The sum of the latter two constitutes the exchange splitting $\delta E_{ex} = \delta E_{d\uparrow} + \delta E_{d\downarrow}$. But if the two valence bands populating the opposite spins do not have much energy overlap at the Fermi level, then the d-band peak energy would be the energy of the filled band and the experimentally observed d-band peak shift $\Delta_\uparrow$ would reflect entirely the upward shift of the filled band, $\Delta_\uparrow = \delta E_{d\uparrow}$. On this basis one obtains from the data of Fig. 5.12, $\delta E_{d\uparrow} = 0.089$ eV. Assuming $\delta E_{d\downarrow} \simeq \delta E_{d\uparrow}$, one derives $\delta E_{ex} \simeq 0.18$ eV. On the other hand, if weighted averages are used, then $\delta E_{d\downarrow} \simeq 2\delta E_{d\uparrow}$, since the density of states for $E_{d\uparrow}$ is twice that of $E_{d\downarrow}$. Then one obtains $\delta E_{ex} \simeq 0.27$ eV as the lower limit of the exchange splitting at optical transition $h\nu \simeq 5$ eV. This experimental result corresponds to an average over the state in the optical transition. For a direct transition this is only a part of the Brillouin zone, but for nondirect transition it encompasses the entire Brillouin zone.

When a difference in the temperature dependence of various EDC peaks exists, dissimilarities in the nature of electron–phonon interaction may be utilized for identification of the states from which different types of photoelectrons originate. For example, a study of silver halides shows that only the hybridized electronic states strongly interact with the lattice. If temperature is varied so that the vibrational energy of the lattice is changed, these special states can be separated out from purer electronic levels.

Temperature dependence of EDCs of AgBr, Ge, LiI are shown in Fig. 5.13. Here Ge has covalent bonding with extended wave functions. The localized valence state orbitals of LiI are quite pure p states and do not have very closely lying energy levels with which p can interact. The reason why AgBr shows such a significant departure in behavior is that $Br(4p)$ and $Ag(4d)$ valence-band wave functions are strongly localized and strongly mix with each other. In AgBr, as the ions vibrate, the wave function mixing undergoes modulation, and fluctuations in hybridization also become temperature-dependent along with the vibrational amplitude.

There are two ways by which this dynamic wave function hybridization becomes manifest. First, with the fluctuation of the valence-state wave functions, the matrix elements of the optical excitation process are also modulated and in turn may lead to energy broadening of the EDC structure. Energy broadening

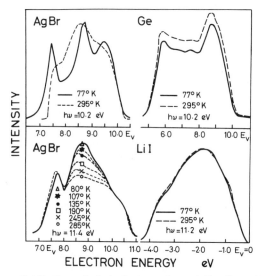

Fig. 5.13. Energy distribution of photoelectrons from AgBr, Ge and LiI at 77°K and 295°K. The electron energy is referred to the valence band maximum, E_V. The distribution from AgBr is also shown at 80°K, 107°K, 135°K, 190°K, 245°K, and 285°K [468].

occurs because, due to the dynamical changes in the matrix elements, amplitudes at all energies change; thus the peak is wider than when the matrix elements and hence also the amplitudes have fixed values. Second, dynamic changes in separation between ions cause a significant modulation of the hybridizing state energies. A change in the lattice constant can be used to simulate the dynamic motion of the lattice. One may then calculate the effect of the static changes on the two-center overlap term as well as on the wave function normalization function to the electronic energies. A change of about 10% in the nearest-neighbor separation changes the overlap integral by 33%, and since the overlap contribution is significant, this means a strong modulation of the electronic energy. Thus it can be seen that the change in EDC due to temperature variation is controlled by the extent of hybridization. In Fig. 5.13, where AgBr results are shown, the peak on the left side of the lower figure is unchanged by cooling, in contrast with transitions that sharpened markedly at $hv = 10.2\,eV$. Such states therefore can be assigned to unhybridized pure states. This would correspond to the uppermost Ag $4d$ states in AgBr. These temperature-dependent UPS studies of AgBr provide strong evidence that silver halide valence band maximum is derived from the halogen p orbitals.

SURFACE AND BULK INTENSITIES, DEPTH PROFILING

In order to be directly useful for binding energy measurements of solids, photoemitted electrons must not undergo any energy loss in the process of emerging from the solid into a vacuum. The *inelastic mean free path* (IMFP) λ, which is a measure of the mean distance traveled by the electron without energy loss, thus also contains information regarding the depth that can be effectively sampled by photoelectron spectroscopy. λ is also called the *escape depth*, and it relates the intensity I_d of the photoelectrons obtained from a layer of material of thickness d, with I_∞, photoelectron intensity from an infinitely thick layer, by the equation $I_d = I_\infty (1 - e^{-d/\lambda})$. By successively putting $d = \lambda$, $d = 2\lambda$, $d = 3\lambda$, it can be seen that this equation implies that 63% of the total signal intensity is due to electrons originating from a layer of thickness λ, 87% from 2λ, and 95% from 3λ. This means that it is quite reasonable to say that most of the photoelectrons originate within a surface thickness of about 3λ.

The effective escape depth of a material can be determined by measuring the intensity of a photoelectron spectral line obtained from the bulk material and then comparing it with the intensity of the same line from a film of known thickness of the same material deposited on a substrate. Escape depth can also be determined by first measuring the intensity of a photoelectron line of a substrate with the film of the sample deposited on it and then comparing it with that of the photoelectron line of the substrate without the film on it.

If a system contains surface overlayers, then the ratio I_0/I_s may be used to denote the ratio of the peak photoelectron intensity I_0 from the overlayer to I_s that due to the substrate. If one writes an expression for the ratio I_0/I_s using the above equation and puts the substrate thickness as infinite, it can be seen that the ratio I_0/I_s may be expected to increase exponentially with the overlayer thickness. This property can be used for probing the nature of overlayers. For example, in many practical applications, a coating of one material is applied on another. A plot of the ratio of photoelectron intensity of a surface constituent to that of a substrate constituent against the percentage of the surface constituent may be used to analyze the nature of surface coating. Whereas an exponential growth of this ratio would be indicative of surface uniformity, a nonexponential growth of the ratio may be taken to indicate islanding of the surface.

In catalyst preparation, metal crystallites are often dispersed over the surface of an inert carrier material. The metal signal intensity depends on the relative magnitudes of particle diameter and λ_M, λ_M representing the inelastic mean free path of photoelectrons of the metal M involved. The signal intensity of the support will depend on the extent of metal coverage, metal particle size, and the inelastic mean free path λ_s of the photoelectrons from the support S. Although parameters like porosity might complicate the analysis, the nature of catalyst dispersion on the surface can be obtained once calibration curves on the intensity ratios against the weight percentage of one component is established.

λ is a function of electron kinetic energy. The nature of its dependence is

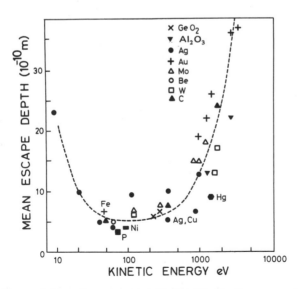

Fig. 5.14. The variation of mean escape depth with kinetic energy of the escaping electron [32].

shown in Fig. 5.14, compiled from data obtained in XPS, UPS, and AES studies of a large number of systems. In the XPS region, where electron kinetic energies are in the 100–1000 eV range, the escape depth increases near linearly with the electron kinetic energy. This implies that in this range, most of the escaping core electrons that have the high binding energy (yielding low kinetic energy electrons) originate at shallow depths. This can be seen in the germanium XPS spectra presented in Fig. 5.15. The spectra show both Ge 3d and Ge 2p peaks, obtained from a disk of pure Ge, having a thin passive oxide overlayer. The binding energy of the Ge 2p electrons is relatively high, therefore the kinetic energy is low. For the same reason, the mean escape depth is also low. While this is adequate for sampling the entire overlayer thickness, very little of the substrate element is sampled, and as a result the Ge 2p peak from the oxide layer is more intense. With Ge 3d electrons, on the other hand, the photoelectron kinetic energy is high and the escape depth is correspondingly high. This means that electrons from greater depths of the substrate element emerge, and the substrate peak dominates in intensity over that from the overlayer. Thus, this method can be effectively used for *depth profiling* of the sample, by choosing the appropriate photoelectron line to derive information about composition of the sample at a certain depth.

Another method by which depth profile data on a solid can be obtained is argon ion etching. Argon ion bombardment at energy < 10 keV can be used to

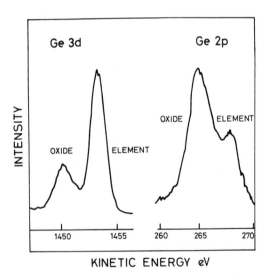

Fig. 5.15. Ge 3d and Ge 2p photoelectron spectra of a disk of pure Ge with a passive oxide overlayer excited by Al K_α, illustrating the relative sampling depths [469].

sputter away material thus exposing the underlying structure. Alternate operations of argon ion etching and XPS measurements can yield information on layer composition, that is, depth profile data. Although this is a fast and simple method for depth profiling, problems may arise because of differential sputtering rates of different atoms, uncertainties caused by the reduction of valence states due to the etching, or subsurface damage caused by beam sputtering. Subsurface damage can affect up to 15 atom layers and can also randomize lattice sites. Another problem is nonuniformity in the etching process itself, which can lead to nonuniform thinning.

The energy inhomogeneity of the argon beam also causes difficulties. With axial energy higher than that at beam periphery, the beam tends to cause pits. Defocusing the beam helps, but only with a concomitant reduction of etching efficiency and a large increase in the neutral content of the beam.

INITIAL OXIDATION OF POLYCRYSTALLINE ZINC

The combination of UPS and XPS constitutes a powerful tool for the study of adsorption phenomena. An interesting recent study of initial oxidation of polycrystalline zinc used UPS and XPS and, in addition, Auger electron spectroscopy. In this particular experiment, the substrate Zn was formed by evaporation of the metal from a tungsten filament at a pressure 10^{-10} torr. Figure 5.16a shows the

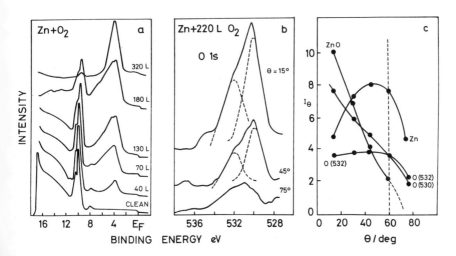

Fig. 5.16. Initial oxidation of zinc. (a) He I spectra with increasing O_2 exposures. (b) Variation of O 1s spectra with the angle of electron emission at 220 L O_2. (c) Variation of peak intensities of photoelectrons from Zn, ZnO, O (530 eV), and O (532 eV) with angle of electron emission ($T \simeq 300°K$) [475].

HeI spectra of the substrate recorded as oxygen was gradually introduced in the system. The valence band spectra of zinc and zinc oxide have two major features: the peak around 4 eV below E_F, which has the characteristics of an O $2p$ peak, and the Zn $3d_{3/2, 5/2}$ peak located at 9.8 eV below E_F. The first peak increases in intensity up to an oxygen exposure of 200 L [1 L (Langmuir) = 10^{-6} torr sec] with a concomitant decrease in Zn $3d$ intensity. Beyond this stage, the overall intensity decreases, although the peak continues to increase with respect to Zn $3d$. By 250 L, the Zn $3d_{3/2, 5/2}$ components become unresolvable and they practically disappear at 320 L. Beyond 250 L, a weak structure appears between E_F and the onset of the peak, which increases in intensity on further exposure. The UPS spectrum of a sample exposed to 320 L reverts to the spectrum observed at 200 L when the sample is pumped for 16 hours at 10^{-10} torr.

The average work function $\bar{\phi}$ of the sample can be evaluated from the relationship $\bar{\phi} = h\nu - (E_T - E_F)$ where $h\nu$ is the incident photon energy and E_T and E_F are respectively the binding energy of the threshold photoelectrons and the energy of the Fermi level, both with respect to the vacuum level. E_T may be interpreted as the photon energy at which photoelectrons obtained from level T have zero kinetic energy. If the energy of the Fermi level is taken as zero, then we can write $\bar{\phi} = h\nu - E_T^F$, where E_T^F is the energy of the level T expressed with respect to the Fermi level. At $h\nu = 21.22$ eV, we thus have $\bar{\phi} = 21.22 - E_T^F$. For a freshly evaporated clean Zn film, $\bar{\phi}$ is found to be 4.2 eV. As oxidation progresses, $\bar{\phi}$ decreases monotonically, reaching a value of 3.9 eV at 40 L and finally 3.45 at 320 L.

In the XPS measurements, right from the initial exposures, two oxygen peaks could be detected (Fig. 5.16b) at 529.8 eV and 532 eV. With progress of oxidation, the 532 eV peak increases at a steady linear rate, but initially at one-third the growth rate of the 530 eV peak. At about 150 L, growth of the 530 eV peak decreases rapidly and thus the O $1s$ peak, at 220 L, almost reaches a plateau. In the associated experiments on Auger spectra, intensity of the most prominent Auger peak Zn $L_3M_{4,5}M_{4,5}$ at 992.3 eV decreases steadily with increase of oxygen exposure. The angular variation of photoelectron peak intensities I_θ of O(530 eV), O(532 eV), ZnO, and Zn, against θ in the range 15–75°, are shown in Fig. 5.16c, where the θ of photoemission is measured from the sample surface.

A number of features in the spectra indicate oxide formation during exposures up to 200 L. The growth, with increasing oxygen exposures, of a broad Auger peak 0–5 eV higher than the Zn $L_3M_{4,5}M_{4,5}$ peak, which corresponds to a chemically shifted ZnO peak, indicates this. The HeI, NeI spectra of a cleaved ZnO crystal are found to be identical with those obtained from the surface oxidation experiments, which also supports this conclusion. Further support comes from the fact that the Zn $2p_{3/2}$ peak splits into two after oxidation, the new component being consistent with a cation formation having a binding energy about 0.3 eV higher than the pure metal component. As regards the O 1s peaks, the 530 eV O 1s peak is due to oxide in bulk ZnO, and the higher binding energy peak at 532 eV is consistent with some form of chemisorbed oxygen having a lower formal negative charge than O^{2-}.

The assignment of O(530 eV) to ZnO and O(532 eV) to chemisorbed O on Zn is clear also from similarities of angular distribution of the two pairs of peaks. It is possible to construct a model of a layer of ZnO on the substrate Zn and evolve a relationship between the ratio R_θ of the ZnO/Zn peak heights as a function of the angle of electron emission θ and the ratio d/λ, d being the layer of thickness and λ the electron escape depth. Dependence of R_θ on escape depth is understandable because the effective depth may be expected to undergo change for electrons emitted obliquely from the surface. The use of experimental data on variations of R_θ and application of the model give a value for $d/\lambda \simeq 0.25$. For the Zn $L_3M_{4,5}M_{4,5}$ electrons, one obtains from the standard λ-electron kinetic energy curve $\lambda \simeq 1.4$ nm, which means that $d = 0.35$ nm. The Zn/O atomic ratios from the Zn $2p_{3/2}$ and O 1s intensities are found to be greater than 1, which indicates that the surface ZnO is in the form of islands. The invariance of O(532 eV) peak to change in θ (Fig. 5.16c) may be taken to indicate that O atoms are embedded in the exposed Zn patches rather than adsorbed on them, that is, the chemisorbed oxygen leads to a reconstructed Zn surface. Overall, it indicates that upon adsorption on Zn, oxygen dissociates into atoms reconstructing the surface, which may or may not encourage nucleation of ZnO islands. This mechanism is consistent with the observed decrease in average work function $\bar{\phi}$.

There have been other studies on oxidation of Zn and Ni films by XPS and UPS, at fixed emission angle 45° and at temperatures 80°K and 300°K. In both cases of oxidation of Zn and of Ni, models have been developed to explain the presence of the O 1s peaks, and in both cases the 530 eV O 1s peak is thought to represent oxygen in oxide islands. In nickel–oxygen systems, the 532 eV peak may be due to chemisorbed oxygen atoms remaining in an overlayer position, with possible d orbital bonding. Angular distribution studies on zinc–oxygen systems indicate that the oxygen atoms are embedded in patches on the metal surface, which means that a 532 eV/530 eV O 1s intensity ratio will increase with low-angle emission. LEED calculations show that oxygen atoms can be very well within 0.09 nm above the plane of the metal surface, whereas ZnO islands can be at least 0.3 nm thick, implying that low-angle enhancement of the 532/530 eV (O 1s) ratio may be due to a shadowing effect as indicated in Fig. 5.17. In another experiment on Ni surface, at a binding energy of 533 eV rather than 530 eV or 532 eV, a third oxygen state has been observed, stable only at the low temperature of 77°K. This may be the true chemisorbed oxygen, a precursor to the O observed in zinc oxidation at 532 eV.

The above model, which invokes the effect of shadows, seems to have the

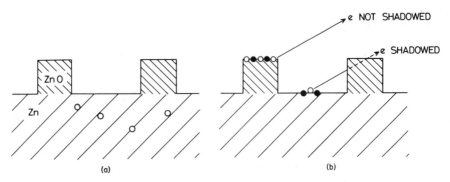

Fig. 5.17. (a) Model of patched oxide surface. (b) Model of patched oxide surface with chemisorbed O in an overlayer position. ○ oxygen atom position, ● zinc atom position [476].

shortcoming that the absolute variation of the intensity of the 532 eV peak in the zinc–oxygen experiment approximates that expected from a homogeneous surface. So, if the above model invoking shadows is correct, then fortuitous compensation would have to operate over a rather large range of angles. A more likely possibility might be that island formation leads to a rougher surface than the evaporated film, which can result in artifacts in the angular dependence curves. Within the escape depths involved, it is possible that the embedded oxygen species forms a concentration gradient into the metal, manifested by a decrease in the O(532)/Zn ratio with increase of θ. But it is difficult to recognize differences between patching of the surface layers and surface roughness.

ADSORPTION STUDIES ON POLYCRYSTALLINE NICKEL

To take another example, let us consider a study of adsorption on polycrystalline Ni films. Nickel is first evaporated at a base pressure of the order of 10^{-10} torr to form a film, which acts as the substrate. Even at such low pressures traces of contamination may be present. A typically clean Ni surface may show the peak intensities Ni(2p):O(1s) 130:1, Ni(2p):C(1s) 260:1. In this experiment with Ni, the substrate is cooled to 77°K. At this temperature, the adsorbate gas is permitted in at 10^{-8} to 10^{-5} torr pressure while photoelectron spectra are taken. Spectra are also taken after warming up and also when the film is subjected to pressures of several torrs. He I and He II spectra showing N_2, CO, and H_2O adsorption on a clean Ni surface are shown in Fig. 5.18.

The spectra of the clean surface are due to a weighted sum of (100), (110), and (111) crystal faces. Contaminants like carbon or oxygen on the evaporated

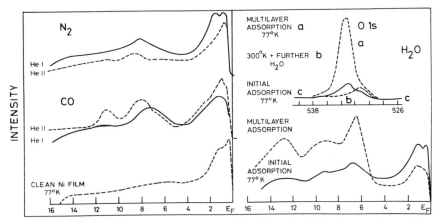

Fig. 5.18. He I and He II spectra of adsorption of N_2, CO, and H_2O on polycrystalline Ni film at $T = 77°K$. Also shown is the O 1s peak from XPS studies of H_2O adsorption at $T = 77°K$ and $300°K$ [477].

Ni film increase the ultraviolet photoelectron spectral intensity both in the 4–7 eV region and the 1.21 eV Ni d-band feature, in comparison with the peak at 0.4 eV in the He I spectrum. This phenomenon can be used to monitor the cleanliness of the surface, the technique being comparable to monitoring the surface by XPS core level spectra. It may be expected that the intensity changes in the spectra due to the adsorbate reflect both the extent of interaction of the adsorbate with the surface and the relative ionization cross sections of the levels. A comparison between He I and He II spectra of the same system provides information on changes due to electron kinetic energy and hence also corresponding changes due to both in escape depths and ionization cross section. Whereas the molecules CO, N_2, and NO show maximum coverage at $77°K$ and 5×10^{-6} torr pressure, which is below their saturation vapor pressure, the molecules H_2O, CO_2, N_2O, H_2S, and SO_2 are probably condensed as multilayers at the same temperature.

In these experiments on polycrystalline Ni film, it is observed from XPS that the binding energy of the Ni $2p_{3/2}$ electron remains practically unaltered although the peak undergoes an intensity decrease and a slight broadening. Only near a grazing electron ejection angle are small shifts of the order of 0.3 eV observed. The difference between the binding energies of a particular type of electron in the gaseous state (E_g) and in the adsorbed state (E_{ads}) can be expressed as a sum of (1) the work function correction ϕ, (2) the initial state chemical shift ΔE^B due to bonding difference, and (3) the difference ΔE^R in the final hole state relaxation energies in the two cases. On this basis, one may write

$$E_g - E_{ads} = \phi + \Delta E^R + \Delta E^B$$

Since the relaxation is greater in the adsorbed solid state, ΔE^R ($\Delta E^R = E_{ads}^R - E_g^R$) will be always positive, whereas ΔE^B can be either positive or negative. In the Ni systems studied, ϕ is of the order of 5–6 eV and ΔE^B is 2–3 eV. Where strong molecule–molecule forces do not exist, as would be the case for condensed organic molecules, ΔE^B is nearly zero, and only the first two terms of the above equation need to be retained. It has been found in a study on the condensation of organic molecules on Ni(111) surface at 77°K that the $E_g - E_{ads}$ values encountered are the same for all valence levels but are larger than ϕ by 1–2 eV. The difference, therefore, must be due to the ΔE^R values for the valence levels that remain nearly constant. This near constancy of $E_g - E_{ads}$, with values greater than ϕ by 1–2 eV, has been found to hold for both core and valence orbitals of condensed (weak physical adsorption) inorganic species at 77°K and can be used for diagnostic tests of molecules not involved in surface bonding. For molecules that strongly interact with the surface this would not be generally valid, but one can find a nonbonding valence orbital and apply the same principle to it, that is, $\Delta E^B = 0$. In fact, using this principle it has been shown that since $E_g - E_{ads}$ increases with increased chemisorption, ΔE^R increases with a changeover from condensed to chemisorbed state. In case of dissociation, the spectral structure due to the original valence level molecular orbital is lost, and shifts occur in the core level lines corresponding to the dissociation products.

In the He I spectrum of CO adsorption on a polycrystalline Ni surface (Fig. 5.18), two adsorbate-induced peaks are observed, one at 7.2 eV and a weak one at 11.1 eV – more clearly seen in a difference spectrum. The former is due to both 5σ and 1π levels of CO and the latter is due to the 4σ level. Actually, the 7.2 eV peak has two components, at 6.3 eV and at 8.1 eV; this is shown in Table 5.3. From Table 5.4, $E_g - E_{ads}$ for O(1s) is 10.9 eV and that for C(1s) is 10.6 eV. Both the differences are large. Since ϕ can only account for about 5–6 eV, and calculations on Ni(CO)$_4$ and CO show a negative ΔE^B value of ≤ 2 eV between the C(1s) and O(1s) values in these two environments, one may conclude that CO is strongly interacting with the surface. By taking differences between the Ni$_{ads}$, He I and free molecule values it can be seen from Table 5.3 that for the valence orbitals 5σ, 1π, and 4σ, the $E_g - E_{ads}$ are 7.7, 8.7 and 8.6 eV respectively. An assumption that the bonding to the surface is primarily through C lone pair 5σ orbital, i.e., $\Delta E_{1\pi, 4\sigma}^B = 0$, implies that $\Delta E_{1\pi, 4\sigma}^R = 8.7 - \phi \simeq 3.7$ eV, using $\phi = 5$ eV. If it is further assumed that the relaxation energy difference $\Delta E_{5\sigma}^R$ for the 5σ orbital also is about 3.7 eV, then $E_{5\sigma}^B = (E_g - E_{ads}) - \phi - \Delta E^R = 7.7 - 5 - 3.7 = -1.0$ eV, which means that the 5σ binding energy increases by 1 eV due to the bonding between Ni metal and free CO. The assumption $\Delta E_{5\sigma}^R \simeq 3.7$ eV is compatible with the results of calculations on end-on interaction of CO through the 5σ orbital. If one further assumes that ΔE_{core}^R also is about 3.7 eV, then it follows that ΔE^B C 1s and ΔE^B O 1s are 1.4 and 1.7 eV

Table 5.3 UPS Valence Level Energies (eV) for Free Molecule and Adsorbed States at 77°K [477]

Molecule	Assignment	5σ	1π	4σ	
CO	Free molecule	14.0	16.8	19.7	
	Ni$_{ads}$ He I	6.3	8.1	–	
	He II	6.3	8.1	11.1	
	Assignment	5σ	1π	4σ	
N$_2$	Free molecule	15.6	16.7	18.8	
	Ni$_{ads}$ He I		8.0	11.8	
	He II		8.0	11.1	
	Assignment	2π	4σ	1π	3σ
N$_2$O	Free molecule	12.9	16.4	18.2	20.1
	Ni$_{cond}$ He I	6.3	9.7	11.3	13.5
	He II	6.3	9.8	11.5	13.3
	Assignment	$1\pi_g$	$1\pi_u$	$2\sigma_u$	$2\sigma_g$
CO$_2$	Free molecule	13.7	17.6	18.0	19.4
	Ni$_{ads}$ He I	7.2	11.2		
	He II	7.4	11.4		12.9
	Ni$_{cond}$ He I	All are ~ 1–1.5 eV higher than for Ni$_{ads}$			
	He II				
	Assignment	$1b_1$	$3a_1$	$1b_2$	
H$_2$O	Free molecule	12.6	14.7	18.3	
	Ni$_{cond}$ He I	6.5	9.2	13.0	
	He II	6.5	9.2	13.0	

respectively, that is, a decrease in the C 1s and O 1s binding energies due to adsorption, implying a net charge transfer to the CO molecule and hence backbonding to CO $2\pi^*$.

If the system is warmed from 77°K to 300°K, changes in XPS intensity ratio Ni(2p):O(1s):C(1s) indicate about 10% desorption, the UPS showing little change. Heating at 400°K in vacuo or at 10^{-4} torr CO produces further desorption and prolonged heating causes a shoulder in the C(1s) spectrum which may be taken to indicate diffusion of carbon to the surface of the nickel film.

Nitrogen adsorbs at 77°K, but it is completely desorbed at 300°K. Two broad peaks are observed in the N 1s spectrum. For the highest binding energy peak, $E_g - E_{ads}$ is 4.2 eV (Table 5.4) which can be accounted for entirely by ϕ. This means the formation of a weakly adsorbed state having little electronic interaction with the surface. For the second peak, $E_g - E_{ads} = 9.3$ eV. There are two possible assignments of these two peaks. Either they represent two different adsorbed states, or alternatively, since the signal intensities are equal, they are two distinguishable atoms of a single state. The energy difference between the

Table 5.4. XPS Binding Energies (eV) for Free Molecule and in Adsorbates [477]

Molecule	Condition		C 1s	O 1s	N 1s
CO	Gaseous		296.2	542.3	
	Ni_{ads}	77°K	285.6	531.4	
		300°K	285.6	531.4	
N_2	Gaseous				409.9
	Ni_{ads}	77°K			400.6, 405.7
		300°K			None
N_2O	Gaseous			541.2	408.5, 412.5
	Ni_{cond}	77°K		534.8	402.1, 406.1
		300°K		530.5	None
CO_2	Gaseous		297.7	541.3	
	Ni_{ads}	77°K	291.1	534.4	
	Ni_{cond}	77°K	~292.0	535.5	
		300°K	None	None	
H_2O	Gaseous			539.7	
	Ni_{cond}	77°K		533.1	
		300°K		531.4	

peaks, 5.1 eV, is quite large for both these assignments. But it is possible that the contribution of ΔE^R is largely determined by the distance of the adsorbed atom from the surface, and a weakly bonded end-on N_2 molecule, with negligible ΔE^B, for example, might cause two different binding energies for two nitrogen atoms. In UPS, no clear orbital picture is discernible, and the question of the nature of the adsorbed state is not resolved. A peak at 8 eV and a somewhat uncertain one at 2.2 eV to the higher binding energy side are observed; they could be due to $5\sigma + 1\pi$ and 4σ respectively. Coadsorption of traces of CO could lead to similar observations. A spectrum at 300°K where N_2 completely desorbs could resolve the problem of CO coadsorption, but such a spectrum was not available in this study.

In the N_2O and CO_2 studies, the spectra, which are not shown here, can be directly correlated with the gas phase situation. For N_2O, although desorption takes place at 300°K, a small residual oxygen is retained at the surface. For CO_2, a multilayer condensation is observed above the saturation vapor pressure with O1s binding energy about 1 eV higher than the adsorption value. In UPS, a second set of molecular orbitals is observed at 1–1.5 eV higher binding energy. Since ΔE^B is expected to be small, the shifts probably represent a difference in ΔE^R. For N_2O valence level $E_g - E_{ads}$ is 6.7 eV and for the core level $E_g - E_{ads}$ is 6.4 eV. For physically adsorbed CO_2, the $E_g - E_{ads}$ value for valence level is

6.3 eV, for C 1s 6.4 eV, and for O 1s 6.9 eV; for condensed CO_2, it is about 5.1 and 5.4 eV for the valence and core levels respectively. ϕ alone can explain the $E_g - E_{ads}$ difference in the condensed situation, and if ΔE^B is taken as about zero, then ΔE^R undergoes an increase from about zero to about 1.5 eV in the condensed state to the physical state transformation.

In studies involving H_2O adsorption at 77°K on Ni surface, desorption occurs on heating at 300°K with a chemisorbed layer left on the surface. Peaks at 6.5 eV, 9.2 eV and 13.0 eV due to the three H_2O molecular orbitals are recognizable in the He I spectrum taken at early stages of adsorption. Both for $3a_1$ and $1b_2$ orbitals $E_g - E_{ads}$ is about 5.5 eV, and for $1b_1$ O lone-pair orbital it is 6.1 eV, indicating that they are accountable by ϕ alone. Ordinarily, for the multilayer situation the orbital binding energies remain at the same values. In H_2O, a significant relative shift of orbitals with respect to the free molecule is observed in the multilayer condensation regime. This is likely to be due to H bonding between H_2O molecules rather than bonding with the surface. A hydrogen bonding effect is also indicated by $E_g - E_{ads}$ for O 1s of 6.6 eV at 77°K, which is significantly different from valence levels. At 300°K only a small O 1s signal is left at 531.4 eV. This is too large to be explained by any possible CO coadsorption, since the C 1s intensity is quite small and therefore most likely represents dissociated H_2O as OH or O atoms at the surface. The He I spectrum at this temperature is consistent with this description, as the spectrum shows a broad band at 6.0 eV characteristic of O atoms and with no correlation to H_2O molecular orbitals.

ADSORPTION OF ORGANIC MOLECULES ON PLATINUM SINGLE CRYSTAL

The preceding two examples involved polycrystalline substances. Adsorption phenomena on a polycrystalline substrate should be easier to understand if the nature of adsorption on the various crystallographic planes of a single crystal of the same substrate is known, although the experiments are considerably more difficult to carry out. Many adsorption studies on single crystals have now been carried out using photoelectron spectroscopy. We consider now an example of the chemisorption studies of organic molecules on a single crystalline metallic surface using both UPS and XPS; it involves the adsorption of vinyl chloride, vinyl fluoride, ethylene, and propene on a Pt single crystal.

Photoemission from associatively chemisorbed ligands should closely relate to that from the molecules in the gas phase both in terms of core binding energies and emission intensity ratios of two or more chemically nonequivalent atoms. The core emission intensity ratios of chlorine and fluorine atoms with that from carbon atom are well represented by appropriate ratios of ionization cross

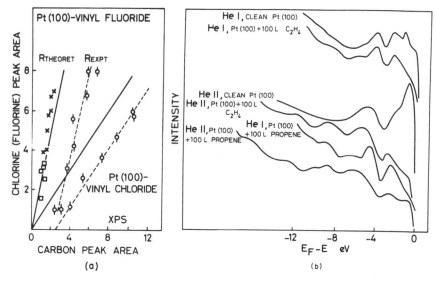

Fig. 5.19. (a) XPS results of the Pt(100)-vinyl fluoride and Pt(100)-vinyl chloride surface. For fluoride, the solid line refers to the ratio $R(F/C) = 2.27$ and corresponds to the gas phase spectrum and also to the ratio of ionization cross sections. For chloride, $R(Cl/C) = 0.78$. Open circles represent adsorption on clean surface, crosses refer to experimental data of fluoride following the preadsorption of 0.3 monolayer of CO, and the open squares refer to the 0.6 monolayer of CO. (b) He I and He II spectra of the Pt(100)-C_2H_4 and Pt(100)-propene surfaces [478].

sections. Fig. 5.19a shows heteroatom-carbon emission intensity ratios $R(X/C)$ for chemisorption of vinyl chloride and vinyl fluoride on a clean Pt(100) surface and when the surface is pretreated with a fraction of a monolayer of carbon monoxide.

He I and He II spectra of Pt(100)-ethylene and Pt(100)-propene surfaces are shown in Fig. 5.19b; the energies are expressed with respect to the Pt Fermi level. The adsorbed state ligand energies E_L may be calculated from the relation $E_L = I_L - E_F - \Delta\phi + R$; I_L is the gas phase ligand ionization potentials and R is a relaxation shift, which has been assumed to be constant (-2.5 eV) for all levels. E_F is 5.5 eV; work function changes are -0.8 eV for C_2H_4 and -1.1 eV for C_3H_6; I_L for C_2H_4 are 10.7, 12.8, 14.8, 16.0, and 19.1 eV; I_L for C_3H_6 are 9.9, 12.1, 13.1, 14.2, 16.0, and 18.1 eV. (Note that though the symbols used here are somewhat different, the relationship for E_L above is similar to that for E_{ads} presented in the previous section.)

CO provides a very small electronic perturbation of the Pt(100) surface with

$\Delta\phi = +0.2$ V; the preadsorption of CO ligands thus does not significantly affect the electrophilicity of surface metal atoms. The fact that the observed ratio of the emission intensities R(X/C), where X is halogen, is much less than that in the gas phase spectra — or in the case of vinyl fluoride much less than the ratio of F to C ionization cross section — indicates that both vinyl fluoride and vinyl chloride are dissociatively chemisorbed on the Pt(100) surface so the original atomic ratio in the molecule is not adhered to. The initial dissociative mode changes to an associative pattern after one-quarter monolayer coverage because the experimental data when the surface is partially covered with CO correspond to theoretical expectations. This changing pattern indicates that only some sites on the metal surface have the geometric or electronic properties needed for initiating surface oxidative addition reactions.

With regard to the low energy photoelectron spectroscopy of metal–alkene and other metal–ligand surfaces, the possibility of at least partial intermolecular dehydrogenation of ethylene on the Pt(100) surface is indicated by the hydrogen halide elimination of vinyl halides. Some data of ionization energies for Pt(100)-alkene surfaces are given in Table 5.5. The difference in photoelectron spectra between the clean Pt surface and the surface of Pt-C_2H_4 indicates the formation of perturbed ligand states. Whereas the unperturbed ionization energies are 10.7, 12.8, 14.8, 16.0, and 19.1 eV, the ligand states on the surface have energies of 3.0, 5.5, 8.6, and 11.5 eV. The additional structure on Pt(100)-propene spectra reinforces the ethylene assignments and, together with the results for trifluoropropene, supports the inference that in alkenes the binding is predominantly associative. At low ethylene coverage a peak is observed that could be due to a molecular orbital built largely from acetylenic π level but it is better explained by shifted ethylene states. Whereas stoichiometric dehydrogenation may be ruled out, low emission intensities of certain bands of C_2H_2 indicate partial dehydrogenation equivalent to dissociative sorption of the haloalkenes.

Table 5.5. Ionization Energies (eV) for Pt(100)–Alkene Surfaces [478]

Molecule								
H_C=C_/H (H/ \\H)	E_{obs} He I	?	5.8		8.5			
	E_{obs} He II	3.0	5.5		8.6 (broad)		11.5	
	E_L	3.5	5.6		7.6		8.8	11.9 ($\Delta\phi = -0.8$ V)
H_3C_C=C_/H (H/ \\H)	E_{obs} He I	2.8	5.0		6.4 8.2	9.3		
	E_{obs} He II	2.6		5.7 (broad)	8.0	9.3	11.0	15.2
	E_L	3.0	5.2		6.2 7.3	9.1	11.2	($\Delta\phi = -1.1$ V)

E_{obs} is relative to Fermi level. E_L is calculated from $E_L = I_L - E_F - \Delta\phi + R$.

174   SPECTRA FROM SOLIDS AND SURFACES

The pseudo DOS of the metal valence band at about 2 eV below the Fermi level undergoes an intensity decrease on adsorption, similar to the Pt(100)-CO surface. It is the doublet component at 2 eV that is modified most. This may be interpreted to indicate back donation of the metal d electrons, which have considerable t_{2g} character, to ligand antibonding states. Other associated data of haloalkanes on Pt(100), Pt(111) show that the more closely packed the surface is, the more complete is the dissociation of vinyl halides.

OXIDATION OF GaAs (110) SURFACE

Let us now look at some surface reaction studies that employ photoelectron spectroscopic techniques with synchrotron radiation as the photon source. We have already noted the characteristic properties of synchrotron radiation: its very high intensity of emitted radiation, its natural collimation, its ability to provide a wide range of photon frequencies with a suitable monochromator, and the natural linear polarization of the radiation in the plane of the electron orbit. The first three properties are used in surface reaction studies. Of particular usefulness is the selectable photon energy, which permits one to either augment or suppress photoionization cross sections and control electron escape depths

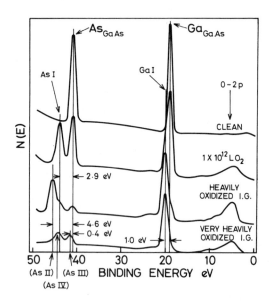

Fig. 5.20. Photoelectron spectra ($h\nu = 100$ eV) of GaAs(110) surface exposed to oxygen [54].

such that the surface photoemission is accentuated. Our first example is on oxidation of the GaAs (110) surface.

The oxidation experiments we discuss here involves gradual exposure of oxygen to an atomically clean GaAs (110) surface obtained by vacuum cleaving of a single crystal at 10^{-10} torr. A photon energy of 100 eV is chosen so that the As $3d$ electrons have a kinetic energy of about 60 eV and the Ga $3d$ electrons about 80 eV, the former at the escape depth minimum and the Ga $3d$ very close to it. At these energies the photoionization cross sections for both As $3d$ and Ga $3d$ are near their maximum values.

Figure 5.20 shows photoelectron spectra under a wide range of oxygen exposures. With the initial exposure of oxygen one observes the appearance of a new As peak at a higher binding energy and its increase in intensity with increasing oxygen exposures. In contrast, the Ga peak does not show any shift. With about 10^{12} L oxygen, which corresponds to about a monolayer coverage, the

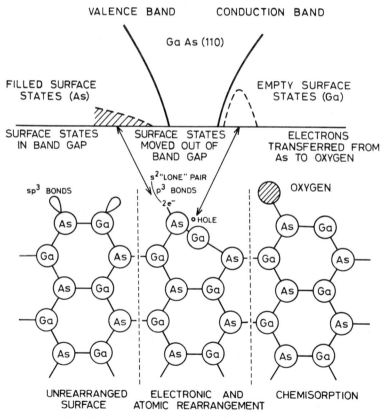

Fig. 5.21. Surface rearrangement of GaAs(110) [54].

two As peaks – the shifted and the unshifted – have about equal intensity, indicating that 50% of the photoelectrons are from the surface. With higher escape depth electrons, such as would have occurred had Al K_α or Mg K_α been used, such fine changes in surface composition would have been quite difficult to observe. Fig. 5.20 gives definite evidence of the removal of electrons from As due to oxygen, or in other words, of chemisorption at As site. If Ga and As were in the free state, then from heat of combustion data one might have expected preferential oxidation of Ga relative to As. But the detailed photoelectron spectral data show the result to be quite to the contrary. We need to note here that the material is GaAs, so thermodynamic data on the elemental solids are not relevant. Oxidation is associated with, first, scission of GaAs covalent bonds. Several conclusions about GaAs (110) surface electronic structure and surface states can be derived from oxygen chemisorption studies. In the simplest possible configuration for GaAs (110), the surface is shown schematically in Fig. 5.21. One bond each of Ga and As on the surface is broken and thus each atom might be considered to be in association with the dangling bond electron, but due to higher electronegativity of As atoms one might expect the electrons to be more localised on the surface As atoms. LEED and other investigations indicate a surface rearrangement with respect to the above structure. A more likely rearranged structure is that in which As atoms move outward and Ga inward – the movements are of the order of 10–20% of cell dimensions, causing a major change in surface potential. Whereas in the bulk the structure is of an sp^3 nature, the formation of a (110) surface along with a minimization of energy involves a p^3 bond formation at the As site along with s^2 lone-pair electrons. The p^3 bonding configuration reduces the bond angle and pushes the As atom out of the surface. For Ga, on the other hand, a minimum energy is found with an sp^2 configuration that increases the bond angle and pulls the Ga atom towards the crystal.

Removal of the s^2 electrons by chemisorbed oxygen can cause change again. The $s^2 p^3$ structure changes more towards p^3, and As atoms are pushed back into the crystal. This rearrangement causes a large strain field, which modifies the idealized arrangement, and the valence band electronic structure at the surface undergoes strong change.

Ordinarily, Ga atoms on the surface are unaffected in spite of large oxygen exposures. But a dramatic change takes place if some excitation source is turned on and can provide the oxygen atoms with some additional energy. Even an ionization gauge out of direct sight from the surface can excite oxygen atoms sufficiently that an activation energy is overcome and oxidation takes place at the Ga site, resulting in a chemically shifted Ga peak in the spectrum. This is presumably due to a long-lived ($\sim 15'$) excited state of oxygen. Under very heavy oxygen exposures and in the presence of an excitation source such as an ionization gauge, several oxides of As are discernible.

CARBON MONOXIDE ADSORPTION ON PLATINUM

Let us consider a final example that further illustrates the power of photoelectron spectroscopic studies based on synchrotron radiation to investigate the nature of the adsorption of gases on solid surfaces. The example concerns CO adsorption on the Pt surface. The CO orbitals are schematically shown in Fig. 5.22a. Here 3σ is the strong bonding orbital between carbon and oxygen; the 1π bonds are significantly weaker. The 4σ and 5σ are the orbitals primarily associated with the oxygen and carbon atom respectively. Because of the difference in

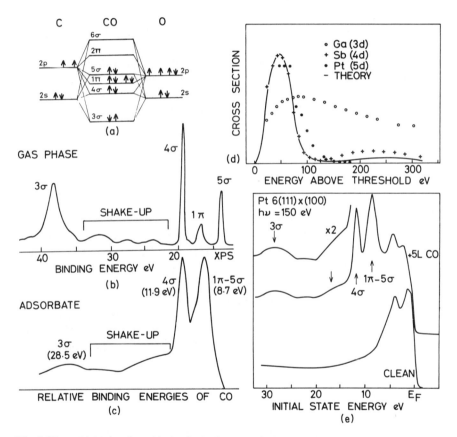

Fig. 5.22. (a) Molecular orbitals of CO. (b) XPS of gas phase CO. (c) The difference curve of CO on Pt, representing the spectrum of CO-covered Pt minus the spectrum of clean Pt. (d) Photoionization cross sections as a function of photon energy above the photoionization threshold for $3d$, $4d$, and $5d$ electrons. (e) Valence band spectra (at $h\nu = 150$ eV) of clean Pt surface and of Pt with adsorbed CO [54, 479, 480, 481].

electronegativities between oxygen and carbon, it may be expected that 4σ will have a higher binding energy than 5σ. The gas phase photoelectron spectrum of CO and the assignment of the peaks to the different CO orbitals are shown in Fig. 5.22b. The 4σ, 1π, and 5σ have binding energies less than 20 eV, and the 3σ has a binding energy of about 38 eV. UPS studies using ordinary ultraviolet radiation, such as those with Ni discussed earlier, show that when CO adsorbs on metal there is a shift of the 5σ peak to higher energy, indicating that a bond formation takes place with the carbon orbital; the 4σ peak, however, does not undergo a shift. In these UPS studies, 3σ levels were not observed for two reasons. First, the matrix element and therefore the cross section for photoionization from 3σ, which determines the photoelectron intensity, is weak in ultraviolet. Second, the secondary electrons generated by photoionization of valence band electrons of the metal, on which adsorption takes place, obscures the 3σ peak. Application of synchrotron radiation circumvents both of these problems. One selects a synchrotron radiation frequency where the matrix element of the relevant orbital of the adsorbed molecule has a much larger value and also where the photoemission from the substrate metal surface is suppressed. That such selective suppression of metal photoemission is possible can be understood by examining the photoionization cross sections for the $3d$, $4d$, and $5d$ levels shown in Fig. 5.22d. Whereas the cross section for $3d$ changes monotonically with photon energy, the cross sections for $4d$ and $5d$ show a strong maximum at about 50 eV and go through a minimum at about 150 eV. This minimum, known as the *Cooper minimum*, is observed when there are nodes in the initial state orbital, as there are in $4d$ and $5d$, and arises due to destructive interference of the initial and final states in the matrix element. The platinum valence bands seen at the photon energy of 150 eV — that is, at the Cooper minimum — from a clean surface as well as after CO adsorption, are shown in Fig. 5.22e. A difference spectrum is shown in Fig. 5.22c. As a result of suppression of Pt photoemission, the CO spectrum dominates that of Pt. As a consequence, emission from 3σ are observed, including shake-up spectra, which is a remarkable improvement over spectra seen with ordinary ultraviolet radiation. The 5σ peak moves upward in binding energy, showing bond formation on the metal surface, and changes in the shake-up structure indicate that metal electrons are involved in the shake-up process.

CHAPTER

6

PHOTOIONIZATION CROSS SECTIONS AND QUANTITATIVE ANALYTICAL CHEMISTRY APPLICATIONS

Any quantitative application of photoelectron spectroscopy based on spectral intensities is directly related to the cross section of the photoionization process itself. Phenomenologically, the *photoabsorption coefficient* α at a particular radiation frequency ν is defined through the relationship

$$I = I_0 e^{-\alpha d} \tag{6.1}$$

where I_0 is the intensity of the photon flux incident on a column of length d cm, and I is the intensity of the transmitted flux. The photoabsorption coefficient α, expressed in cm^{-1}, is a characteristic constant of the sample and is a function of ν. The microscopic *photoabsorption cross section* σ_p is then defined by the equation $\alpha = \sigma_p N$, with N representing the number density of molecules in the gas, in cm^{-3}. Therefore, σ_p has dimensions of cm^2. The validity of Eq. (6.1) requires radiation to be monochromatic so that the Lambert–Beer law is obeyed, that is, a decrease in intensity dI of the radiation as it penetrates a distance dx in the sample is proportional to I.

The process of photoabsorption encompasses a large number of component processes. They include processes in which electrons are not given off, such as simple excitation to upper bound states and excitations leading to predissociation, and processes where one or more electrons are ejected, such as direct excitation to ionization continua, autoionization, and dissociative ionization. The measured photoabsorption coefficient represents a sum total of probabilities of all these processes.

In general, the total photoabsorption cross section σ_p of a molecule has several components and may be written $\sigma_p = \sigma_b + \sigma_s + \sigma_i + \sigma_f$, where σ_b represents excitation to a normal bound state, σ_s excitation to a superexcited neutral state located above the ionization limit, σ_i ionization to a stable molecular ion without fragmentation, σ_f ionization to a fragment ion. The total *photoionization cross section* can be expressed as $\sigma = \sigma_i + \sigma_f + \sigma_s [k_+/(k_+ + k_n)]$. k_+ represents the rate of decay of the superexcited state by autoionization

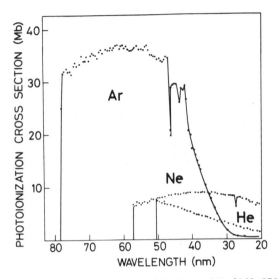

Fig. 6.1. Photoionization cross sections of Ar, Ne, and He [569, 570, 571, 577].

and k_n is the total probability for nonionic radiative and collision-induced nonradiative decay processes. The ratio of total photoionization cross section to total photoabsorption cross section, σ/σ_p, is called the photoionization efficiency. Figure 6.1 shows the photoionization cross section of He, Ne, and Ar

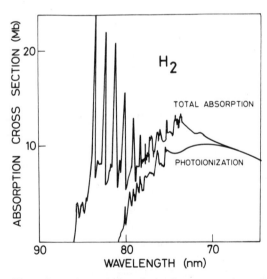

Fig. 6.2. Photoabsorption and photoionization cross sections of H_2 [584].

involving photoionization of the most loosely bound electron: $1s$ for He, $2p$ for Ne, and $3p$ for Ar. The details of the cross section curve as a function of energy depends on the initial state of the neutral and also on the nature of the final states of the ions formed, which in the molecular case may have contributions from rotational and vibrational components. Figure 6.2 shows the relationships between the photoionization and photoabsorption cross sections of H_2, and the higher energy threshold of the former is clearly seen.

EXPERIMENTAL DETERMINATION OF PHOTOIONIZATION CROSS SECTIONS

For the experimental determination of photoionization cross sections one needs to make quantitative measurements of the intensities of ejected photoelectrons at known sample density and known flux of incident photons. The photoabsorption coefficient α and therefore the photoabsorption cross section σ_p can be determined from a knowledge of N, d, and a measurement of the relative intensity I/I_0; there is no need of any absolute intensity measurement. A conventional apparatus for determining the photoabsorption cross section of gaseous samples consists of an intense continuum photon source, a monochromator, a sample chamber, and a photoelectric detector.

The need for a monochromator arises because one is interested in a photoabsorption cross section as a function of radiation frequency. The relative placement of the absorption cell and the monochromator is important. If the absorption cell is placed outside the exit slit of the monochromator and relatively distant from the source, one has the advantage that chances of photolysis of the sample molecules and consequent complications in interpretation of photoabsorption data are reduced. The advantage of placing the absorption cell between the source and the monochromator, however, is that the entire absorption spectrum of the absorber molecules can be photographed in one exposure.

To compensate for any intrinsic variation of the light source intensity, double beam methods have been used. A simple contraption uses a set of grid wires sensitized with a coating of sodium salicylate and placed in path of the light beam. When these wires are radiated by ultraviolet radiation, they scatter enough fluorescence radiation in the visible region for a photoelectric detector, placed at right angles to the primary beam, to be used as a monitor for the ultraviolet light source intensity.

For photoionization cross section determination, the number of ions generated due to a given incident photon flux needs to be measured. To collect all the photoions produced in any ionization that takes place in the absorption cell, a pair of parallel metal plates are placed above and below the light beam in the absorption cell. The potential difference applied between the plates is such that it is adequate to collect all the photoions produced, yet is not so large as to cause secondary ionization.

In such an experimental setup, the photoionization cross section can be determined in the following manner. First, an atomic gas having a low ionization potential is taken into the cell such that the entire photoabsorption, which amounts to $I_0 - I$, leads to ionization. If i_s represents the saturation ion current collected by the plates in the ionization chamber, and e the electronic charge, then one can write

$$\frac{i_s}{e} = I_0 - I \tag{6.2}$$

Using Eq. (6.1), one obtains

$$\frac{i_s}{e} = I_0(1 - e^{-N\sigma_p d}) \tag{6.3}$$

or

$$I_0 = \frac{i_s/e}{1 - e^{-N\sigma_p d}} \tag{6.4}$$

If σ_p of the same system is already known from relative intensity measurements of photoabsorption, N and d are also known, and i_s is measured, then the absolute flux of incident radiation I_0 can be obtained by using Eq. (6.4). If for measurement of radiation intensity a simple photoelectric detector such as a platinum plate is used, then one may define the photoelectric yield γ as

$$\gamma = \frac{i_p^0}{I_0 e} \tag{6.5}$$

where i_p^0 is the photoelectric current emitted by the detector plate in the absence of an absorbing gas. Once γ and I_0 are determined by using a rare gas as absorber the following relationship can be used to determine the photoionization cross section σ:

$$\frac{\sigma}{\sigma_p} = \frac{i_g}{eI_0(1 - e^{-N\sigma_p d})} = \frac{i_g \gamma}{i_p^0 - i_p} \tag{6.6}$$

where i_g is the ion current collected by the parallel plates and i_p represents the corresponding photoelectric current emitted by the detector plate when the sample is in the absorption cell at a concentration of $N \, \text{cm}^{-3}$. Here, we use

$$I_0 - I = I_0(1 - e^{-N\sigma_p d}) = \frac{i_p^0 - i_p}{\gamma e} \tag{6.7}$$

Absolute measurement of the flux can also be made by using a double chamber filled with a rare gas. In this arrangement, two plates, one following the other, of diameter d are used to collect the ion currents i_1 and i_2 in the two chambers respectively, one following the other. If I_0 is the radiation flux from the monochromator and L_1 is the distance between the monochromator exit slit and the entrance aperture of the first ionization chamber, then the flux entering and the flux leaving the first chamber are respectively $I_0 e^{-\kappa L_1}$ and $I_0 e^{-\kappa L_1} e^{-\kappa d}$, where κ is the photoionization coefficient. If all those photons which are lost in the first chamber are used up in ionization, then

$$\frac{i_1}{e} = I_0 e^{-\kappa L_1}(1 - e^{-\kappa d}) \tag{6.8}$$

Similarly, if the distance between the monochromator exit slit and the entrance aperture to the second chamber is L_2, then equating the number of photons lost and the number of ions collected in the second chamber, we have

$$\frac{i_2}{e} = I_0 e^{-\kappa L_2}(1 - e^{-\kappa d}) \tag{6.9}$$

From Eq. (6.8) and Eq. (6.9) the two unknowns, the absorption coefficient κ and the flux I_0, can be determined.

If $L_1 \to 0$ and $L_2 \to d$, we have a modified version of the double ionization chamber, and I_0 is given by the expression

$$I_0 = \frac{i_1^2/e}{i_1 - i_2} \tag{6.10}$$

For measurement of absolute cross sections, an intense radiation source having a closely spaced line spectrum is used. Hydrogen glow discharge is often used for wavelengths above 100 nm. Other discharge sources are used for shorter wavelengths, down to about 15 nm (e.g., condensed discharge argon continuum source can be used in the region 60–100 nm). Synchrotron radiation sources are also used. The intensity of the radiation passing through the absorption chamber is measured either by platinum photocathode or photomultiplier coated with sodium salicylate. The detectors are calibrated by a single-ionization chamber using a rare gas.

In recent determinations of photoionization cross sections, the ejected electrons are also energy analyzed using the cylindrical mirror analyzer or one of the other analyzers described in Chapter 2. This permits the measurement of photoelectron flux intensity originating from various sublevels of the sample atom or the molecule, and thus *partial photoionization cross sections* of various levels can be determined. Figures 6.3 and 6.4 show total photoabsorption and partial photoionization cross section data for various electronic states of oxygen obtained in a similar manner.

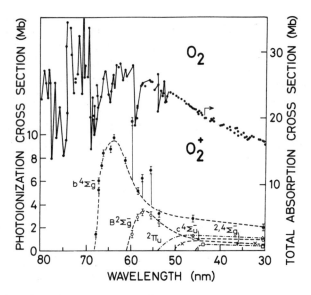

Fig. 6.3. Photoabsorption cross section of O_2 and partial photoionization cross sections for production of excited states of O_2^+ [596].

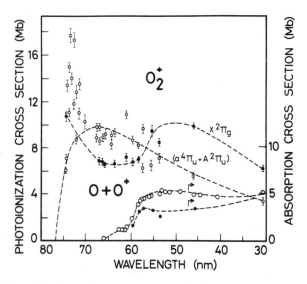

Fig. 6.4. Partial photoionization cross sections of O_2^+ states and photoabsorption cross section for the dissociative ionization $O + O^+$ [596].

THEORETICAL CALCULATION OF PHOTOIONIZATION CROSS SECTIONS

Photoionization cross sections are also obtained by theoretical calculations. Besides their contribution to understanding the details of the photoionization process, theoretical results are particularly useful for species where experiments are difficult to carry out. In photoelectron spectroscopic studies, the photoelectron intensities are directly dependent on the quantitative aspects of the photoionization process. The angle-dependent properties of photoelectron ejection also form an integral part of photoionization cross section studies; but since the work on angular distribution of photoelectrons has received somewhat special attention, we consider this aspect separately in the next chapter. Here we first sketch a few methods of theoretical calculation of photoionization cross section and then consider some quantitative analytical applications of photoelectron spectroscopy.

The photoionization cross section of a process in which the system in the initial state i having a total energy E_i absorbs an unpolarized photon of energy $h\nu$ and produces a final state f of total energy E_f, consisting of a photoelectron of kinetic energy ϵ and an ion in the jth state, is given by

$$\sigma_{ij}(\epsilon) = \frac{4\pi^2 \alpha a_0^2}{3g_i} (\epsilon + I_{ij}) |M_{if}|^2 \qquad (6.11)$$

Here α is the fine structure constant ($\alpha = e^2/\hbar c = 1/137.037$), a_0 is the Bohr radius of hydrogen atom (5.2917×10^{-9} cm), and g_i is the statistical weight of the initial state. The factor $4\pi^2 \alpha a_0^2/3$ has a value of 2.689×10^{-18} cm^2. The difference $E_f - E_i$ is equal to $I_{ij} + \epsilon$ (expressed in Rydbergs Ry $= e^2/2a_0$), that is, the sum of ionization potential I_{ij} and the photoelectron kinetic energy ϵ. The expression M_{if} is the *matrix element* of the transition. The square of the matrix element is given by

$$|M_{if}|^2 = \frac{4}{(I_{ij} + \epsilon)^2} \sum_i \sum_f \left| \int \psi_f^* \sum_\mu e^{i\mathbf{k}_\nu \cdot \mathbf{r}_\mu} \nabla_\mu \psi_i d\tau \right|^2 \qquad (6.12)$$

$|M_{if}|^2$ in Eq. (6.11) is expressed in Rydberg atomic units: length in terms of a_0, wave number in terms of $1/a_0$, and energy in Rydbergs; $|M_{if}|^2$ therefore has units of Ry^{-1}. ψ_f^* is the complex conjugate of the final-state wave function and ψ_i is the initial-state wave function, $\mathbf{k}_\nu$ is the propagation vector of the radiation of frequency ν, $\mathbf{r}_\mu$ is the position coordinate of the μth electron, ∇_μ is the vector differential operator, and $d\tau$ is the volume element. The summations are over all possible initial and final states.

Equation (6.12), in a simplified notation, can be written

$$|M_{if}|^2 = \frac{4}{(I_{ij}+\epsilon)^2} \sum_{i,f} \left| \left\langle f \left| \sum_\mu e^{i\mathbf{k}_\nu \cdot \mathbf{r}_\mu} \nabla_\mu \right| i \right\rangle \right|^2 \quad (6.13)$$

When the magnitude of $\mathbf{k} \cdot \mathbf{r}$ is small, the term $e^{i\mathbf{k}\cdot\mathbf{r}}$ in Eq. (6.12) can be approximated by unity. This is called the *dipole approximation* or *neglect of retardation*. Under this approximation the expression for the matrix element reduces to

$$|M_{if}|^2 = \frac{4}{(I_{ij}+\epsilon)^2} \sum_{i,f} \left| \left\langle f \left| \sum_\mu \nabla_\mu \right| i \right\rangle \right|^2 \quad (6.14)$$

The form in which the matrix element is expressed in Eq. (6.14) is called the *dipole velocity* form. Two other equivalent ways of expressing the matrix element are the *dipole length* form and *dipole acceleration* form. The dipole length form of the matrix element is given by

$$|M_{if}|^2 = \sum_{i,f} \left| \left\langle f \left| \sum_\mu \mathbf{r}_\mu \right| i \right\rangle \right|^2 \quad (6.15)$$

and the dipole acceleration form by

$$|M_{if}|^2 = \frac{16Z^2}{(E_f-E_i)^4} \sum_{i,f} \left| \left\langle i \left| \frac{\mathbf{r}_\mu}{r_\mu^3} \right| f \right\rangle \right|^2 \quad (6.16)$$

where Z is the nuclear charge. As mentioned earlier, $(E_f - E_i) = I_{ij} + \epsilon$.

We note from the above equations that a basic necessity in the calculation of photoionization cross sections is the availability of dependable initial bound state and final continuum state wave functions.

HYDROGENIC SYSTEMS

For hydrogen like systems, the wave functions can be obtained in exact form and exact analytical expressions for the matrix elements are available. The fin cross section then turns out to be

$$\sigma_\nu(n) = \frac{G \, 64\pi^4 e^{10} m Z^4}{3\sqrt{3} \, ch^6 \, \nu^3 g_n n^3} \quad (6.1)$$

G, the bound-free Gaunt factor, is close to unity and is given by

$$G = [1 + 0.1728 b^{1/3} - 0.0496 b^{2/3}]$$

$$+ \frac{1}{n^2} \left[-\frac{0.3456}{b^{2/3}} + \frac{0.0333}{b^{1/3}} \right] - \frac{0.0333}{n^4 b^{4/3}} \quad (6.1)$$

$b = \nu/\nu_H$, and $h\nu_H = 13.6$ eV represents the ionization threshold of atomic hydrogen.

Although in some atomic calculations the results of one-electron systems can be applied to complex systems by replacing Z with some effective nuclear charge, the extension of the above result in a similar manner to photoionization of multielectron systems fails. This is so even in the systems containing a single electron outside a closed core, where the approximation, it would seem, is most likely to work. The approximation fails because in the calculation of the transition moment integral, the bound-state and the final-state wave functions require the use of different Z values, and the use of a mean Z does not give the correct degree of overlap between the wave functions. As a consequence, such approximation leads to photoionization cross sections that give errors in both the magnitude and the nature of variation with incident photon energy.

QUANTUM DEFECT METHOD

In the dipole length formulation, contributions to the integrals arise primarily from large distances, where the effective field may be taken to be approximately coulombic. With a knowledge of experimentally observed energies and the requirement that the wave functions approach zero as $r \to \infty$, it is possible to derive the bound-state wave functions at large values of r. This is the basis of the quantum defect method. In order to derive the radial wave function, a radial differential equation with the radial function $R(\epsilon, l, r)$ is constructed (l is the orbital angular momentum and ϵ is the electron energy) that uses the potentials $V(r) = -2Z/r$ (in Rydbergs) at small distances and $V(r) = -2Z'/r$ at large distances, where Z is the nuclear charge and the effective charge is $Z' = Z - N + 1$, the N representing the number of atomic electrons. For discrete bound states, that is, photoelectron kinetic energy $\epsilon < 0$, the eigenvalues are equated with the experimentally observed series by $\epsilon_{nl} = -Z'^2/\nu_{nl}^2$ where ν_{nl} is an effective quantum number. The difference $n - \nu_{nl} = \mu_{nl}$, called the quantum defect, provides a measure of departure from the hydrogenic state. The μ_{nl} tends to vanish with increasing angular momentum of the electron. In practice, therefore, one considers only nonhydrogenic $l \leq 2$ or 3 cases. For $\epsilon > 0$, the photoelectron kinetic energy for the final-state electron can be represented by k^2, and the continuum wave function, the solution for the final state $R(k^2, l, r)$, has an asymptotic form that is bounded at the origin and depends on k, l, Z', and a phase angle δ_l, the latter accounting for the departure from a pure coulombic field. This method has been applied to a number of cases, such as the photoionization of He, Na, Mg, Si, K, Ca, for the determination of bound-state and continuum-state wave functions and the evaluation of dipole matrix elements, hence cross sections.

CENTRAL FIELD CALCULATIONS

In the central field approximation, each electron moves in a certain effective field created by the nucleus and all other remaining electrons. In photoionization cross section calculations involving an N electron system using the central field approximation, the initial-state and final-state wave functions ψ_i and ψ_f are taken as antisymmetrized products of N one-electron wave functions and ψ_i and ψ_f are approximated by single Slater determinants. Out of these N one-electron wave functions, $N-1$ are taken to be exactly the same for initial and final states, and it is assumed that in the photoionization process, all electrons except the active electron remain unaffected. Since the passive electrons remain "frozen," a consequence of this assumption is that no core relaxation or rearrangement is accounted for in the calculations, and multiple transitions are also excluded. It is further assumed that like the bound-state wave functions, the one-electron continuum wave functions are also eigenfunctions of the same effective central potential. Since the field remains unchanged in the ionization process, all other orbitals due to the passive electrons remain unchanged and integrate out to unity in the dipole matrix element. As a result the photoionization cross section $\sigma_{nl}(\epsilon)$ for an nl electron, where n and l are principal and azimuthal quantum numbers, depends on the bound-state wave function of the active electron and the corresponding wave function for the continuum state. σ_{nl} is given by

$$\sigma_{nl}(\epsilon) = \frac{4\pi^2 \alpha a_0^2}{3} \frac{N_{nl}(\epsilon - \epsilon_{nl})}{2l+1} [l\mathscr{R}_{l-1}(\epsilon)^2 + (l+1)\mathscr{R}_{l+1}(\epsilon)^2] \quad (6.19)$$

Here ϵ is the photoelectron kinetic energy, $-\epsilon_{nl}$ represents the binding energy of an nl electron, and N_{nl} is the occupation number of the nl subshell. The $\epsilon - \epsilon_{nl}$ in Eq. (6.19) is equivalent to the factor $I + \epsilon$ of Eq. (6.11). The two terms inside the square brackets of Eq. (6.19) arise due to two possibilities of angular momentum change in the photoionization process, $l-1$ and $l+1$, the l being the initial-state value. The dipole matrix elements $\mathscr{R}_{l\pm 1}$ are defined as

$$\mathscr{R}_{l\pm 1}(\epsilon) = \int_0^\infty P_{nl}(r) r P_{\epsilon, l\pm 1}(r) dr \quad (6.20)$$

where $P_{nl}(r)$ and $P_{\epsilon, l\pm 1}(r)$ are respectively the bound-state and free-state normalized radial functions of the active electron. If we write

$$\frac{P_{nl}(r)}{r} = R_{nl}(r) \quad \text{and} \quad \frac{P_{\epsilon, l\pm 1}(r)}{r} = R_{\epsilon, l\pm 1} \quad (6.21)$$

where $R_{nl}(r)$ and $R_{\epsilon, l\pm 1}(r)$ are the normalized radial functions of the final states of the active electron, then in terms of $R_{nl}(r)$ and $R_{\epsilon, l\pm 1}(r)$ the dipole matrix elements can be written

$$\mathcal{R}_{l\pm 1}(\epsilon) = \int_0^\infty R_{nl}(r) r^3 R_{\epsilon, l\pm 1}(r) \, dr \tag{6.22}$$

The radial wave functions are solutions to the one-body Schrödinger equation with a potential $V_{nl}(r)$ and energy ϵ_{nl}

$$\left[\frac{d^2}{dr^2} - \frac{l(l+1)}{r^2} + V_{nl}(r) + \epsilon_{nl} \right] P_{nl}(r) = 0 \tag{6.23}$$

Eq. (6.23) has been written for the bound-state wave function $P_{nl}(r)$. Continuum-state radial wave functions $P_{\epsilon l}(r)$ are obtained by solving Eq. (6.23) with $P_{nl}(r)$ replaced by $P_{\epsilon l}(r)$ and ϵ_{nl} replaced by ϵ.

The normalized continuum functions (normalization in the energy scale $\epsilon = \frac{1}{2} k^2$, $k \propto \epsilon^{1/2}$) have the form

$$P_{\epsilon l}(r) \xrightarrow[r \to \infty]{} \pi^{-1/2} \epsilon^{-1/4} \sin \left[\epsilon^{1/2} r - \tfrac{1}{2} l\pi - \epsilon^{-1/2} \ln 2\epsilon^{1/2} r \right.$$
$$\left. + \sigma_l(\epsilon) + \delta_l(\epsilon) \right] \tag{6.24}$$

In Eq. (6.24), $\sigma_l(\epsilon) = \arg \Gamma(l + 1 + i\epsilon^{-1/2})$ is the complex phase of the Γ function, and $\delta_l(\epsilon)$ is another shift with respect to coulomb waves.

The simplest form in which the central potential can be represented is $V(r) = -2Z_{\text{eff}}/r$, where Z_{eff} is a fixed effective nuclear charge. The principal shortcoming of this potential is that it is too small for small r, where an electron actually experiences a full nuclear charge Z without screening and is thus subject to a potential $-2Z/r$. For sufficiently large r, on the other hand, an atomic electron should experience the screening due to $Z - 1$ electrons and this should lead to an effective Z of unity. The resulting potential should then be $-2/r$. To this should be added contributions from shell effects and electron exchange. But since an exchange is a noncentral nonlocal interaction, approximations are required to incorporate the effect of the exchange into the central potential. The Slater approximation, which uses the exchange potential of a free electron gas, has the form

$$V(r) = -6 \left[\frac{3}{8\pi} |\rho(r)| \right]^{1/3} \tag{6.25}$$

where ρ is the total charge density of electrons of both spins.

On incorporation of this exchange potential into a Hartree SCF calculation followed by a self-consistent solution of the resulting equations, Hartree–Slater (HS) central-field wave functions and potential are obtained [they are also referred to as Hermann–Skillman (HS) wave-functions and potential]. The HS approximation provides the best central field for photoionization cross section calculations. As mentioned earlier, the multiplet structure cannot be explained from central field calculations, which can be quite important for open-shell atoms. Although the exchange forces are quite often attractive, cases are known where they have been found to be repulsive, which would be in variance with the above model. Nonetheless, HS calculations have been quite useful in obtaining theoretical atomic photoionization cross sections.

Recently some central field calculations of atomic photoionization cross sections have found considerable use in analytical applications of XPS. In this work the electrons are treated relativistically in a Hartree–Slater central potential–relativistic considerations are required because of the high photoelectron kinetic energies involved when X-ray sources are used. The effective central potential that has been used, which also approximates the effect of exchange, is

$$V(r) = -1.5 \left[\frac{3}{\pi} |\rho(r)| \right]^{1/3} \qquad (6.26)$$

where ρ is the electron charge density. Some numerical results of this cross section calculation for the X-ray region are given in Table 6.1.

We present here also the results of central field calculations on Ne, Ar, and Kr, along with experimental results, and show the way the wave function and matrix elements contribute to the spectral shape. The theoretical results are only for outer shell ionization which are experimentally observed; inner shell ionization would involve processes such as $2s \to \epsilon p$ and $3s \to \epsilon p$ (a change of l by only $+1$, from $l = 0$ to $l = +1$; a change by -1 is not possible). The results show that the theory predicts the cross section to the correct order of magnitude; spectral shapes of the cross section are also in fair agreement. Neon results agree reasonably well; also, the observed drop of the Ar cross section at threshold is qualitatively predicted by theory in a correct manner. Consideration of correlation effects may be expected to smoothen the rapid decrease.

A consideration of the energy dependence of the dipole matrix integral helps in understanding the spectral shape of cross sections. The ground state radial function $P_{nl}(r)$ ($2p$, $3p$, $4p$) and also the $P_{\epsilon l}(r)$ for the d wave ($l = 2$) at zero energy ($\epsilon = 0$) are presented in Fig. 6.5 for Ne, Ar, and Kr. In a p electron ionization both s and d wave matrix elements contribute, but for Ne, Ar, Kr at all energies the d wave matrix element is several times larger than the s wave matrix element. Therefore, its spectral behavior determines the overall spectral behavior of the cross section. The calculated values of the dipole matrix elements are shown in Fig. 6.5. We note that at $\epsilon = 0$ the d wave matrix element is small and positive for Ne but large and negative for Ar and Kr. The results of calculation show that with an increase in electron energy the d waves move towards the origin. As a consequence, for Ne this results at first in an increase in cross section followed by a decrease when the first node of the d wave has moved

Table 6.1. Theoretical Photoionization Cross Sections (in Barns) at 1000 and 2000 eV Photon Energy [604]

$1s_{1/2}$	1000	2000	$1s_{1/2}$	1000	2000
Li	2.6536 03[a]	3.0365 02	O	1.1553 05	1.7476 04
Be	8.7130 03	1.0706 03	F	1.7817 05	2.8496 04
B	2.1103 04	2.7303 03	Ne	2.3215 05	3.9194 04
C	4.2045 04	5.7308 03	Na	0	5.4247 05
N	7.3239 04	1.0516 04	Mg	0	7.2229 04

$2s_{1/2}$	1000	2000	$2s_{1/2}$	1000	2000
Na	1.4701 04	2.7478 03	K	6.6596 04	1.6693 04
Mg	1.9466 04	3.8110 03	Ca	7.4392 04	1.9222 04
Al	2.4855 04	5.0835 03	Sc	8.2051 04	2.1891 04
Si	3.0801 04	6.5575 03	Ti	8.9688 04	2.4675 04
P	3.7279 04	8.2237 03	V	9.6843 04	2.7534 04
S	4.4168 04	1.0076 04	Cr	1.0339 05	3.0487 04
Cl	5.1430 04	1.2109 04	Mn	1.1016 05	3.3448 04
Ar	5.8891 04	1.4315 04	Fe	1.1695 05	3.6465 04

$2p_{3/2}$	1000	2000	$2p_{3/2}$	1000	2000
Na	6.6200 03	6.2900 02	V	2.6138 05	3.6426 04
Mg	1.1084 04	1.0979 03	Cr	3.1063 05	4.4409 04
Al	1.7374 04	1.7946 03	Mn	3.6259 05	5.3408 04
Si	2.5849 04	2.7786 03	Fe	4.1803 05	6.3649 04
P	3.6899 04	4.1122 03	Co	4.8067 05	7.5058 04
S	5.0955 04	5.8598 03	Ni	5.5143 05	8.7747 04
Cl	5.8294 04	8.0944 03	Cu	6.3893 05	1.0182 05
Ar	8.9311 04	1.0895 04	Zn	0	1.1691 05
K	1.1434 05	1.4353 04	Ga	0	1.3349 05
Ca	1.4383 05	1.8551 04	Ge	0	1.5138 05
Sc	1.7819 05	2.3595 04	As	0	1.6993 05
Ti	2.1761 05	2.9531 04	Se	0	1.9003 05

$3p_{3/2}$	1000	2000	$3p_{3/2}$	1000	2000
Ga	6.9729 04	1.3936 04	As	8.5195 04	1.8084 04
Ge	7.7286 04	1.5926 04	Se	9.3431 04	2.0398 04

$3d_{5/2}$	1000	2000	$3d_{5/2}$	1000	2000
Ga	3.3370 04	3.0613 03	Sr	1.3843 05	1.5343 04
Ge	4.2891 04	4.0503 03	Y	1.6192 05	1.8377 04
As	5.4018 04	5.2508 03	Zr	1.8777 05	2.1810 04
Se	6.6868 04	6.6885 03	Nb	2.1594 05	2.5664 04

Table 6.1. (*Continued*)

$3d_{5/2}$	1000	2000	$3d_{5/2}$	1000	2000
Br	8.1578 04	8.3874 03	Mo	2.4679 05	2.9980 04
Kr	9.8313 04	1.0372 04	Te	2.8024 05	3.4778 04
Rb	1.1721 05	1.2681 04	Ru	3.1649 05	4.0084 04
Rh	3.5537 05	4.5928 04	La	1.0129 06	1.6765 05
Pd	3.9721 05	5.2315 04	Ce	1.0970 06	1.8344 05
Ag	4.4162 05	5.9303 04	Pr	1.2038 06	1.9986 05
Cd	4.8907 05	6.6896 04	Nd	1.1220 06	2.1677 05
In	5.3911 05	7.5156 04	Pm	0	2.3447 05
Sn	5.9069 05	8.4067 04	Sm	0	2.5348 05
Sb	6.4602 05	9.3695 04	Eu	0	2.7359 05
Te	7.0327 05	1.0409 05	Gd	0	2.9407 05
I	7.5973 05	1.1515 05	Tb	0	3.1492 05
Xe	8.1545 05	1.2709 05	Dy	0	3.3599 05
Cs	8.7788 05	1.3988 05	Ho	0	3.5611 05
Ba	9.4349 05	1.5330 05	Er	0	3.7957 05

$4d_{5/2}$	1000	2000	$4d_{5/2}$	1000	2000
Te	7.6604 04	1.3101 04	Eu	1.7637 05	3.8720 04
I	8.6287 04	1.5087 04	Gd	1.8603 05	4.1794 04
Xe	9.6346 04	1.7218 04	Tb	1.9113 05	4.4067 04
Cs	1.0692 05	1.9524 04	Dy	1.9808 05	4.6817 04
Ba	1.1794 05	2.1999 04	Ho	2.0474 05	4.9611 04
La	1.2928 05	2.4628 04	Er	2.1109 05	5.2440 04
Ce	1.3553 05	2.6436 04	Tm	2.1708 05	5.5300 04
Pr	1.4407 05	2.8756 04	Yb	2.2270 05	5.8187 04
Nd	1.5243 05	3.1146 04	Lu	2.2984 05	6.1502 04
Pm	1.6062 05	3.3605 04	Hf	2.3711 03	6.4945 04
Sm	1.6861 05	3.6130 04			

$4f_{7/2}$	1000	2000	$4f_{7/2}$	1000	2000
Hf	1.9689 05	2.1001 04	Pb	5.2789 05	6.8887 04
Ta	2.2277 05	2.4283 04	Bi	5.7009 05	7.5779 04
W	2.5028 05	2.7865 04	Po	6.1385 05	8.3105 04
Re	2.7931 05	3.1761 04	At	6.5926 05	9.0867 04
Os	3.0989 05	3.5979 04	Rn	7.0594 05	9.9097 04
Ir	3.4186 05	4.0510 04	Fr	7.5420 05	1.0779 05
Pt	3.7577 05	4.5427 04	Ra	8.0333 05	1.1695 05
Au	4.1126 05	5.0694 04	Ac	8.5422 05	1.2664 05
Hg	4.4844 05	5.6360 04	Th	9.0612 05	1.3877 05
Tl	4.8740 05	6.2416 04	Pa	9.5738 05	1.4731 05

[a] Read 2.6536 03 = 2.6536 × 10³ barns. 1 barn = 10^{-24} cm²; 1 megabarn (Mb) = 10^{-18} cm²

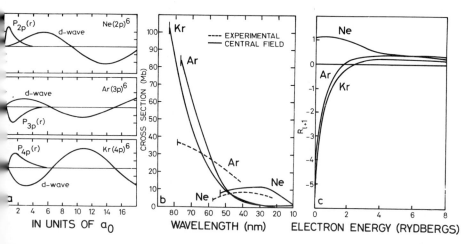

Fig. 6.5. (a) Outer subshell radial wave functions and d waves for Ne, Ar, and Kr at zero photoelectron kinetic energy. (b) Calculated and experimental photoionization cross sections for Ne, Ar, and Kr. (c) Matrix elements for $p \to d$ transitions in Ne, Ar, and Kr [561].

close enough to the origin, that is, at higher photoelectron energy, when negative contributions to the integrand become important. We see that calculations in the energy range 0–10 Ry show that the d-wave matrix element is positive. Since the matrix element must be positive in the high energy limit, a reversal of sign at higher energies is not expected. (This is because at higher photoelectron kinetic energies the oscillation of the continuum wave function is of higher frequency, resulting in an increased positive contribution to the matrix element from the bound-state wave function.) For Ar and Kr, on the other hand, at $\epsilon = 0$ the matrix element is negative; with increasing energy its magnitude decreases and becomes zero when positive and negative portions of the dipole matrix element integrand are equal. This occurs at about $\epsilon = 2$ Ry for Ar, and at about $\epsilon = 2.4$ Ry for Kr. At higher energies the matrix element is positive and further sign reversals are unlikely.

HARTREE–FOCK CALCULATIONS AND OTHER METHODS

In the Hartree–Fock method, the simplicity of wave functions consisting of single Slater determinants is maintained while still treating exchange correctly. The HF wave function is the most accurate independent particle wave function possible, as it is obtained via the variation method. For the initial bound state, the steps involve first setting up for the system a wave function ψ that is an antisymmetric product of one-electron functions $R_{nl}(r)Y_l^m(\theta,\phi)$ where $Y_l^m(\theta,\phi)$ are spherical harmonics. A linear combination of such products may also be used so as to correctly represent the angular momentum couplings of an open shell many-electron system. On the basis of the exact nonrelativistic Hamiltonian $\mathcal{H}$, the energy functional $\langle\psi|\mathcal{H}|\psi\rangle$ is then constructed subject to the constraints of the orthonormality of the one-electron functions; the variation principle is then applied for minimization of the energy functional. This process yields a set of self-consistent coupled integrodifferential equations for the $P_{nl}(r)$, which are treated as unknowns, and a solution of these equations leads to the explicit form of the wave function.

For the final continuum state, first HF procedure is carried out for the residual ion core less the photoelectron. This yields the final-state radial functions $P_{nl}^f(r)$. After this, the core orbitals are frozen and the HF procedure is carried out with the core plus the photoelectron final state, with only the radial part of the continuum orbital, $P_{\epsilon l}(r)$, unknown.

Within the framework of the HF approximation for the wave function, the general expression for photoionization of an electron from an initial state described by $(nl)^q$ ^{2S+1}L to an $\{[(nl)^{q-1}$ $^{2S_c+1}L_c]$, $(\epsilon l')\}$ $^{2S+1}L'$ final state is given by

$$\sigma_{nl}(LS, L_cS_c, \epsilon l'L') = \frac{4\pi^2 \alpha a_0^2}{3g_i}(I_{ij}+\epsilon)\frac{1}{4l_>^2 - 1}$$
$$\zeta(LS, L_cS_c, l'L')\,\gamma\,[\mathcal{R}_{l'}(\epsilon)]^2 \qquad (6.27)$$

where q is the number of electrons in the atom prior to ionization, l and $l'(=l\pm 1)$ are respectively the angular momenta of the active electron before and after photoionization, $l_>$ is the greater of l and l', ζ is the relative multiplet strength, L and S are respectively the total orbital and spin angular momenta for the initial state, L_c and S_c the same for the residual ion core, and L' is the total orbital angular momentum of the ion plus photoelectron final state. ^{2S+1}L, $^{2S_c+1}L_c$, and $^{2S+1}L'$ are the respective term symbols. The overlap integral γ is given by

$$\gamma = \prod_{\substack{\text{passive}\\\text{electrons}}}\left|\int_0^\infty P_{nl}^i(r)P_{nl}^f(r)\,dr\right|^2 \qquad (6.28)$$

where i represents the initial state and f the final state. The dipole matrix element $\mathcal{R}_{l'}(\epsilon)$ is expressed as $\mathcal{R}_{l'}(\epsilon) = \int_0^\infty P_{nl}^i(r)\,rP_{\epsilon l'}^f(r)\,dr$. Since $P_{nl}^i(r)$ and $P_{nl}^f(r)$ are not identical, γ represents the effect of core relaxation that was not included in central field calculations. The cross section for a particular photoionization channel $i(LS) \to j(L_cS_c)$ is a sum over all possible l' and L' values:

$$\sigma_{ij}(\epsilon) = \sum_{L'}\sum_{l'}\sigma_{nl}(LS, L_cS_c, \epsilon l'L') \qquad (6.29)$$

None of these calculations, however, take into consideration resonances in photoionization cross section due to autoionization as well as those populating the Auger states. These appear only in a range of about a few eV above each ionization threshold, but when they do appear, the cross sections are affected drastically. For an adequate treatment of such situations one needs to go beyond HF approximation.

For many-electron atoms the possibility of electron exchange needs consideration. When wave functions with exchange are used, the effect of the exchange terms is to pull in all the nodes of the wave function towards the nucleus, which is equivalent to the presence of an attractive force field. The extent of this effect depends on the atom being considered. For example, in photoionization of Ca the difference between calculated cross sections with and without exchange is significant, whereas for the photodetachment of H$^-$ it is not important.

When excessive cancellation occurs within the matrix element, it is no longer possible to derive wave functions of the active electron by application of the Hartree or Hartree–Fock methods. The accuracy then can be improved by incorporating a correlation between the active electron and the passive electrons by adding to the self-consistent potential energy a term, $V_p(r)$, that represents the attractive force between the active electron and the dipole moment induced by it on the residual system. The $V_p(r)$ has a form $V_p = \frac{1}{2}p/(r_c^2 + r^2)^2$ where r_c represents the mean radius of the core and p is its polarizability.

There are now methods that permit the calculation of bound-state wave functions to any desired degree of accuracy. The problem is to obtain accurate continuum wave functions. In a method that uses configuration interaction in the continuum wave function, a complete set of solutions for the final state is used; it includes discrete states as well. When such complete sets of solutions are used, the effect of autoionizing states on the photoionization cross section comes out directly. In the one-particle approximation to the total Hamiltonian, which is the Hartree–Fock case, solution of the Schrödinger equation yields a spectrum of discrete and continuum eigenstates. Ordinarily, a consideration of only the discrete states is sufficient for bound states. But for the final-state wave function in the photoionization process — which corresponds to a parallel situation where the interaction potential between an incident electron and an excited state of the target is strong enough to support a bound state — continuum states mix with the discrete states and therefore need to be taken into consideration for a satisfactory description of the system. Two efficient methods by which the continuum states can be incorporated into the total system description are the *close coupling scheme* and the *many-body perturbation theory* (MBPT).

The close coupling scheme is based on an expansion of the total wave function in terms of the eigenstates of the core Hamiltonian. The Schrödinger equation describing the scattering of an electron by an atomic system with $N - 1$ electrons can be written

$$(\mathcal{H}_N - E)\, \psi\, (x_1, x_2, \ldots, x_N) = 0 \qquad (6.30)$$

where $\mathcal{H}_N$ is the N-electron Hamiltonian, and x_i represents the space and spin coordinates of the ith electron. In the close coupling method, the final-state continuum wave function is written as $\psi(x_1, x_2, \ldots, x_N) = \sum_i a_i \phi_i(x_1, x_2, \ldots, x_{N-1}) F_i(x_N)$. The summation is over the complete set of eigenstates ϕ_i, including the continuum, of the core Hamiltonian (i.e., without the Nth electron). The F_i in the equation describes the motion of the scattered electron in the ith channel. Coupled second-order integrodifferential equations are then derived for the radial parts of the wave function F_i by the following procedure. By left-multiplying Eq. (6.30) with each ϕ_i and integrating, the eigenvalue expressions are obtained, which are then minimized variationally with respect to the expansion coefficients a_i. From this, coupled equations of the following type are obtained:

$$\left(\frac{d}{dr^2} - \frac{l(l+1)}{r^2} + \epsilon\right) F_i(r) = \sum_j V_{ij}(r) F_j(r) \qquad (6.31)$$

The potential matrix V (with elements V_{ij}) on the right side of Eq. (6.31) is in general nonlocal, with long-range asymptotic inverse r behavior. With F_i's and a_i's known, the final-state wave function can be obtained and the dipole matrix elements evaluated.

Another way of improving on the HF method is by application of MBPT, which also attempts to simultaneously improve both initial-state and final state wave functions of the exact Hamiltonian itself. For an N-particle problem the exact Hamiltonian is decomposed into a one-body part $\mathcal{H}_0$ and a correction $\mathcal{H}'$ that has both one- and two-body components such that

$$\mathcal{H} = \mathcal{H}_0 + \mathcal{H}', \quad \mathcal{H}_0 = \sum_\mu (T_\mu + V_\mu), \quad \text{and} \quad \mathcal{H}' = \sum_{\mu<\mu'} v_{\mu\mu'} - \sum_\mu V_\mu \qquad (6.32)$$

where T_μ is the kinetic energy operator, V_μ a one-body operator, and $v_{\mu\mu'}$ a two-body operator. The solutions to $\mathcal{H}_0$, the one-body approximation to the exact Hamiltonian, are obtained (as is done in the HF case, for example) and the residual term $\mathcal{H}'$ is treated as a perturbation. This procedure can be adopted for obtaining both the correlated initial-bound and the final-continuum states; the expectation value of the transition moment operator can be calculated as in the usual procedure. Alternatively, a direct calculation for the transition moment expectation values with the many-body effects incorporated through wave functions can be performed. Unlike normal perturbation theory, where limitation on keeping track of terms restricts the perturbation expansion to a few orders, in MBPT it is possible to perform summation to infinite order of selected terms. Physically meaningful arguments can be used for the choice of such terms. The efficacy of the MBPT method rests entirely on the choice of partitioning of the Hamiltonian, such as HF and HFS, and the ensuing set of single-particle eigenstates of $\mathcal{H}_0$. Notwithstanding complications, the methods that improve both initial and final states simultaneously yield the best results for photoionization cross sections. They also have the advantage of clearly showing which interactions are important for evaluation of the dipole matrix element.

CROSS SECTIONS FOR MOLECULES

The major difficulty in calculating photoionization cross sections for molecules is in constructing sufficiently accurate wave functions for the final state, the state containing the photoelectron. For the bound-state wave function, SCF molecular orbitals can be used.

In the photoionization cross section calculations in molecules, simple *plane waves* (PW), $e^{i\mathbf{k}\cdot\mathbf{r}}$, have been used to represent the photoelectron. Cross sectional studies of π-electron ionization in planar hydrocarbons, where plane wave approximations have been used, show that in the low photon energy range (~ 40 eV) increasing the number of nodes in a molecular orbital (MO) tends to reduce the photoionization efficiency of the orbital. All trans-polyenes are found to yield cross sections smaller than the corresponding cis- and cyclic polyenes. The calculations show no clear relationship, however, between the magnitude of the cross section and the size of the molecule. With low photoelectron kinetic energy, as would be the case in UPS, simple plane wave turns out to be a poor approximation for the photoelectron wave function. It is a better approximation when the kinetic energy of the ejected electron is much greater than the ionization potential.

A better set of continuum wave functions is plane waves that have been orthogonalized by the Schmidt process to all of the occupied valence shell molecular orbitals. We consider the *orthogonalized plane wave* (OPW) method in some detail in Chapter 7. Such an orthogonalization incorporates the influence of the attractive molecular potential. In this approximation different continuum functions in general are no longer orthogonal to each other, but this factor may not affect the bound-free transition probabilities. The OPW simulates best a short range rather than a long range attractive potential. The method is thus relatively more suitable for describing photodetachment processes like $M^- \rightarrow M^0 + e$ than for describing ionization of a neutral $M^0 \rightarrow M^+ + e$.

For H_2, OPW results show good agreement for $h\nu \geqslant 30$ eV. At low photon energies, however, the agreement is poor. For example, OPW results show a peak in photoionization cross section at 22.5 eV with $\sigma_{tot}^{OPW} = 3.2 \times 10^{-18}$ cm^2. An experimental peak at 17 eV has $\sigma_{tot} = 10 \times 10^{-18}$ cm^2. At lower energies near the photoionization threshold the agreement is worse, as much as an order of magnitude or more.

The final state should not only be orthogonal to all the bound states but also to other continuum states. In an extended OPW method, a minimal number of additional orthogonalities for some excited states involving virtual molecular orbitals are included. A basic limitation of the OPW method is that whereas it takes into account the attractive potential of the molecular ion, long-range effects of the coulomb potential are not accounted for. Additional and more realistic virtual orbitals to improve the basis set could improve the cross sections obtained by the OPW method.

Like OPW, several other kinds of unbound orbitals may be used, such as *spherical waves* (SW), $e^{i\mathbf{k}\cdot\mathbf{r}}/r$, which can be orthogonalized to occupied orbitals. Spherical waves and plane waves have similarities as they both describe a particle in the region of constant potential. Another alternative is the use of *coulomb waves*, unbound states of the hydrogen atom, or rather, *orthogonalized coulomb waves* (OCW), which provide asymptotic description of the final states of any molecular ionization. In practice, however, OPW has been a frequently used method for calculation of molecular photoionization cross sections. Results of photoelectron angular distributions obtained by the application of the OPW techniques and some details of the OPW formalism are presented in Chapter 7.

Among other approaches, a pseudopotential model has been used for photoionization cross section of N_2. It assumes the core region to be spherically symmetric with the center at the midpoint of the molecule; the phase shifts are obtained from quantum defect values of the molecular Rydberg states. For H_2, a single-center approximation, including exchange and screening effects, has been used. The theoretical cross section is in good agreement with experimental results. For polyatomic molecules, one-center SCF functions have been applied. A common origin is used for the SCF wave functions of the bound electrons and for the photoelectron represented by a coulomb wave continuum function. The use of a single-center formulation for a semiquantitative understanding of photoelectron angular distribution is discussed in the next chapter.

The PW may be expected to constitute a better approximation in photoionization using X-rays than in vacuum ultraviolet photoionization. The results of relative intensities of several free molecules of XPS $h\nu = 1254\,\text{eV}$, however, do not show any marked changes in results when OPW instead of PW approximation is used. For example, for N_2 and H_2O the relative intensities are presented in Table 6.2. Another PW application has been in the calculation of photoionization cross section of H_2 at high energies. At 5.41 and 8.39 keV of photon energy, where experimental data are available, the cross sections are found to be in good agreement. Experimental values at these energies for $\frac{1}{2}\sigma^{MO}$ are 54.0 ± 2.9 and 12.0 ± 0.6 Mb/atom respectively. Corresponding PW results are 54.7 and 12.0 Mb/atom.

Table 6.2. Relative Intensities in Photoionization [555]

	Molecular Orbital	Plane Wave	Orthagonalized Plane Wave	Experimental Peak Area [4]
N_2	$2\sigma_g$	1.00	1.00	1.00
	$2\sigma_u$	1.02	0.94	1.11
	$1\pi_u$	0.04	0.05	0.11
	$3\sigma_g$	0.19	0.18	0.22
H_2O	$2a_1$	1.00	1.00	1.00
	$1b_2$	0.02	0.02	0.08
	$3a_1$	0.23	0.22	0.27
	$1b_1$	0.05	0.04	0.08

Besides this, there have been several qualitative and semiquantitative models on photoelectron intensities. Two of these are presented in the following sections.

CROSS SECTION CHARACTERISTICS AND NODAL PROPERTIES OF ORBITALS

A qualitative idea about the magnitude of photoionization cross section may be obtained from the dimensions and nodal characteristics of the orbital involved as well as the wavelength associated with the ejected photoelectron. The latter is known from the relationship between the wavelength of the photoelectron and its kinetic energy, and hence it indirectly depends on the energy of the incident radiation. The dipole matrix element involved in the cross section expression depends on both the probability distribution and the nodal properties of the initial state and the wavelength and phase of the photoelectron in the final state. The cross section becomes maximum when all these terms contribute positively to the integral. This is illustrated in Fig. 6.6. For continuum states, plane waves are used, de Broglie wavelength decreasing with increasing photoelectron kinetic energy. Qualitatively, the photoionization cross section can be expected to maximize when the half-width of orbital spread is about $\lambda/4$, where λ is the photoelectron wavelength. This may be visualized from the overlaps of the various constituents of the dipole matrix element integral shown in Fig. 6.6. It explains how, near the threshold of ionization using low photon energy, the cross sections of molecular orbitals built from p atomic orbitals are greater than those constructed from atomic s orbitals. At the threshold of ionization of p electrons by vacuum ultraviolet radiation, the photoelectron

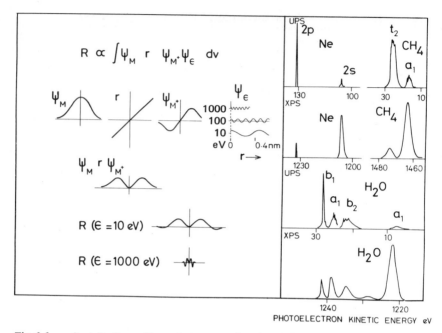

Fig. 6.6. Contributions of bound-state wave function and continuum-state wave function with photoelectron kinetic energy ϵ to the transition moment integral. Examples of change in relative intensities of bands from p- and s-type orbitals for ionizing photons of low and high energy [29, 564].

kinetic energy is less, therefore the corresponding wavelength is large. The p orbital being relatively more spread out, the matching is better with a longer wavelength, thus yielding a larger cross section. At high photoelectron energy the situation reverses, since with the shorter wavelength of photoelectron, small radial distances are emphasized, where an s electron contribution is greater. This explains why the cross section for the p orbital is large in UPS and low in XPS, as illustrated in the observed spectra of Ne, CH_4, and H_2O.

A MODEL FOR APPROXIMATE PHOTOIONIZATION CROSS SECTIONS

Since cross sections determine the intensities of photoelectron spectral peaks, it would certainly be useful if simple correlations between peak intensities and orbital properties were available, even if they were approximate. We have seen that in X-ray photoionization the valence electron cross sections depend largely on the inner parts of the orbitals, those in the near core region. In this region the valence molecular orbitals have steep slopes in order to be orthogonal to core molecular orbitals, which are built from core electron wave functions of the constituent atoms. On this basis it should be possible to express the photoionization cross section of a valence molecular orbital j as a sum of atomic contributions from each atomic center of the molecule: $\sigma_j^{MO} = \sum_A \sigma_{Aj}$. The σ_{Aj}'s can be further partitioned as

$$\sigma_{Aj} = \sum P_{A\lambda j}\sigma_A^{AO} \qquad (6.33)$$

where σ_A^{AO} is the photoionization cross section of the atomic $A\lambda$ subshell, and $P_{A\lambda j}$ denotes the net atomic population on atom A from the atomic $A\lambda$ orbital in the jth molecular orbital. The summation is carried out over all atomic subshells $\lambda(ns, np, \ldots)$. For photoelectrons emitted at an angle $\pi/2$ from a source of unpolarized photon beam, the intensity I_j^{MO} of the jth molecular orbital, using Eq. (6.33) can be written

$$I_j^{MO} \propto (2 + \tfrac{1}{2}\beta) \sum_{A,\lambda} P_{A\lambda j}\sigma_A^{AO} \qquad (6.34)$$

where β is the asymmetry parameter. This model has been quite successful in explaining the prominent features of X-ray photoelectron spectra of valence electrons of many molecules: CF_4, C_3O_2, C_4H_4O, C_4H_4S, C_6H_6, SF_6, to name a few.

The XPS valence electron spectrum for NNO is shown in Fig. 6.7. The

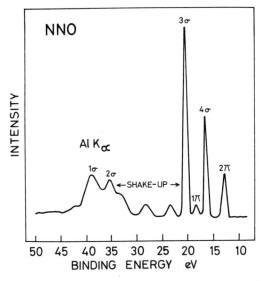

Fig. 6.7. XPS valence electron spectrum of N_2O [36].

ground state configuration for NNO is $1\sigma^2 2\sigma^2 3\sigma^2 1\pi^4 4\sigma^2 2\pi^4$, and these electron peaks are seen in the spectrum. σ orbitals have higher intensity because of their $2s$ character. On the basis of atomic orbital cross sections, the 2π nonbonding orbital has 40% N $2p$ and 60% O $2p$ character whereas the strongly bonding orbital 1π has 74% N $2p$ and 26% O $2p$ character. Since the O $2p$ cross section is about twice that of N $2p$, the 2π peak is more intense. For the 1π bonding orbital most of the electron density, in comparison with 2π, is in the interatomic regions and not near the nucleus, which results in a low cross section. The net atomic populations of 1π and 2π orbitals have been calculated to be 1.48 and 2.12 respectively, and this is compatible with the spectral intensities.

We should now have a fair idea about both experimental and theoretical aspects of photoionization cross sections. The next step concerns more practical considerations, such as applications of photoionization cross sections to analytical chemistry.

APPLICATIONS TO ANALYTICAL CHEMISTRY

The potential of photoelectron spectroscopy for quantitative analytical applications stems from the following factors. (1) The method is sensitive to practically all the elements in the periodic table. (2) The core binding energies lead directly to qualitative identification of the atomic species present in the sample. (3) The XPS chemical shifts, obtained as primary data, yield additional information on the chemical nature of the analyte. (4) The technique can be used to examine the surface region of solids. (5) The method is practically nondestructive and has high absolute sensitivity.

For quantitative analysis, the photoelectron line intensity is the key observable. This depends on the photon flux incident on the sample, the photoionization cross section, the space-integrated escape probability of the photoelectrons in the direction of the entrance slit of the analyzer, and the sensitivity of the instrument's electron detector. Since the photons at the energies commonly used in XPS have a mean free path of about hundred times that of the photoelectrons, the former may be taken to act unattenuated and therefore uniformly over the entire sample. Photoionization cross sections obtained from theoretical calculations are used in quantitative applications; reliability of the available cross sections is thus limited by the accuracy of the theoretical methods used in the calculations. The electron escape probability poses no less challenging a problem, and when experimental values are not available, one often needs suitable theoretical methods to obtain estimates of this quantity. With regard to the sensitivity of the detector, the detected electron current has been found to be inversely proportional to the kinetic energy of the photoelectrons. Thus at several stages, approximations as well as appropriate corrective factors need to be applied.

APPLICATIONS TO ANALYTICAL CHEMISTRY

As we have seen in Table 6.1, the photoionization cross section at the photon energies used for XPS experiments is strongly influenced by the angular momentum quantum number. With increasing atomic number, the binding energy of electrons from the given shell increases. It is interesting that when such binding energies become very high — higher than the photon energies from commonly available photon sources — a new series becomes accessible with ionization from the shell of next higher principal quantum number, and therefore with lower binding energy. The latter also results in strong photoelectron lines; the lines have high intensities and are also otherwise suitable for use in quantitative analysis. The effect of angular momentum is dominant. For a given principal quantum number, the strong photoelectron lines have the highest angular momentum. Except for the first members of each series, ionizations that provide strong lines are $1s$ for elements through $Z = 12$, $2p$ or $2p_{3/2}$ for $Z = 14-33$, $3d$ or $3d_{5/2}$ for elements $Z = 36-68$, and $4f$ or $4f_{7/2}$ for elements $Z = 72-92$.

Since one of the major application areas of analytical XPS is the analysis of solids, the electron escape probability in the solid sample constitutes essential information. A peculiar situation in solid analysis with XPS is that not all the atoms in the sample contribute equally to the observed photoelectron intensity; the relative contribution from those near the surface with respect to those in the bulk depends on the inelastic scattering cross section of the electrons or their mean free path λ in the solid. Uncertainty in this mean free path λ also limits the application of XPS to analytical chemistry. Accuracies in λ are not better than 20%, and λ for different elements vary, as a function of electron kinetic energy, within 50%.

Chemical composition is also an important constraint on quantitative analysis. One of the very early quantitative applications was measurement of relative sensitivities of elements up to $Z = 92$ in fluorine-containing compounds with fluorine used as the internal standard. It was noted that the relative intensities depend on the chemical composition of the material studied. A difference in composition means a difference in the nature of chemical bonding, and the latter affects one or more of the factors that contribute to the sensitivity. Chemical bonding is unlikely to affect the cross section of core ionization significantly; more likely it affects the escape probability of the photoelectrons through the inelastic scattering cross sections of the photoelectrons. The variability of response among compounds is thus a major concern for routine quantitative application. Quantitative analysis using XPS, till now, is thus not as routinely and conveniently usable as some other more established analytical methods. As a surface analysis method, however, its application has been much more successful. Ordinarily, without using special conditions to improve the bulk sensitivity (as discussed in Chapter 5), the latter is limited to about 0.1% based on bulk percentage and it is thus not a particularly sensitive technique for analysis of bulk constituent. In contrast, it has been possible to detect as low as 0.2% of a monolayer of heavy metals, which amounts to about 10^{12} atoms and hence approximately picogram material. If these atoms can be scavenged from 100 ml of a solution, a sensitivity of about 2 parts per trillion is indicated.

In the following sections we discuss two different types of models that have been used in quantitative analysis using XPS. Thereafter, we consider some examples of trace metal analysis and matrix dilution technique. Finally, we summarize the present status of the field.

A MODEL FOR QUANTITATIVE APPLICATIONS

In order to evolve a simple model for use in quantitative analysis, one may proceed as follows. If σ is the photoionization cross section for a given subshell of an element, n is the concentration of the element in terms of atoms per unit volume, and I_0 is the incident photon flux, then it is reasonable to assume that N_0 the number of photoelectrons ejected per unit volume from a given subshell is given by $N_0 = \sigma n I_0$. Since the detector looks at a definite angle the nature of angular distribution matters, but at high energies the corrections do not vary much from one subshell to another, and if one is ultimately interested in relative intensities, then the effect is less. It is possible to realize a further simplification by measuring the ejected photoelectron intensity at an angle of 54.73°, where the intensity does not depend on the asymmetry parameter β. This can be seen by a consideration of Eq. (1.2) in Chapter 1. We discuss this aspect further in the next chapter.

The probability of escape of the electrons from the solid material must also be considered. The photoelectron must emerge without any energy loss due to inelastic collisions, must pass through the spectrometer, and must be detected. The number of electrons originating on a vertical line segment dx located at a depth x below the surface is given by

$$dN = N_0 e^{-x/\lambda} S \, dx \tag{6.35}$$

where $1/\lambda$ is the reciprocal of the mean free path, equivalent to the inelastic scattering cross section of electrons in the solid, and S is a spectrometer factor. The total signal intensity to be expected is obtained by integrating the above equation from the surface to infinite depth, and one obtains

$$N = \int_0^\infty N_0 e^{-x/\lambda} S \, dx = N_0 \lambda S = \sigma n I_0 \lambda S \tag{6.36}$$

The relative intensity for two different photoelectron peaks from the same sample is therefore given by

$$\frac{N_1}{N_2} = \frac{\sigma_1 n_1 \lambda_1 S_1}{\sigma_2 n_2 \lambda_2 S_2} \tag{6.37}$$

The spectrometer factor may depend on several factors, one being the electron energy, but all of them are determinable. If the two peaks are similarly affected, then it is possible to write $S_1 = S_2$. In the energy range 100–1500 eV, theoretical and experimental results show that the mean free path λ and the electron kinetic energy ϵ approximately satisfy a relationship $\lambda \propto \epsilon^{0.5}$. On the basis of this (although some experimental results show somewhat different dependence) it is possible to write $\lambda \propto \sqrt{h\nu - E_B}$, since $\epsilon = h\nu - E_B$, E_B being the binding energy of the atomic orbital of the sample from which the photoelectron is ejected. In order to obtain an expression for relative intensities when concentrations of the two atomic species are identical, that is, $n_1 = n_2$, one may write, using C 1s as the reference:

$$\frac{N_{Z,nl}}{N_{C,1s}} = \frac{\sigma_{Z,nl} \sqrt{h\nu - E_B(Z, nl)}}{\sigma_{C,1s} \sqrt{h\nu - 284}} \tag{6.38}$$

where 284 eV represents the C 1s binding energy in the solid, and $h\nu$ and E_B are expressed in eV. If λ dependence on ϵ is somewhat different, appropriate modifications are to be incorporated in Eq. (6.38).

As mentioned earlier, $\sigma_{Z,nl}$ values for elements have been calculated based on relativistic Hartree–Slater wave functions and partly shown in Table 6.1. The values are dependable so long as the photon energy is not too close to the

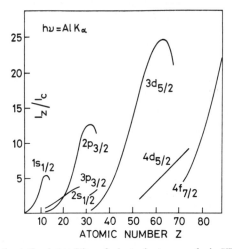

Fig. 6.8. Calculated relative intensities of photoelectron peaks in XPS of solids for Al K_α radiation [607].

threshold of photoionization. Figure 6.8 gives the relevant ratio for Al K_α radiation on the basis of Eq. (6.38) with subshells that have potential XPS applications; the cross sections for these states were presented in Table 6.1; the comparison of relative intensities for photoelectron peaks in solids $h\nu = \text{Al}\,K_\alpha$ is presented in Table 6.3. As mentioned earlier, these often are the subshells that have the highest angular momentum for a given principal quantum number. Vacancies in these orbitals cannot be filled by Coster–Kronig transitions, which can drastically shorten the half-life of such states and thus the photoelectron peak; this makes their choice very desirable in quantitative photoelectron spectroscopy. The results of actual applications show that agreement between theory and experiment are quite satisfactory. Photoelectron peaks of the same element separated by chemical shifts sometimes do not show intensities that would be expected from the stoichiometric formula. This could be due to differences in the degree of electron shake-up and shake-off for the different atoms.

In the special situation where a two-component homogeneous material of known concentration has a coating of a contaminant layer, it is possible to determine the layer thickness by quantitative measurements of photoelectron intensities. If N_1' and N_2' are the respective intensities from the two constituents in the absence of the contaminant layer, then the presence of an overlayer of thickness d will alter the intensities by an exponential factor as given in Eq. (6.35). If a ratio of the intensities is taken, then one obtains:

$$\frac{N_1}{N_2} = \frac{N_1' e^{-d/\lambda_{1'}}}{N_2' e^{-d/\lambda_{2'}}} = \frac{N_1'}{N_2'} e^{d(1/\lambda_{2'} - 1/\lambda_{1'})} \tag{6.39}$$

where λ_1' and λ_2' are respectively the mean free paths of electrons of kinetic energy ϵ_1 and ϵ_2 from the two components in the contamination layer and the

Table 6.3. Comparison of Relative Intensities of Photoelectron Peaks in Solids $h\nu = \text{Al}\,K_\alpha$ [607]

Ratio	Experiment				Theory [607]
	$I_{[606]}$	$I_{[607]}$	$I_{[608]}$	Average I	
C 1s/F 1s	0.24	0.29	0.24	0.26	0.277
O 1s/Na 1s	0.61	0.53	0.35	0.50	0.522
Na 1s/F 1s	2.09	1.44	1.89	1.80	1.320
Si $2p_{3/2}$/F 1s	0.17	0.23	0.15	0.18	0.161
P $2p_{3/2}$/Na 1s	0.26	0.18	0.12	0.19	0.167
S $2p_{3/2}$/Na 1s	0.33	0.30	0.18	0.27	0.232
Cl $2p_{3/2}$/Na 1s	0.46	0.43	0.25	0.38	0.312
K $2p_{3/2}$/F 1s	0.85	1.03	0.83	0.90	0.723
Ca $2p_{3/2}$/F 1s	1.01	1.06	0.98	1.02	0.903
Na 2s/Na 1s	0.065	0.145	0.077	0.096	0.0915

ratio N_1'/N_2' is that expected for a homogeneous substance. If $(1/\lambda_2' - 1/\lambda_1')$ is known or can be estimated, the thickness of the contamination layer can be derived from Eq. (6.39). This method is suitable when the contamination layer thickness is comparable to the smaller of the mean free paths, and its effectiveness increases with increasing difference in energy of the photoelectrons being observed. Naturally, this can be usefully applied to the determination of a metal and its oxide in the surface layer, using the two chemically shifted metal lines.

If the mean free path depends not only on photoelectron kinetic energy but also on the nature of atoms present in the solid, then Eq. (6.37) will need modification. The photoelectron inelastic scattering cross section may be taken as the reciprocal of the mean free path. Additivity of inelastic scattering cross sections further implies that the total inelastic scattering cross section $1/\lambda^{ik}$ of the photoelectron from level i of atom k can be written

$$\frac{1}{\lambda^{ik}} = \frac{n_k}{\lambda_k^{ik}} + \sum_{l \neq k} \frac{n_l}{\lambda_l^{ik}} \qquad (6.40)$$

where n_k and n_l denote respectively the concentration of k atoms and all atoms of a type other than k in the solid; $1/\lambda_k^{ik}$ and $1/\lambda_l^{ik}$ represent inelastic scattering cross sections for k- and l-type atoms. By incorporating this in the expression for the ratio of intensities applicable to k- and m-type atoms (in the same sample), one obtains

$$N_R = \frac{N^{ik}}{N^{jm}} = \left[\frac{n_k}{n_m}\right]\left[\frac{\sigma^{ik}S^{ik}}{\sigma^{jm}S^{jm}}\right]\left[\frac{n_m/\lambda_m^{jm} + \sum_{p \neq m} n_p/\lambda_p^{jm}}{n_k/\lambda_k^{ik} + \sum_{l \neq k} n_l/\lambda_l^{ik}}\right] \qquad (6.41)$$

The terms in the first two sets of square brackets are identifiable on the basis of Eq. (6.37), as they apply to photoelectrons from i level of k atom and j level of m atom. The terms in the third square brackets evolve out of the application of Eq. (6.40) to both types of photoelectrons, that is, to the mean free paths λ^{ik} and λ^{jm}. By writing K for the terms in the second square brackets, n_R for the terms in the first square brackets, and dividing both the numerator and the denominator by n_m, one obtains

$$N_R = K n_R \frac{1/\lambda_m^{nm} + \sum_{p \neq m} n_R^p/\lambda_p^{jm}}{n_R/\lambda_k^{ik} + \sum_{l \neq k} n_R^l/\lambda_l^{ik}} \qquad (6.42)$$

where $n_R = n_k/n_m$, $n_R^l = n_l/n_m$, and $n_R^p = n_p/n_m$. For a binary system, one has to consider only one term each under the two summation signs, that is, for the second species. Under such conditions $\sum_{p \neq m} n_R^p/\lambda_p^{jm} = (n_k/n_m) 1/\lambda_k^{jm} = n_R/\lambda_k^{jm}$

and $\sum_{l \neq k} n_R^l / \lambda_l^{ik} = (n_m/n_m) \, 1/\lambda_m^{ik} = 1/\lambda_m^{ik}$. That means, Eq. (6.42) takes the form

$$N_R = Kn_R \frac{n_R/\lambda_k^{jm} + 1/\lambda_m^{jm}}{n_R/\lambda_k^{ik} + 1/\lambda_m^{ik}} \qquad (6.43)$$

When the inelastic scattering cross section does not depend on the nature of atomic constituents in the solid, $1/\lambda_k^{ik} = 1/\lambda_m^{ik}$, and Eq. (6.43) reduces to Eq. (6.37). Plots of experimental intensity ratios versus atomic ratios in Au–Cu, Ag–Au, CoO–MgO, and CuO–MgO systems are presented in Fig. 6.9. The curvature in plots of observed intensity ratios against atomic ratios may be explained by application of the above equations (6.42 and 6.43).

In a recent study on the feasibility of XPS for quantitative analysis using solid solution phases, Cu–Ni, Cu–Au, Au–Ag alloys, mixed oxides such as $Cu_xMg_{1-x}O$ (CuO–MgO), $Ni_xMg_{1-x}O$ (NiO–MgO), $Co_xMg_{1-x}O$(CoO–MgO), $Cr_{2x}Al_{2-2x}O_3$ ($Cr_2O_3-Al_2O_3$), $Cr_{2x}Fe_{2-2x}O_3$ ($Cr_2O_3-Fe_2O_3$), and also $Co_{3x}Mn_{3-3x}O_4$ ($Co_3O_4-Mn_3O_4$) compounds, one finds that plots of intensity ratios against atomic ratios show departures from a slope of unity, and a curvature results such as that shown in Fig. 6.9. Calculated intensity ratios assuming that the mean free path is proportional to $\epsilon^{0.5}$ are in considerable variance from the experimental curves. Measurements of relative mean free paths with respect to that of Cu $2p_{3/2}$, that is, $\lambda/\lambda_{Cu\,2p_{3/2}}$, and plotted as a function of energy, show that the best fitting curve is obtained where the mean free path varies more as $\epsilon^{0.7}$, as is also obtained in some other studies.

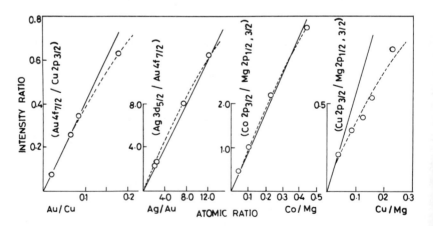

Fig. 6.9. Intensity ratios of photoelectron peaks versus atomic ratios for Au–Cu, Ag–Au, CoO–MgO, and CuO–MgO systems. Dashed lines represent experimental values and continuous lines calculated values, assuming cross sections from [604] and $\lambda \propto \epsilon^{1/2}$ [611].

Whereas standard samples are needed for quantitative analysis, the preparation of standard samples for quantitative analysis of surfaces is very difficult with regard to their homogeneity, contamination, chemical state, absorption, and other properties. Thus an analytical procedure using Eq. (6.37), $\lambda \propto \epsilon^{0.5 - 0.7}$, relevant photoionization cross sections and measured intensities, is a suitable alternative for the quantitative estimation of surface composition.

If the probable error in N_R is about 20–30% in Eq. (6.43), the intensity-ratio/atomic-ratio curves may be considered to be linear. Under such conditions a relationship between the mean free path and kinetic energy may be adopted from empirical equations and experimental data, the λ_s can be obtained as such, and then from a knowledge of spectrometer factors and photoionization cross section, quantitative estimation of solid surfaces can be carried out without using standards. For the measurement of a metal having its oxide on the surface layer, this method can be usefully applied using the chemically shifted lines.

Electron mean free paths play a crucial role in quantitative applications of photoelectron spectroscopy. Quantitative applications to surface analysis is restricted to two major areas: determination of impurity concentrations on a surface when the composition of the host material is known, and determination of the surface composition of an alloy or a compound. Complications arise because the energy dependence of the mean free path needs to be explicitly incorporated into the analysis. In the following section we examine a method of quantitative analysis that uses mean free path estimates from a theoretical model.

MEAN FREE PATH ESTIMATES

If I_0 represents the incident X-ray flux, n_i the density of impurity atoms i in the host h, $\lambda_T(\epsilon_i)$ the mean free path of the photoelectrons of energy ϵ_i originating from impurity atoms, σ_i the photoionization cross section, and $S(\epsilon_i)$ the fraction of the electrons detected by the analyzer (equivalent to the spectrometer factor of the previous section), then as in Eq. (6.36) it is possible to write

$$I_i = I_0 n_i \sigma_i \lambda_T(\epsilon_i) S(\epsilon_i) \tag{6.44}$$

where I_i is the signal strength due to electrons of energy ϵ_i. The absolute values of I_0 and S are rather difficult to determine. To circumvent this, one can resort to a ratio representation of intensities from impurity and host atoms. Using Eq. (6.44) for the two species, the equation can be written

$$\frac{n_i}{n_h} = \frac{I_i \lambda_T(\epsilon_h) \sigma_h S(\epsilon_h)}{I_h \lambda_T(\epsilon_i) \sigma_i S(\epsilon_i)} \tag{6.45}$$

The subscript h represents the host atoms. Thus if it is somehow possible to determine the ratio $S(\epsilon_h)/S(\epsilon_i)$ in the above equation, then, from available calculated values of σ (e.g., Table 6.1), a knowledge of λ_T's as a function of ϵ, and from experimental values of I_i and I_h, n_i/n_h can be determined.

If measurements are carried out in geometrically identical samples using separately pure host and impurity material, it is possible to obtain the quantity $\sigma_h S(\epsilon_h)/\sigma_i S(\epsilon_i)$ using the following equation, derived by transposing Eq. (6.44) and applying it to the pure samples.

$$\frac{\sigma_h S(\epsilon_h)}{\sigma_i S(\epsilon_i)} = \frac{n_i' I_h' \lambda_T^{(i)}(\epsilon_i)}{n_h' I_i' \lambda_T^{(h)}(\epsilon_h)} \tag{6.46}$$

Here n_i' and n_h' are known densities of impurity and host atoms in respective pure samples, I_h', I_i' the respective signal intensities, and $\lambda_T^{(i)}(\epsilon_i)$, $\lambda_T^{(h)}(\epsilon_h)$ are the corresponding mean free paths. The ratio $\sigma_h S(\epsilon_h)/\sigma_i S(\epsilon_i)$ then can be used in Eq. (6.45) for finding n_i/n_h. In Eq. (6.45) one can further assume $\lambda_T^{(h)}(\epsilon_h) \simeq \lambda_T(\epsilon_h)$, because the impurity concentration is usually small in a sample under study.

This method can be directly used for the determination of alloy composition. The equation that would be applicable to the two components 1 and 2 of a binary alloy is

$$\frac{n_1}{n_2} = \frac{I_1 \lambda_T(\epsilon_2) \sigma_2 S(\epsilon_2)}{I_2 \lambda_T(\epsilon_1) \sigma_1 S(\epsilon_1)} \tag{6.47}$$

where n_1/n_2 represents the relative concentration of the two types of atoms, and I_1 and I_2 are the photoelectron signal strengths due to the two components. The ratio $\sigma_2 S(\epsilon_2)/\sigma_1 S(\epsilon_1)$ can be determined by appropriate measurements involving pure metals constituting the alloy. Actual values of n_1 and n_2 can then be determined by the application of an iterative method.

Let us assume that $\lambda_T(\epsilon_2)/\lambda_T(\epsilon_1)$ can be calculated if the composition is known. One may then start with an approximate alloy composition n_1/n_2, on the basis of which $\lambda_T(\epsilon_2)/\lambda_T(\epsilon_1)$ can be calculated. The latter, on insertion into Eq. (6.47), leads to an improved estimate of n_1/n_2; absolute values of n_1 and n_2 can then be derived from an independent measurement of the sample density. From a knowledge of this n_1 and n_2, a final value of λ_T can be obtained. The process of iteration can continue until, in an iteration cycle, no further changes are obtained between the starting value of n_1/n_2 and the final result. It is necessary, however, that λ_T dependence on n_1 and n_2 be sufficiently weak, so that the initial value of n_1/n_2 obtained from Eq. (6.47) is close ($\lesssim 15\%$) to the corresponding value obtained in several iterations and extensive iterations are thus unnecessary.

The mean free paths can be calculated as follows. For materials that exhibit well defined plasmons with energies close to the free electron value $\hbar\omega_p/2\pi = \hbar(4\pi n e^2/m)^{1/2}$, which includes all elements except the transition and inner transition metals, the mean free path due to valence and core electron excitation is to a good approximation given by

$$\lambda_T(\epsilon) = \frac{\epsilon}{a(\ln \epsilon + b)} \tag{6.48}$$

where a and b are functions of both electron concentration of the host material and of core levels of the constituent host atoms. λ_T has a substantial contribution from the valence electrons. In terms of contributions from valence and core electrons to the total mean free path λ_T, it is possible to write $\lambda_T^{-1} = \lambda_v^{-1} + \lambda_c^{-1}$; the major contribution is from the first term, due to the valence electrons. For λ_v and λ_c, one may write equations similar to Eq. (6.48), such that $a = a_v + a_c$, $b = (a_v b_v + a_c b_c)/(a_v + a_c)$. The $\lambda_T(\epsilon)$ can be calculated with the help of the theoretical values of a_v, a_c, b_v, and b_c. Such values for some elements are shown in Table 6.4. In practical applications the ratios of λ_T's are involved, which are more accurate than the absolute values, and this permits application of the technique even to transition and inner transition metals. The results given in Table 6.4 agree to within 20% of a statistical model for the calculation of λ_T.

Application of this model can be illustrated by an example. For a compound of the type X_iY_j, the valence electron density n_e can be expressed as $n_e = [6.02 \times 10^{23} N_v(A/\rho)]$ cm^{-3}, where N_v is the number of valence electrons, ρ the density in g cm^{-3}, and A is the formula weight in g, A/ρ represents the volume associated with one formula weight. In the case of Al_2O_3, with $A = (2 \times 26.98) + (3 \times 16) = 101.96$ g, $N_v = (2 \times 3) + (3 \times 6) = 24$ and ρ being equal to 3.7 g cm^{-3}, one obtains $n_e = 5.24 \times 10^{23}$ cm^{-3}. The values a_v, b_v are given in terms of r_s in Fig. 6.10, where r_s is defined as

$$r_s = \left(\frac{3}{4\pi n_e}\right)^{1/3} \frac{1}{a_0} \tag{6.49}$$

where a_0 is the Bohr radius. The value of r_s corresponding to the n_e as calculated above for Al_2O_3 turns out to be 1.45. From this value of r_s, the values of a_v and b_v from Fig. 6.10 are $a_v = 15.75$ and $b_v = -2.73$.

The expressions for a_c, b_c, which represent the core contribution are

$$a_c = 3.92 \times 10^2 \sum_i \left(\frac{N_i}{\Delta E_i}\right) \frac{\rho}{A} \tag{6.50}$$

$$b_c = -\frac{\sum_i (N_i/\Delta E_i) \ln \frac{1}{4} \Delta E_i}{\sum_i (N_i/\Delta E_i)} \tag{6.51}$$

where N_i is the number of electrons from the highest atomic core level of type i (usually the contribution from other levels is negligible), and ΔE_i is the

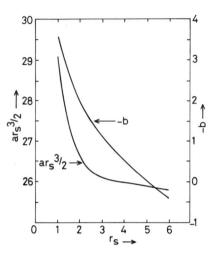

Fig. 6.10. Plots of $ar_s^{3/2}$ and $-b$ versus r_s [610].

corresponding excitation energy. This excitation energy is estimated from atomic binding energies E_B; if $E_B < 70\,\text{eV}$ then $\Delta E = 2 E_B$, and if $E_B > 70\,\text{eV}$ then $\Delta E = E_B + 70\,\text{eV}$. For Al, $N = 6$ (6 $2p$ electrons) and $\Delta E = 143\,\text{eV}$. For oxygen, $N = 0$, since, in this application, there is no important core level. With $N_{\text{Al}_2} = 2 \times 6 = 12$ and $\Delta E = 143\,\text{eV}$, one obtains from Eq. (6.50) and Eq. (6.51) $a_c = 1.19$ and $b_c = -3.58$. By combining these a_c, b_c values with a_v, b_v values derived earlier, one obtains $a = a_v + a_c = 16.94$ and $b = (a_v b_v + a_c b_c)/(a_v + a_c) = -2.79$.

An extension of Table 6.4 that includes all the elements shows that for ϵ values in the domain 200–2400 eV, (i.e., $5.3 \leqslant \ln \epsilon \leqslant 7.8$), the absolute value of b is in range 1–3, and the b term in Eq. (6.48), compared with $\ln \epsilon$, is small. Using an intermediate value of $b = -2.3$, it is thus possible to write an expression like

$$\frac{\lambda_T(\epsilon_h)}{\lambda_T(\epsilon_i)} = \frac{\epsilon_h}{\epsilon_i} \frac{\ln \epsilon_i - 2.3}{\ln \epsilon_h - 2.3} \quad (6.52)$$

from which approximate values of $\lambda_T(\epsilon_h)/\lambda_T(\epsilon_i)$ can be obtained with a typical error of 5% and a maximum possible error of 14%. For ϵ_h, $\epsilon_i \geqslant 200\,\text{eV}$, the expression becomes essentially material independent. In expressions where dependence on the parameter a remains because of the mean free path in two different substances, such as in Eq. (6.45) and Eq. (6.46), $a^{(i)}$ and $a^{(h)}$ need to be calculated, as has been shown above for free-electron-like materials. For transition and noble metals, dependable experimental measurements on λ_T are not available, and therefore the validity of this theoretical model is less certain.

Table 6.4. Total Mean Free Paths for the Elemental Solids Calculated from Eq. (6.48) λ (nm) = $\epsilon/[a(\ln \epsilon + b)]$ for ϵ Values from 200 to 2400 eV [610]

Z:	3 Li	6 C	12 Mg	13 Al	26 Fe	29 Cu	30 Zn	47 Ag	50 Sn	79 Au
a:	4.97	18.3	7.74	10.2	21.1	23.6	21.1	19.5	12.9	19.6
b:	−1.23	−2.95	−1.77	−2.16	−3.00	−3.21	−3.10	−2.95	−2.04	−2.95
eV										
200	0.989	0.464	0.732	0.620	0.410	0.405	0.430	0.436	0.474	0.432
400	1.690	0.717	1.220	1.010	0.631	0.608	0.654	0.674	0.782	0.667
600	2.330	0.949	1.670	1.370	0.833	0.796	0.861	0.892	1.060	0.884
800	2.950	1.160	2.100	1.720	1.020	0.974	1.050	1.090	1.330	1.080
1000	3.540	1.370	2.510	2.050	1.200	1.140	1.240	1.290	1.580	1.280
1200	4.120	1.580	2.910	2.370	1.380	1.300	1.420	1.480	1.830	1.470
1400	4.680	1.770	3.300	2.680	1.550	1.460	1.590	1.670	2.070	1.650
1600	5.230	1.970	3.680	2.980	1.720	1.620	1.770	1.850	2.310	1.830
1800	5.780	2.160	4.060	3.280	1.880	1.770	1.930	2.030	2.540	2.010
2000	6.310	2.340	4.430	3.570	2.050	1.920	2.100	2.200	2.770	2.180
2200	6.840	2.520	4.790	3.860	2.210	2.070	2.260	2.370	3.000	2.350
2400	7.360	2.700	5.150	4.150	2.360	2.220	2.420	2.540	3.200	2.520

AN ALTERNATIVE APPROACH TO QUANTITATIVE XPS

In an alternative method, the equation that is applied to the quantitative application of XPS is derived on the basis of a consideration of X-ray fluorescence and then appropriate changes are incorporated to make the method applicable to XPS. We consider first a situation where a parallel beam of X-rays impinges on the sample surface at an angle α to the surface normal and illuminating a surface area A'. The detector that measures the X-ray fluorescence is positioned so that it looks at a surface area A'' and the beam that reaches the detector from the surface makes an angle γ to the surface normal. The area A is common to both A' and A'' and the detector subtends a solid angle Ω at the sample surface area A. We are interested in the number of photons emitted from a layer dt located at a depth t from the surface. For this we need to consider the extent of penetration of the incident X-ray flux to the depth t then follow the outgoing beam of fluorescence photons in the process of emergence through the surface and finally reaching the detector. If the incident X-ray beam has n number of quanta crossing per unit time through a plane perpendicular to the beam direction, then the number of quanta eventually registered by the photon detector is given by

$$\left[(nA\cos\alpha) \left(\exp \frac{-\bar{\mu}_{ac}t}{\cos\alpha} \right) \left(\frac{\bar{\mu}_{ac}dt}{\cos\alpha} \right) \left(\frac{\bar{\tau}_{aj}}{\bar{\tau}_{ac}} \frac{d_j}{dt} \right) \left(\frac{S_{qj}-1}{S_{qj}} \right) \right]$$
$$\left[(p_{qj}^r)(W_{qj}^r) \left(\exp \frac{-\bar{\mu}_{jc}t}{\cos\gamma} \right) \left(\frac{\Omega}{4\pi} \right) (\kappa_j) \right]$$
(6.53)

where the terms in the first set of square brackets represent the process of the incident photons reaching the layer dt and the terms in the second set of square brackets represent the passage of the outgoing photons towards the detector. The terms in the first set of parentheses stand for the quantum rate of impinging of X-ray photons on the area A; the radiation is treated as a monochromatic parallel beam. The exponential factor that follows represents the attenuation of the incident beam before it reaches the layer dt, $\bar{\mu}_{ac}$ being the total linear absorption coefficient in cm^{-1} of the sample c with respect to the radiation signified by a. The factor inside the third set of parentheses represents the attenuation in the layer dt. Further, if we assume that in the layer dt the effective thickness of the jth element is d_j, the $\bar{\tau}_{aj}$ and $\bar{\tau}_{ac}$ respectively stand for linear photoabsorption coefficients in cm^{-1} owing to the element j and the total linear photoabsorption coefficients in the sample c, and we also assume that the photoelectric component is the most significant one in the total absorption coefficient — the others being coherently and incoherently scattered components — then the overall factor inside the fourth set of parentheses stands for the partial photoabsorption due to d_j relative to the total photoabsorption. Finally, the factor $(S_{qj} - 1)/S_{qj}$ accounts for the fact that of the total photoabsorption by the element j only that component of radiation is significant which falls on a particular energy level; this is described by the absorption edge jump S_{qj}. Since $\bar{\tau}_{ac}$ is the major component of $\bar{\mu}_{ac}$, the former can replace the latter in the third set of parentheses. Among the terms that describe the emergence of the fluorescence photons, p_{qj}^r is the fluorescence transition probability of the $r \to q$ transition of the jth element, W_{qj}^r is a correction factor because part of the radiation is lost due to Auger effect, the exponential term that follows is due to the attenuation suffered by the outgoing photon beam ($\bar{\mu}_{jc}$ being the linear total attenuation coefficient of the j radiation in the sample c), the $\Omega/4\pi$ factor accounts for the fact that whereas the emission takes place in all 4π steradians the detector intercepts only Ω of it, and κ_j is the wavelength-dependent sensitivity factor of the detector.

The expression d_j/dt represents the ratio of the effective width d_j which can be ascribed to the element j in the layer dt to the width of the overall layer dt. The higher the weight fraction c_j of the element j in the sample, the greater becomes the effective width d_j. The width is also inversely proportional to the relative density (ρ_j/ρ_c) of the pure jth element with respect to the density of the alloy. Therefore, one can write

$$\frac{d_j}{dt} = \frac{c_j \rho_c}{\rho_j} \qquad (6.54)$$

The linear and the mass attenuation coefficients are related by equations of the type $\bar{\mu}_{ac} = \rho_c \mu_{ac}$, $\bar{\tau}_{ac} = \rho_c \tau_{ac}$, where $\bar{\mu}_{ac}$ and $\bar{\tau}_{ac}$ are in cm^{-1} and the mass attenuation coefficients μ_{ac} and τ_{ac} are in cm^2 g^{-1}, the ρ_c being in g cm^{-3}. Incorporating these substitutions in Eq. (6.53) and simplifying, one obtains

$$nA \exp\left(\frac{-\bar{\mu}_{ac} t}{\cos \alpha}\right) \rho_c c_j \tau_{aj} \frac{S_{qj} - 1}{S_{qj}} p_{qj}^r W_{qj}^r \frac{\Omega}{4\pi} \exp\left(\frac{-\bar{\mu}_{jc} t}{\cos \gamma}\right) \kappa_j \, dt \qquad (6.55)$$

To modify this relationship, which involves the detection of quanta of photons, to the detection of photoelectrons emitted by the solid, a few changes need to be made. The n, A, ρ_c, c_j, τ_{aj}, S_{qj}, $\bar{\mu}_{ac}$, α, and γ would retain their meaning. Since we are no longer considering X-ray fluorescence, the optical transition probability p_{qj}^r and the correction factor W_{qj}^r are not necessary, therefore both may be replaced by unity. The photoionization angular distribution and polarization may affect the number of photoelectrons received by the detector. This can be accounted for by incorporating a factor $\phi(\theta, \psi)$ besides $\Omega/4\pi$. The sensitivity factor κ_j of the photoelectron detector may depend on electron kinetic energy. Since the latter for photoionization from a given level depends on the energy of radiation used, κ_j may thus depend on the nature of radiation;

we indicate this by κ_j^a. The photoelectrons that finally emerge through the surface also experience scattering and hence attenuation. This is accounted for by the linear inelastic scattering coefficient $\bar{\sigma}_{jc}^a$, and a corresponding exponential term replaces the $\bar{\mu}_{jc}$ term of Eq. (6.55). So we have as photoelectron counting rate:

$$n_j^a = nA \exp\left(\frac{-\bar{\mu}_{ac}t}{\cos\alpha}\right) \rho_c c_j \tau_{aj} \frac{S_{aj}-1}{S_{aj}} \frac{\Omega}{4\pi} \phi(\theta,\psi) \exp\left(\frac{-\bar{\sigma}_{jc}^a t}{\cos\gamma}\right) \kappa_j^a \, dt \qquad (6.56)$$

To apply this to a solid sample, we integrate the above equation between the limits $t = 0$ and $t = \infty$. If we represent by N_{jc}^a the number of photoelectrons registered by the detector from the jth element of an alloy c owing to photoionization by radiation a, and represent the number owing to the pure elemental solid as N_{jj}^a, then we have

$$N_{jc}^a = \left[nA c_j \tau_{aj} \frac{S_{aj}-1}{S_{aj}} \frac{\Omega}{4\pi} \phi(\theta,\psi) \kappa_j^a \right] \frac{1}{(\mu_{ac}/\cos\alpha) + (\sigma_{jc}^a/\cos\gamma)} \qquad (6.57)$$

For the pure element N_{jj}^a, all the terms would be the same as in Eq. (6.57) except $c_j = 1$, and μ_{ac} and σ_{jc}^a are replaced respectively by μ_{aj} and σ_{jj}^a. Since $\sigma_{jj}^a \gg \mu_{jj}$ (for Mg K_α radiation, Au, and Au_{N_7} level $\mu_{aj} = 10^4$ cm^2 g^{-1} and $\sigma_{jj}^a = 10^5$ cm^2 g^{-1}), we neglect μ_{aj}. Applying this approximation also to N_{jc}^a, we have from Eq. (6.56):

$$N_{jc}^a = \mathscr{C} \frac{\cos\gamma}{\sigma_{jc}^a} \qquad (6.58)$$

where $\mathscr{C}$ represents all the terms inside the square brackets of Eq. (6.57).

For thin films, we need to integrate Eq. (6.56) between the limits $t = 0$ and $t = t'$, where t' is the film thickness. We can neglect $\bar{\mu}_{ac}$ with respect to $\bar{\sigma}_{jc}^a$, and $\bar{\mu}_{aj}$ with respect to $\bar{\sigma}_{jj}^a$, use the relationships $\bar{\sigma}_{jc}^a/\rho_c = \sigma_{jc}^a$ and $\bar{\sigma}_{jj}^a/\rho_c = \sigma_{jj}^a$, and replace $t'\rho_c$ with m/F (mass and area of the film are m and F respectively) in g cm^{-2} as mass per unit area of the film. Then we have

$$n_{jc}^a = \mathscr{C} \frac{1 - \exp\left[(-\sigma_{jc}^a/\cos\gamma)(m/F)\right]}{\sigma_{jc}^a} \cos\gamma \qquad (6.59)$$

and

$$n_{jj}^a = \mathscr{C} \frac{1 - \exp\left[(-\sigma_{jj}^a/\cos\gamma)(m/F)\right]}{\sigma_{jj}^a} \cos\gamma \qquad (6.60)$$

The relationship between the mean free path λ discussed in the earlier sections and the inelastic mass scattering coefficient σ_{ik} of the i photoelectrons due to element k is:

$$e^{-t/\lambda} = e^{-\sigma \cdot m/F} \qquad (6.61)$$

where t is the absorber thickness.

The above equations can be employed in several ways in quantitative analysis. For example, the use of photoelectron counting-rate ratios of two samples rather than the absolute counting-rate from one sample has the obvious advantage of eliminating instrumental parameters. In an analysis of a multicomponent sample consisting of n elements, the counting-rate ratio involving the jth element in the alloy and the pure element using Eq. (6.57) for N_{jc}^a, and the corresponding equation for N_{jj}^a with $c_j = 1$, can be written:

$$r_{jc}^a = \frac{N_{jc}^a}{N_{jj}^a} = \frac{c_j \sigma_{jj}^a}{\sigma_{jc}^a} = \frac{c_j \sigma_{jj}^a}{\sum_{k=1}^{p} c_k \sigma_{jk}^a} \qquad (6.62)$$

Here the effect of inelastic scattering of photoelectrons from the jth element by all atoms in the sample has been made explicit; σ_{jc}^a has been expressed in terms of a summation of inelastic mass scattering coefficients σ_{jk}^a of the constituent elements weighted by the respective concentrations (in weight fraction) c_k. This is along the same vein in which the mean free path was expressed in Eq. (6.40). By rewriting Eq. (6.62) we see that it results in p equations of the type

$$r_{jc}^a \sum_{k=1}^{p} c_k \sigma_{jk}^a - c_j \sigma_{jj}^a = 0 \qquad (6.63)$$

Solving these p equations, p concentrations can be determined. To this, although it is not essential, one can add the constraint $\sum_{k=1}^{p} c_k = 1$ to make $p + 1$ equations, from which the p c_k's can be determined. This procedure gives better accuracy in an iterative solution in comparison with a uniquely solvable system of p equations.

Alternatively, one can also write Eq. (6.62) in the form

$$r_{jc}^a = \frac{c_j}{\sum_{k=1}^{p} (c_k \sigma_{jk}^a / \sigma_{jj}^a)} \qquad (6.64)$$

where now only a knowledge of ratios of the inelastic scattering coefficients is needed. Again, with p equations of the type

$$r_{jc}^a \sum_{k=1}^{p} c_k \frac{\sigma_{jk}^a}{\sigma_{jj}^a} - c_j = 0 \tag{6.65}$$

and the constraint $\sum_{k=1}^{p} c_k = 1$, all the p concentrations can be determined.

One could also plot a calibration curve of the photoelectron counting rate from the jth element against the concentration c_j of a binary system. Then from the experimental count rate for the jth element from a sample of unknown composition, the concentrations of the constituents can be determined.

Further, we may rewrite Eq. (6.65) in the following manner:

$$\sum_{k=1}^{p} c_k \frac{\sigma_{jk}^a}{\sigma_{jj}^a} - \frac{c_j}{r_{jc}^a} = 0 \tag{6.66}$$

which, for $j = 1, 2, \ldots$, will be given by

$$c_1 \left(1 - \frac{1}{r_{1c}^a}\right) + c_2 \frac{\sigma_{12}^a}{\sigma_{11}^a} + \cdots + c_p \frac{\sigma_{1p}^a}{\sigma_{11}^a} = 0 \tag{6.67}$$

$$c_1 \frac{\sigma_{21}^a}{\sigma_{22}^a} + c_2 \left(1 - \frac{1}{r_{2c}^a}\right) + \cdots + c_p \frac{\sigma_{2p}^a}{\sigma_{22}^a} = 0 \tag{6.68}$$

For a p component system there will be p such equations, each having $(p-1)$ coefficients of the type $\sigma_{12}^a/\sigma_{11}^a$, that is, a total of $p(p-1)$ such coefficients. From $(p-1)$ known samples of different composition of the same p-component system, these $p(p-1)$ coefficients can be determined. Thus, one needs one calibration sample for the analysis of a two-component system, two calibration samples for a three-component system, and so on.

The influence of surface roughness can be circumvented by considering the counting-rate ratio of two elements of one and the same sample. Using Eq. (6.58), one obtains

$$r_{ijc}^a = \frac{N_{ic}^a}{N_{jc}^a} = \frac{\tau_{ai}(1 - 1/S_{qi}) \kappa_i^a c_i \sum_{k=1}^{p} c_k \sigma_{jk}}{\tau_{aj}(1 - 1/S_{qj}) \kappa_j^a c_j \sum_{k=1}^{p} c_k \sigma_{ik}} \tag{6.69}$$

where, because it is the same sample, roughness-dependent shading effects cancel out. For a binary system, the above equation assumes the form:

$$r_{12c}^a = \mathscr{K} \frac{1 + \mathscr{L}(c_1/c_2)}{1 + \mathscr{M}(c_2/c_1)} \tag{6.70}$$

where

$$\mathcal{K} = \frac{\tau_{a1}(1 - 1/S_{q1})\kappa_1^a \sigma_{22}^a}{\tau_{a2}(1 - 1/S_{q2})\kappa_2^a \sigma_{11}^a} \qquad (6.71)$$

$$\mathcal{L} = \frac{\sigma_{21}^a}{\sigma_{22}^a}, \quad \text{and} \quad \mathcal{M} = \frac{\sigma_{12}^a}{\sigma_{11}^a} \qquad (6.72)$$

The constants $\mathcal{K}$, $\mathcal{L}$, and $\mathcal{M}$ can be determined by measurement of three different alloys from the two-component system i, j.

A basic requirement for application of the above methods to quantitative studies is the availability of either values of the absolute inelastic scattering coefficients or their ratios. We describe here a few methods for determining inelastic scattering coefficients.

First, if we measure the photoelectron counting rate from a pure j element thin film, and separately from a pure j element solid, then we have from Eqs. (6.57) and (6.60):

$$\frac{n_{jj}^a}{N_{jj}^a} = 1 - \exp\left(-\frac{\sigma_{jj}^a}{\cos\gamma}\frac{m}{F}\right) \quad \text{or} \quad \sigma_{jj}^a = \frac{\cos\gamma}{m/F} \ln\frac{N_{jj}^a}{N_{jj}^a - n_{jj}^a} \qquad (6.73)$$

Besides the experimentally measured values of N_{jj}^a and n_{jj}^a, if the emergent angle γ and the mass per unit area of the film m/F are known, σ_{jj}^a can be determined.

For determination of inelastic scattering coefficients of the type σ_{ij}^a, that is, the scattering of i photoelectrons by the element j, the number of photoelectrons from the substrate needs to be measured once without any film and once with a film of the j element on it. The photoelectron count rate in the first case, N_{ic}^a, is given by an expression like Eq. (6.57). When a film of thickness t' of the j element is on the substrate, the i photoelectron count rate through the film $\bar{N}_{ic}^a$ is given by Eq. (6.57) with a multiplicative factor $\exp(-\bar{\sigma}_{ij}^a t'/\cos\gamma)$ where $\bar{\sigma}_{ij}^a$ is the linear inelastic scattering coefficient. The ratio of $\bar{N}_{ic}^a$ and N_{ic}^a is therefore given by

$$\frac{\bar{N}_{ic}^a}{N_{ic}^a} = \exp\left(-\frac{\sigma_{ij}^a}{\cos\gamma}\frac{m}{F}\right) \quad \text{or} \quad \sigma_{ij}^a = \frac{\cos\gamma}{m/F} \ln\frac{N_{ic}^a}{\bar{N}_{ic}^a} \qquad (6.74)$$

since $\bar{\sigma}_{ij}^a/\rho_j = \sigma_{ij}^a$ and $t'\rho_j = m/F$. It follows that the last two methods for the determination of inelastic scattering coefficients may be combined to yield the scattering coefficient ratios $\sigma_{ij}^a/\sigma_{jj}^a$.

Figure 6.11 shows the calibration curve for the binary system Ag–Au alloy with a plot of $N_{\text{Au}}^{\text{Mg}K\alpha}$ against composition. It also shows the calibration curves obtained from Cu–Ni alloy using valence band Cu–Ni$_{M_{2,3}}$ levels. Theoretical curves are also shown.

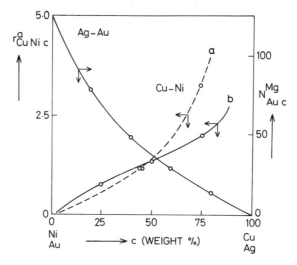

Fig. 6.11. Calibration curves for the binary systems Ag-Au (Au$_{N_7}$) and Cu-Ni. (a) Cu-, Ni-valence band, (b) Cu-, Ni-$M_{2,3}$ level [614].

On the basis of equations derived above, one can proceed to calculate elemental sensitivities. In Eq. (6.59), if we replace nA (in the constant $\mathscr{E}$) with I_0, which is the intensity of radiation falling on the surface of a sample film, and write $\mathscr{G}$ for the factors $(\Omega/4\pi)\phi(\theta,\psi)\kappa_j^a$, then the photoelectron peak intensity I which is proportional to the number of photoelectrons registered by the detector would be given by

$$I = \mathscr{G}\frac{I_0 \tau_{aj}(1-1/S_{qj})c_j}{\sigma_{jc}^a}\left[1-\exp\left(\frac{-\sigma_{jc}^a}{\cos\gamma}\frac{m}{F}\right)\right]\cos\gamma \quad (6.75)$$

If we further assume that the sample is a film of a pure element ($c_j = 1$, and we replace σ_{jc}^a by σ_{jj}^a) with a mass of 1 μg cm^{-2} and the angle of emergence $\gamma = \pi/4$, then we have the following equation and we call the I under these conditions the *sensitivity index* (SI);

$$\text{SI} = \mathscr{G}'\frac{I_0 \tau_{aj}\left(\dfrac{S_{qj}-1}{S_{qj}}\right)}{\sigma_{jj}^a}[1-\exp(-1.4\times 10^{-6}\sigma_{jj}^a)] \quad (6.76)$$

The factor $\mathscr{G}$ changes to $\mathscr{G}'$ as the value of $\cos\pi/4$ is now incorporated into it. In order to derive quantitative estimates of SI as a function of the atomic number Z, we need knowledge about the dependence of the photoelectric cross section τ_{aj}, the absorption edge jump S_{qj}, and coefficient σ^a_{jc} on the atomic number Z. It turns out that these are known only for a narrow range of Z values, much less than would be of interest for quantitative analysis. Extrapolations are necessary to obtain estimates of values beyond the known range. This constitutes a major uncertainty in the application of the XPS technique to quantitative analysis. Further, τ_{aj} and S_{qj} not only depend on the atomic number, but also, for any given incident energy, they depend strongly on the actual level (K, L_3, M_5, etc.) from which the electron is ejected.

To obtain values of cross sections where results of actual theoretical calculations are not available, extrapolation can be resorted to, and where possible, linear regressions may be evolved. For example, for Al K_α excitation, relationships such as $\log \tau_j(L) = -0.73 + 3.02 \log Z$ (experimental data in the range $\log Z = 1.1-1.5$), $\log \tau_j(L_3) = -0.98 + 3.02 \log Z$ (experimental data in the range $\log Z = 1.5-1.6$), $\log \tau_j(M) = -2.46 + 3.55 \log Z$ (experimental data in the range $\log Z = 1.6-1.8$), and $\log \tau_j(M_5) = -2.78 + 3.55 \log Z$ (experimental data in the range $\log Z = 1.8 - 1.9$), which relate cross sections of respectively L, L_3, M, M_5 lines to the atomic number, give fairly good fits to known $\tau_j(Z)$ data. Extrapolations have also resorted to $\log S - \log Z$ plots and the linear regressions obtained, such as $\log S_K = -0.44 \log Z + 1.54$ (experimental data in the range $\log Z = 1.1-2.0$), $\log S_{L_3} = -1.4 \log Z + 2.42$ (experimental data in the range $\log Z = 1.5-1.8$), and $\log S_{M_5} = 1.91 \log Z + 4.07$ (experimental data in the range $\log Z = 1.8-2.0$).

If one of the lines used is a standard line, a *relative sensitivity index* (RSI) can be defined as

$$\text{RSI} = \frac{\tau_{aj}(1 - 1/S_{qj})\,\sigma^a_{rr}[1 - \exp(-1.4 \times 10^{-6}\,\sigma^a_{jj})]}{\tau_{ar}(1 - 1/S_{qr})\,\sigma^a_{jj}[1 - \exp(-1.4 \times 10^{-6}\,\sigma^a_{rr})]} \qquad (6.77)$$

where the subscripts r and j refer respectively to the reference element and the element of interest. For infinitely thick samples, the above relationship takes the following form as the exponential terms vanish:

$$\text{RSI}_\infty = \frac{\tau_{aj}(1 - 1/S_{qj})\,\sigma^a_{rr}}{\tau_{ar}(1 - 1/S_{qr})\,\sigma^a_{jj}} \qquad (6.78)$$

Once this is experimentally determined or theoretically calculated, the RSI for thin samples can be calculated for any desired concentration. The RSI_∞ for a particular series, if plotted without $\sigma^a_{rr}/\sigma^a_{jj}$ against the atomic number, shows that it increases monotonically without any discontinuities. The nature of this variation is shown in Fig. 6.12.

Not much quantitative data is available for $\sigma^a_{rr}/\sigma^a_{jj}$ either. In a given series such as K, L, M involving the ejection of an electron from a particular shell, the ionization energy increases and the kinetic energy of photoelectrons decreases with increasing Z. These low kinetic energy electrons are susceptible to more

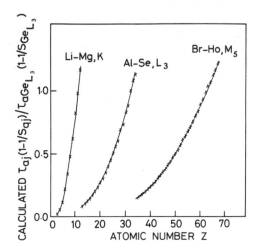

Fig. 612. Calculated curves for relative sensitivity index for infinitely thick samples (RSI$_\infty$) referred to Ge$_{L_3}$. Effects of electron absorption coefficients are not included. Adapted from M. Janghorbani et al. [616] (copyright 1975, American Chemical Society).

scattering, and therefore an increase of σ_{jj}^a takes place. This in turn implies a decrease of RSI with increasing Z. Besides electron energy, σ_{jj}^a also depends on the nature of the absorber material, and knowledge about the extent of such dependence is at present scant.

Absolute photoelectron line intensities are now experimentally known for a large number of atomic levels. Shown in Table 6.5 are a few such values of absolute line intensities of prominent lines of some elements, expressed by using the equation *Absolute intensity* $= 1.064 \times h \times W_{1/2}$, where h is the experimental peak height, count sec^{-1}, and $W_{1/2}$ = FWHM eV. Peak intensities, however, are affected by both X-ray tube and analyzer parameters. Peak intensity increases rather linearly with increasing anode emission current in the range 5–40 mA. In the analyzer, it is affected by analyzer hemisphere voltage, slit widths, electron multiplier high voltage, scan rate, and rate meter time constant. Reproducibility of results obtained by various workers, however, is not satisfactory. Part of the reason may be that some are on pure elemental solids and others are on compounds.

Studies have also been made to calculate as well as experimentally determine intraelemental relative sensitivities — relative intensities between two photoelectron lines of the same element. From Eq. (6.75) we can see that with varying sample thickness, it is the exponential factor that may be expected to cause major variations in relative intensities of photoelectron lines of widely differing kinetic energy.

Table 6.5. Absolute Photoelectron Line Intensities of Some Elements (count/sec eV × 10³) [616]

Element	Z			Element	Z		
Na	11	K	40.4, 0.3[a]	Ag	47	M_5	9.9, 0.2
Mg	12	K	16.4, 1.6	Cd	48	M_5	24, 2
Al	13	L_1	11, 1	In	49	M_5	10.4, 0.2
Ca	20	L_3	2.3, 0.1	Sn	50	M_5	11.7, 0.3
V	23	L_3	4.6, 0.3	Sb	51	M_5	12.3, 0.2
Mn	25	L_3	9.8, 0.6	Ba	56	M_5	18.5, 0.3
Ni	28	L_3	9.3, 0.5	La	57	M_5	25, 1
Cu	29	L_3	21, 1	Ta	73	N_5	4.0, 0.1
Zn	30	L_3	31, 1	W	74	N_5	3.7, 0.1
Ge	32	L_3	52, 3	Pt	78	N_5	4.4, 0.3
Se	34	M_3	1.1, 0.1	Au	79	N_5	6.5, 0.2
Y	39	M_3	3.1, 0.0	Pb	82	N_5	5.1, 0.2
Mo	42	M_5	6.8, 0.5	Bi	83	N_7	5.1, 0.1

[a]The first entry is the absolute intensity for an average of three to six measurements; the second entry is the standard deviation of the individual measurement.

On the basis of Eq. (6.75) a ratio of two photoelectron lines of the same element can be written

$$R = \frac{I_1}{I_2} = \frac{(1-E_1)\sigma_{2j}^a}{(1-E_2)\sigma_{1j}^a} \tag{6.79}$$

where $E = \exp[-(\sigma_{jj}^a/\cos\gamma)m/F]$, and the subscripts refer to two photoelectron lines of the same element. At infinite thickness, R approaches the value $\sigma_{2j}^a/\sigma_{1j}^a$. By expanding the exponential term we can see that it is equal to unity at the other extreme of thickness, regardless of the photoelectron kinetic energy of the lines. A plot of ratio R of two photoelectron lines having kinetic energies 72 and 1403 eV for a gold matrix as a function of ρt, that is, m/F, is given in Fig. 6.13. The nature of the plot implies that a low kinetic energy line having intensity I_1 at low concentrations, when compared with a high kinetic energy line I_2, exhibits relatively low intensity at higher sample thickness. Therefore, for the selection of lines to be used in quantitative analysis, one needs to take into consideration the effect of sample thickness besides effective cross section $\tau_{aj}(1 - 1/S_{qj})$ when two photoelectron lines of significantly different kinetic energy are considered. One needs to know the inelastic scattering coefficients for a detailed evaluation of this effect.

Besides the methods described in the first part of this chapter, photoelectron cross sections have also been calculated from semiempirical X-ray absorption data.

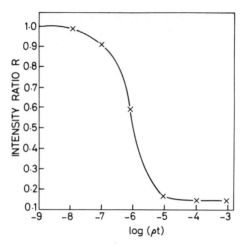

Fig. 6.13. Effect of concentration on the intensity ratio R for two photoelectron lines with 1403 eV and 72 eV kinetic energy in gold [617].

The relationships between photoelectric cross sections σ, absorption edge jump S, and X-ray absorption coefficients τ are of the following types:

$$\sigma_K = \left(1 - \frac{1}{S_K}\right) \tau_K(\epsilon_{h\nu})$$

$$\sigma_{L_1} = \left(1 - \frac{1}{S_{L_1}}\right) \tau_L(\epsilon_{h\nu})$$

$$\sigma_{L_2} = \left(1 - \frac{1}{S_{L_2}}\right) \frac{\tau_L(\epsilon_{h\nu})}{S_{L_1 L_1}}$$

$$\sigma_{L_3} = \left(1 - \frac{1}{S_{L_3}}\right) \frac{\tau_L(\epsilon_{h\nu})}{S_{L_1 L_2}}$$

(6.80)

Based on known data of S's and τ's, and assuming continuous behavior of all the S-values and τ's as a function of Z, S's and τ's have been fitted to expressions as a function of Z. Using such functional dependence, S and τ values for other Z have been calculated and then σ's obtained using the above equations. Some of these values are in good agreement with photoionization cross sections (obtained with the Hartree–Slater relativistic calculations mentioned earlier in the chapter) relative to the cross section of Ca $2p_{3/2}$ for several series of elements.

The intensity for an infinitely thick sample can also be written as

$$I_{ej} = \sigma_j G_0 f'_b g(\theta, Z) \rho \lambda_e(\epsilon, Z) \quad (6.81)$$

where G_0 is a geometric factor in the instrument maintained constant for all samples; f'_b is the gain in sensitivity due to preretardation, which can be experimentally determined as a function of the electron kinetic energy ϵ; ρ is the sample density; and λ_e is the electron mean free path, which also can be described parametrically as a function of ϵ and Z and fitted with known data. A ratio of two experimental I_{ej}'s of two samples will not be equal to the ratio of two σ's. If one wants to compare the photoionization cross section between two samples, the experimental intensity ratios ought to be corrected by the ratios of $g(\theta, Z)$, that is, the asymmetry factor, ρ, and $\lambda_e(\epsilon, Z)$ of two samples.

These corrected experimental results for various elements relative to the cross section of Ca $2p_{3/2}$ have been plotted in Fig. 6.14 and are compared with the cross sections, obtained both from semiempirical absorption data and HS calculations. The agreement between the theory and experiment is good. If the uncorrected experimental results are compared with theoretical results the differences are smaller in the first and second series but are appreciable in the third and fourth series. However, Fig. 6.14 shows that there are differences even on using

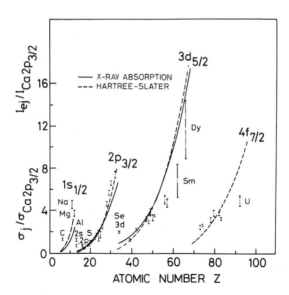

Fig. 6.14. A comparison of corrected experimental peak intensities (relative to Ca$_{2p_{3/2}}$, that is, Ca$_{L_2}$) with cross sections from X-ray absorption and Hartree–Slater central–field calculations [620].

the corrected experimental data. In the first region only the intensity of Mg $1s$ line compares well with the cross-section while C and Na deviate appreciably. Matching is better in the second region. In the third and fourth regions there is good agreement for $34 \leq Z \leq 50$ and $Z \sim 80$. The higher intensity of Na as compared with Mg is probably due to the accumulation of several effects such as surface structure, plasmon losses, and mean free paths of photoelectrons in oxidized metal layers. The large relative errors in Sm and Dy could be due to energy losses in rapid oxidation. Although a better criterion for checking the theoretical cross section is to compare it with the experimental absolute peak intensities, these data indeed give some idea about the proximity of experimental and theoretical cross sections.

TRACE METAL ANALYSIS BY EXTRACTION ONTO SOLID SURFACE

Besides the direct but rather difficult applications of XPS method to quantitative analysis of solids, many innovations have been made in the application of XPS to analytical problems. For example, it is known that in trace metal analysis, techniques to separate metals using ion exchange method, liquid chromatography, and even thin layer chromatography sometimes result in considerable diffusion of adsorbed species below the surface. In such a situation XPS studies become difficult. If trace metals in solution can be suitably extracted onto a nonabsorbing solid surface, then the surface can be subjected to quantitative XPS analysis for indirect determination of trace metals in solution.

The basic technique common to all the examples to be presented here is to suitably scavenge and concentrate the trace metal onto a solid surface, which can then be conveniently subjected to XPS analysis. An innovation employing fiberglass surfaces has recently been used to circumvent the problem of diffusion of adsorbed species below the surface. The fiberglass surface is first treated with a silylating agent containing an amino function that places the amino group on the surface. This is followed by a treatment with CS_2 and a base, which results in the formation of a dithiocarbamate. This material is then used to react with metal ions in solution, and as a result, the metal is scavenged onto the glass surface ready for XPS measurement.

Using this technique and a small disk of dithiocarbamate-coated glass, it has been possible to detect Pb in as dilute a solution as 20 ppb lead. Different kinds of chelating glasses having a variety of functional groups can be employed for the analysis of a wide variety of metals and also for simultaneous multielement analysis.

TRACE METAL ANALYSIS BY EXTRACTION ONTO SOLID SURFACE 227

glass $\begin{array}{l}\text{—OH}\\ \text{—OH}\\ \text{—OH}\end{array}$ + $(CH_3O)_3Si(CH_2)_nCH_2NH_2$

↓

glass $\begin{array}{l}\text{—O}\\ \text{—O}\\ \text{—O}\end{array}\!\!>\!\!Si(CH_2)_nCH_2NH_2$

↓ CS_2 + base

glass $\begin{array}{l}\text{—O}\\ \text{—O}\\ \text{—O}\end{array}\!\!>\!\!Si(CH_2)_nCH_2NHC\!\!<\!\!\begin{array}{l}S\\ S\end{array}\!\!\left(-Na^+\right)$

↓ M^+

glass $\begin{array}{l}\text{—O}\\ \text{—O}\\ \text{—O}\end{array}\!\!>\!\!Si(CH_2)_nCH_2NHC\!\!<\!\!\begin{array}{l}S\\ S\end{array}\!\!\left(-\dfrac{M^{n+}}{n}\right)$

Here is another example of this kind of extraction of metal ions from solution for quantitative analysis. In cases where a 1:1 metal–site complex is formed, the equilibria that control the extraction are

$$\text{SOH} \rightleftharpoons \text{SO}^- + \text{H}^+ \quad K_a = \frac{[\text{H}^+][\text{SO}^-]}{[\text{SOH}]}$$

$$\text{SO}^- + \text{M}^+ \rightleftharpoons \text{SOM} \quad K_f = \frac{[\text{SOM}]}{[\text{SO}^-][\text{M}^+]}$$

where S represents the solid support, M^+ the metal ions in solution. The fraction θ of surface sites covered by the metal, V the volume of solution which interacts with the surface, C_M^0 the total metal concentration in the solution, are related by the following equation:

$$\theta^2 - \theta\left[\frac{NV[\text{H}^+]}{K_a K_f S^0} + \frac{NVC_M^0}{S^0} + 1\right] + \frac{NVC_M^0}{S^0} = 0 \qquad (6.82)$$

where S^0 is the number of sites available, and N is Avogadro's number. The metal XPS signal from the surface will be proportional to the surface coverage θ, so that one may write $I_{XPS} = k\theta$.

Plots of θ versus C_M^0 for 8-hydroxyquinoline complexes of Fe^{2+}, Mg^{2+} and Cu^{2+} are shown in Fig. 6.15. The curves are linear at low concentration, log–linear in moderate concentration, and independent of C_M^0 at high concentration. That the curves are mostly nonlinear indicates a disadvantage where differences in small concentrations are to be determined, but has the advantage of potential use where large changes in concentrations are involved.

At constant pH, the location of the curves depends on the formation constant of the complex, and for a given ligand, the location depends on pH. With stronger acids the curves as function of pH are steeper with a limiting steepness. This is characteristic of ligands forming strong complexes.

The θ–pH plot with C^0 as a parameter shows that in the case of Fe^{2+} 8-hydroxyquinoline complex the threshold position of the curve does not change in the C_M^0 range $10^{-10} - 3.17 \times 10^{-9}$ M. At high pH, when a reasonable equilibrium is reached, θ values show that at these low concentrations there are more sites than the metal ions can fill and even a maximum coverage is incomplete. The nature of this coverage depends on pH. Three orders of magnitude from $\theta = 1 - 10^{-3}$ can be analyzed by XPS. The log–linear region can be used

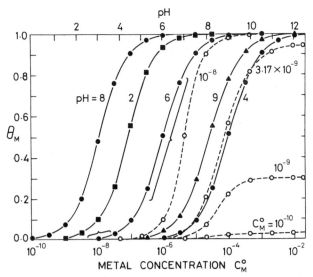

Fig. 6.15. Fraction of metal sites filled, θ_M, as a function of total metal present for an ML_1 complex (continuous line), and as a function of pH (dashed line). ● 8-hydroxyquinoline with Fe^{2+}, $K_f = 6.8 \times 10^9$, ▲ 8-hydroxyquinoline with Mg^{2+}, $K_f = 2.7 \times 10^5$, ■ 8-hydroxyquinoline with Cu^{2+}, $K_f = 1.0 \times 10^{15}$ [628].

to extend it to five orders of magnitude. The concentration range of analysis would be $3.17 \times 10^{-9} - 10^{-10}\ M$ at pH 10; at pH 7, it would be $10^{-5} - 10^{-9}\ M$.

TRACE ANALYSIS BY VOLATILIZATION TECHNIQUE

Another recent example of trace analysis by XPS is the determination of arsenic by a volatilization technique. By coupling XPS with volatilization technique a sensitive method for trace analysis has been obtained; it has been shown that it is possible to detect As in solution in the parts per trillion range. In this method, AsH_3, which is produced by $NaBH_4$ reduction, is trapped as an arsenide of Hg on $HgCl_2$ impregnated paper. A calibration curve was obtained linear up to 650 ppb and the relative standard deviation of the method was found to be 10% when measured at the 100 ppb and 500 ppb levels. Results have been found to agree well with values for NBS reference material. Simultaneous detection of As, Se, Sn, and Sb each at 100 ppb in a 1 ml sample has also been performed. The black precipitate sometimes observed with the $NaBH_4$ method has been determined to be elemental arsenic.

The XPS method gives a detection limit of 3 ng; this compares with the arsenic detection limits of atomic absorption (flame) 4 ng, atomic absorption (flameless) 0.08 ng, atomic fluorescence 2 ng, atomic emission 1 ng, neutron activation 15 ng, anodic stripping voltammetry 0.1 ng, and GC/MS 10 ng.

The advantages of volatilization are as follows. It removes the analyte from the potential interference of its matrix, and it is selective since volatilization removes only certain elements. It is possible to make an estimate of the detection limit in this method. Twenty-five ng of As is equal to 3.3×10^{-10} mole or about 2×10^{14} atoms. If a monolayer consists of 10^{15} sites per cm^2, since the sample area is $1\ cm^2$, the results indicate that a detection of 0.2 of a monolayer takes place.

MATRIX DILUTION TECHNIQUE

Applicability of the matrix dilution technique to XPS chemical analysis has also been examined. The experiments involved a study of the effect of an inert matrix on Pb $4f$ intensities from the three lead salts $PbSO_4$, $PbCl_2$ and PbI_2; MoO_3 was used as the internal standard and graphite as the matrix. Binary mixtures consisting of lead salt and molybdenum oxide have also been studied. Plots of relative intensities of Pb $4f$/Mo $3d$ against the atomic ratios Pb/Mo for both the binary mixture and the mixture with the matrix show that the scatter of data is less in plots involving the binary mixtures. Also, such plots for the three salts mentioned show different slopes, which could be due to changes in photoionization cross section arising from different anions of the lead salts. That this is not so, however, is shown by a comparison of the ratios Pb $4f$/Pb $4d$ with PbI_2 and $PbCl_2$, which are found to be constant within experimental error. In that case the differences in slopes in the three salts may be ascribed to differences in mean escape depths. This implies that a knowledge about the nature of the compound would be necessary before any quantitative application of photoelectron spectroscopy.

This problem has been circumvented by subjecting the lead compound to a $BaSO_4$ fusion at 400°C, which changes it into lead sulphate; $BaSO_4$ used in the fusion then acts as the internal standard. A number of lead compounds have been studied to test the applicability of this method. A plot of the ratio of the intensities $I(Pb\,4f)/I(Ba\,4d)$ against the ratio (Pb atoms)/(Ba atoms) using 11 different lead salts show a linear relationship, with a standard deviation of ± 3%.

ANALYSIS OF ORGANIC POLYMERS

Both qualitative and quantitative analysis of organic polymers have been carried out with XPS. Core binding energies may be used for characterization of the atoms present and the chemical shifts may be used to decide on the nature of chemical environment. Results of the analysis of some organic polymers are presented in Table 6.6.

A STATUS SUMMARY AND COMPARISON OF XPS WITH OTHER ANALYTICAL METHODS

We may summarize now the applicability of the photoelectron spectroscopic method as an analytical tool as it presently stands and provide a brief comparison with other surface analytical methods.

The core binding energies are characteristic of elements and therefore ideally suited for qualitative identification of the elements. For multielectron atoms a series of core electrons get photoionized, and the XPS method thus provides a number of additional checks for elemental identification. In XPS instruments containing more than one photon source, Auger lines are easily identified and sorted out because Auger lines appear at fixed energies, unlike the kinetic energy of photoelectrons, which depend on the photon energy.

An inherent additional advantage of XPS is that while the characteristic core binding energies permit qualitative analysis, the chemical shifts at the same time provide direct information on the electronic environment, that is, on the chemical state of the atoms. Such a powerful probe for structure elucidation is not provided by some otherwise very useful analytical methods, such as mass

Table 6.6. Analysis of Ethylene–Tetrafluoroethylene Comonomer Incorporations [630]

Sample	Composition of Monomer Mixture (mol%) C_2F_4	Predicted from Monomer Reactivity Ratios	Copolymer Composition (mol% C_2F_4)				
			Calculated from C Analysis	Calculated from F Analysis	Calculated from Area Ratio C 1s peak, F 1s peak	Calculated from C 1s (CH$_2$ peak)	Calculated from C 1s (CF$_2$ peak)
1	94	63	61	61	63	62	
2	80	53	52	54	52	52	
3	65.5	50	49	48	47	46	
4	64	50	47	45	44	45	
5	35	45	41	40	42	40	
6	15	36	—	—	32	31	

231

spectrometry, nor by any other surface analytical method. XPS is also the only method other than secondary ion mass spectrometry (SIMS) that permits the analysis of solid organic samples, including polymers, yielding information on both composition and structure. The high-energy electron beam used in Auger electron spectroscopy (AES) causes radiation damage and charging. Because of a superior peak-to-background ratio in photoionization in comparison with electron bombardment methods, XPS is as sensitive as or sometimes more sensitive than the AES technique. Unlike other surface analytical methods such as SIMS, AES, and the ion bombardment-induced light emission (BLE) technique, the XPS method is nondestructive. Also, the analysis of insulators is far more straightforward in XPS than it is with AES, SIMS, BLE, low energy ($<$ 10 keV) ion-scattering spectroscopy (LEIS), or high energy (0.5–3 MeV) ion-backscattering spectroscopy (HEIS). For qualitative and semiquantitative analysis of low atomic weight elements, XPS is not the most desirable method. The information depth for XPS is 0.5–3 nm, and for AES 0.3–2.5 nm; they are about the same. In the photoelectron spectroscopic technique, depth profiling can be done without sputtering by comparing photoelectron lines that have different energies, as has been indicated in the germanium example in Chapter 5. With sputtering, XPS can be applied for depth profiling, as is done in SIMS and AES.

When a scanning mode is used to map the elemental composition of a solid surface, the width of the primary beam is of critical significance. If the projectile is a charged particle beam, much finer focusing is possible. The lateral resolution for XPS is about 1 mm, which is the present limit to which an X-ray beam can be focused. The new transmission scanning instruments might bring this limit down. This lateral resolution of XPS compares with the lateral resolutions of AES at about 100 nm, SIMS 1 μm–100 nm, LEIS 100 nm, HEIS 100 μm, and BLE 1 μm. For fast- and high-resolution surface scanning, therefore, XPS is still not the best method.

The lack of information on the photoelectron mean free path (inelastic electron scattering cross sections) and the photoionization cross sections remain two limiting factors in applications of photoelectron spectroscopy to quantitative analysis of solids and surfaces. Chemical nature of the solid and method of sample preparation also contribute to uncertainties. It has been convincingly demonstrated, however, that with the use of suitable calibration techniques, quantitative analysis using XPS is a practicable method. Although there are still several problems to be solved before XPS becomes a routine tool for quantitative measurements, the advantages are clear: scope of simultaneous multielement analysis, negligible radiation damage, high surface sensitivity, and nondestructiveness. Looking at it another way, there is no other method for solid and surface analysis with fewer problems, and the refinement of techniques in XPS analytical method is fast evolving.

CHAPTER

7

ANGULAR DISTRIBUTION OF PHOTOELECTRONS

The angular distribution of photoelectrons constitutes an integral part of the study of photoionization cross sections. Although logically a discussion of such properties belongs with the material covered in the previous chapter, we treat them separately because of the exceptional character of the experimental techniques involved and also because of some unusual applications of angular distribution data.

Interestingly, the nature of the information that one can derive from angular distribution studies of photoelectrons is drastically different for gaseous and solid samples. As we discussed in the first chapter, a quantitative study of photoelectron angular distribution for gaseous samples can be expressed in terms of the total cross section and a characteristic parameter that describes the angular distribution. This parameter is dependent on the nature of the atom or the molecule as well as on the orbital of the electron from which it is ejected. Since it also depends on the photoelectron kinetic energy, it has an indirect dependence on the incident photon energy and the ionization potential of the sample. From the point of view of theory, one is interested in calculating both the magnitude of the total cross section and the magnitude of this characteristic parameter, the latter determining the physical pattern of the photoelectron angular distribution.

A study of the angular distribution of photoelectrons from solids leads to entirely different kinds of information. With proper experimental techniques, it can be used to determine the electron mean free path in solid samples. From angle-resolved photoemission data it is possible to derive information on the distribution of electron momentum in solids. It is also possible to derive from such data two-dimensional band structure of surfaces and adsorbates. Angular distribution information, in fact, permits the study of surface and adsorption phenomena in a number of ways. When adsorption takes place, angle-resolved photoemission intensities, such as at grazing emergence, lead to information on the nature of surface islanding and permit sputtering-free depth profiling of atoms undergoing photoemission in the solid. The angular anisotropy of photoemission also depends on the nature of orientation of the adsorbate molecules on the surface. Yet another method of studying surface phenomena in minute detail is the use of synchrotron radiation to study angle-dependent diffraction phenomena and backscattering of adsorbate photoelectrons by substrate atoms.

In this chapter, we first consider photoelectron angular distribution from gaseous samples and then from solids.

ANGULAR DISTRIBUTION FROM GASEOUS SAMPLES: ASYMMETRY PARAMETER β

For polarized radiation, the angular dependence $I(\theta)$ of photoelectron intensity, or, identically, the angle-dependent differential cross section of photoionization $d\sigma(\theta)/d\Omega$ in cm^2 steradian^{-1}, is given by

$$I(\theta) = \frac{d\sigma}{d\Omega} = \frac{\sigma_{tot}}{4\pi}\left[1 + \frac{\beta}{2}(3\cos^2\theta - 1)\right] \quad (7.1)$$

where σ_{tot} is the total photoionization cross section and θ is the angle between the direction of photoelectron ejection and the electric field vector of the polarized light. Equation (7.1) is derived using dipole approximation on a single-electron model and assuming random orientation of target particles. Retardation effects are neglected. The β in Eq. (7.1) is called the *asymmetry parameter*, because only when $\beta = 0$ is $I(\theta)$ independent of θ and the photoelectrons isotropically disposed. A coordinate system that shows the interrelationships between various parameters involved in photoelectron angular distribution is presented in Fig. 7.1. It can be seen from the figure that $\cos\theta = \sin\alpha\cos\phi$. Therefore, to obtain a relationship for the differential cross section when unpolarized radiation is used, it is necessary to substitute $\cos\theta$ in Eq. (7.1) by

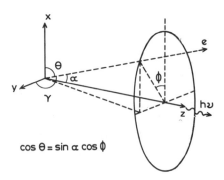

Fig. 7.1. Coordinate system in photoelectron angular distribution.

sin α cos φ and integrate over all φ, as under this condition the direction of the electric field vector is no longer along the x-direction but points to all possible directions on the xy plane. One then obtains, for the differential cross section in the unpolarized case:

$$I(\alpha) = \frac{d\sigma}{d\Omega} = \frac{\sigma_{tot}}{4\pi}\left[1 - \frac{\beta}{4}(3\cos^2\alpha - 1)\right] \quad (7.2)$$

α being the angle between the propagation direction of photons and the direction of photoelectron ejection. It is interesting that for α = 54.73°, the so-called *magic angle*, the $3\cos^2\alpha - 1$ term is equal to zero, hence $I(54.73°)$ — which is the intensity of the photoelectrons at 54.73° with respect to the photon propagation direction — provides a direct measure of the absolute value of the total cross section, σ_{tot}.

Synchrotron radiation, which also can be used for photoelectron spectroscopy, is elliptically polarized. Considering the electric vector components along the major and minor axes of the ellipse that characterize the elliptically polarized beam as equivalent to two perpendicular incoherent electric vector components of a partially polarized beam, the differential cross section is given by

$$I(\alpha, \phi) = \frac{d\sigma}{d\Omega} = \frac{\sigma_{tot}}{4\pi}\left[1 - \frac{\beta}{2}\left(\frac{3\cos^2\alpha - 1}{2}\right) + \tfrac{3}{4}\beta\left(\frac{I_y - I_x}{I_y + I_x}\right)(\cos^2\theta - \cos^2\gamma)\right]$$

(7.3)

where I_x and I_y are radiation intensities respectively along the x and y axes. With $I_y = I_x$, Eq. (7.3) reduces to Eq. (7.2).

The asymmetry parameter β determines the nature of angular distribution displayed by the photoelectrons. For example, when β = 0, both $I(\theta)$ and $I(\alpha)$ are independent of φ the photoelectron ejection angle, and hence the distributions are isotropic. Figure 7.2 shows a polar plot of $I(\alpha)$ for unpolarized radiation, as the latter is often the experimental case, for β = −1, 0, 1, and 2. Note that for β = −1 the photoelectrons are ejected preferentially in the direction of photon propagation, whereas for β = 2, the ejection is skewed at right angles to it. The requirement of nonnegativity of the differential cross section constrains β to values only in the range from −1 to +2. The β for an atom is a function of both the angular momentum quantum number l of the orbital from which the electron is ejected and of the photoelectron kinetic energy. Photoelectrons originating from the s subshell ($l = 0$) leave the atom with angular momentum (l) equal to 1, and in that case β has a value of +2 at all photoelectron energies. If the initial state of the electron is not an s state, both $\Delta l = \pm 1$ are possible, and interference occurring between these two channels results in the deviation of β from +2.

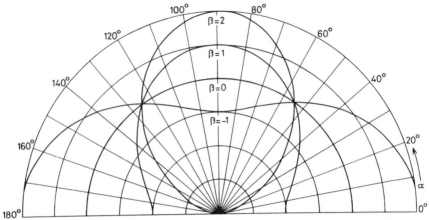

Fig. 7.2. Photoelectron angular distribution $I(\alpha)$ at $\beta = 2, 1, 0,$ and -1. Note the intersection of the curves at $\alpha = 54.73°$.

For unpolarized radiation, Eq. (7.2) can also be written in the form

$$d\sigma/d\Omega = A + B \sin^2 \alpha \tag{7.4}$$

with $A + B = 100$ and $\beta = 4B/(3A + 2B)$. At high photon energies, when retardation effects are not negligible, an approximate expression for the differential cross section is

$$\frac{d\sigma}{d\Omega} = A + B\left[1 + \frac{(4+l)v}{c} \cos \alpha\right] \sin^2 \alpha \tag{7.5}$$

where v is the velocity of the photoelectron and c the velocity of light. Under such conditions the maximum of the photoelectron distribution becomes skewed in the direction of photon propagation.

It is clear from Eqs. (7.2) and (7.4) that if experimental measurements are made of photoelectron intensity at any two angles, then by taking the ratio of the intensities σ_{tot} is eliminated, and one can determine β. As an alternative procedure, lines corresponding to photoelectrons ejected from a subshell ns and from a subshell nl can be measured at two different angles, for instance at $54.73°$ and $90°$, and the β_{nl} value deduced from the following relationship:

$$\frac{\sigma_{ns}}{\sigma_{nl}} = \left.\frac{(d\sigma/d\Omega)_{ns}}{(d\sigma/d\Omega)_{nl}}\right|_{54.73°} = \frac{2(1 + \beta_{nl}/4)}{3} \left.\frac{(d\sigma/d\Omega)_{ns}}{(d\sigma/d\Omega)_{nl}}\right|_{90°} \tag{7.6}$$

This is obtained by applying Eq. (7.2) to photoelectron ejection from both the shells ns and nl, at the angles $\alpha = 54.73°$ and $\alpha = 90°$, and taking $\beta_{ns} = 2$. From Eq. (7.2) it can be seen that since the intensity distributions have a cosine square dependence, sensitivity is greater for smaller angles.

It is apparent from Eqs. (7.1) through (7.5) that the magnitude of the differential photoionization cross section depends on the angles as well as on the asymmetry parameter β. Since the photoionization cross section is determined basically by the nature of the initial orbital before ionization and the final orbital after the electron has been ejected, it is reasonable to expect that the β values would not only be characteristic of the chemical species undergoing photoionization but would be the same wherever the nature of initial and final orbitals are identical. However, since the cross section also depends on the energy of the ejected photoelectron, equality of β values would require similar photoelectron energies. While such relationships permit estimates of β and hence the nature of photoelectron angular distribution for a species when information on other species having similar properties is available, a far more important and powerful application of the orbital–β correlation would be to apply it in reverse, that is, to determine from experimental β values the nature of the orbitals involved in ionization. One could then determine in a routine manner from angular distribution studies of a photoelectron spectral peak the nature of the orbital involved in photoionization, and one could determine from binding energy studies the orbital energetics. The two complementary techniques could then be of immense value in the study of atomic and molecular electronic structure. Such an objective is far from realized at the present stage of development; however, the importance of angular distribution measurements partly arises because of its potential for being of significant assistance in the assignment of orbitals involved in photoionization.

Correlations between orbitals and β require considerable data on angular distribution as a function of the energy of the ejected photoelectron. This means that photoionization studies are needed using either a continuously tunable intense monochromatic light source, such as synchrotron radiation, or a wide range of photon sources from few eVs to energies in the keV range. The initial angular distribution studies were made using UV photon sources, and β values for some simple molecules are shown in Table 7.1. Because β is a function of photoelectron kinetic energy, these values are at one value of energy, as indicated in the table. In Table 7.2, β values are given for Ne $2p$ at different photon energies.

Table 7.3 lists β values of some halomethanes. Although the photoelectron energies associated with the same type of orbitals are different, certain correlations can be made. In general, the highest β value is associated with the $np\pi$ lone-pair electrons or $e(X)$ nonbonding orbital (X stands for halogen), an intermediate value is associated with the bonding orbital $e(CH_3)$, and the lowest β value is associated with the bonding orbital $a_1(CX)$. Further, the β value of a given orbital increases with the atomic number Z. A comparison of β values obtained from atomic calculations and presented in the same table shows that the β values for the lone-pair halogen electrons follow the results of atomic

Table 7.1. Asymmetry Parameter β for H_2, N_2, O_2, H_2O, and H_2S at Low Photoelectron Energies ($h\nu = 21.22$ eV) [644]

Molecule	Orbital	Photoelectron Energy (eV)	β	Molecule	Orbital	Photoelectron Energy (eV)	β
H_2	$\sigma_g\ 1s$	5.7	+1.75	H_2O	$1b_1$	8.6	+1.0
	$\sigma_g\ 1s$	1.1[a]	+2.0		$2a_1$	7.5	+0.3
N_2	$\sigma_g\ 1s$	5.6	+0.5		$1b_2$	4.0	−0.1
	$\sigma_g\ 2p$	1.3[a]	+1.2	H_2S	$1b_1$	10.7	+1.6
	$\pi_u\ 2p$	4.5	+0.3		$2a_1$	8.4	+1.1
	$\sigma_u\ 2s$	2.4	+1.25		$1b_2$	6.5	+0.9
O_2	$\pi_g\ 2p$	9.1	−0.3				
	$\pi_g\ 2p$	4.8[a]	−0.55				
	$\pi_u\ 2p$[b]	5.1	+0.4				
	$\pi_u\ 2p$[c]	4.2	+0.3				
	$\sigma_g\ 2p$	3.0	+0.7				

[a] $h\nu = 16.9$ eV.
[b] $\pi_u\ 2p = a^4\Pi_u$.
[c] $\pi_u\ 2p = A^2\Pi_u$.

calculations. Since HCl has a purer form of lone-pair, the value for HCl is higher. The reason why fluoromethanes do not show good correlations is partly because the β value for the p electron is lower in C whereas for Cl, Br, and I it is higher. Also, compared with other halomethanes, fluorine electrons have greater bonding characteristics and therefore act less like lone-pair electrons.

Angular distributions of photoelectrons that arise out of transitions to various vibrational states of the resulting photoions have also been studied, although mainly for small molecules. For most of these molecules, the relative photoelectron intensities due to various vibrational transitions do not undergo any

Table 7.2. Asymmetry Parameter β of Photoelectrons from Ionization of Ne $2p$ Electrons at Various Photon Energies [645]

X-ray Line	$h\nu$ (eV)	Photoelectron Energy (eV)	β_{2p}
Be K	108.9	87.3	1.35
Y $M\zeta$	132.3	110.7	1.41
Zr $M\zeta$	151.4	129.8	1.49
Mg K_α	1253.6	1232.0	0.85
Al K_α	1486.6	1465.0	0.76

change with the observational angle, and β remains unchanged. This may be expected if the Born–Oppenheimer approximation is valid and the wave function due to the electrons can be factored out from that due to the nuclear motion. Under this condition, the Franck–Condon factors are independent of the angle of photoelectron ejection, and therefore relative photoelectron intensities remain independent of the direction of observation.

In a few cases, however (e.g., N_2, O_2, CO), β's have been experimentally found to be dependent on the vibrational state of the ion. In O_2 this angular dependence is believed to be due to contributions from autoionization states. In such a situation the Franck–Condon factor contributions are different depending on whether the photoionization is direct or through the autoionizing state, and this results in changes in β. It is possible that for N_2 and CO, changes in β depending on the vibrational state are due to a breakdown of the Born–Oppenheimer approximation but this is not fully established.

The β values have also been measured in transitions involving rotational state changes. In one such measurement, using NeI radiation doublet of 74.4 and 73.6 nm wavelength, the hydrogen transition $H_2(^1\Sigma_g^+) \to H_2^+(^2\Sigma_g^+)$ with ($v'' = 0) \to (v' = 0, 1)$, and rotational quantum number changes $\Delta N = 0$ and $\Delta N = 2$, have been studied. The spectrum shows partially resolved rotational lines, among others, due to Q branch $\Delta N = 0$ and S branch $\Delta N = +2$. From the same initial state, different rotational states of different angular momenta are populated, and differences in β values may be expected. The β values are found to be $\beta_Q(\Delta N = 0) = 1.75 \pm 0.03$ and $\beta_S(\Delta N = +2) = 0.85 \pm 0.14$.

In a recent work on CO, β values of twenty-four $X^2\Sigma^+, A^2\Pi, B^2\Sigma^+$ molecular ionic vibrational states have been measured. For the $X^2\Sigma^+$ band an unusually rapid variation of β has been observed as the electron energy changes from 5.9 to 7.2 eV. For the $A^2\Pi$ state a similar nonmonotonic variation in an energy range of 2.9–4.1 eV has been observed. A plot of β against electron energy using both HeI and NeI ionization shows that a straight line can be drawn with 14 points; β's that correspond to $v' = 0, 4, 8, 9, 10$ are significantly off the line. A similar range of variation is observed for the $B^2\Sigma^+$ band. Here, except for the $v' = 3$ point, a straight line can be drawn through the points with a slope of about 1 eV^{-1}.

Careful measurements rule out this large variation of β as arising from contributions from overlapping features resulting from the 53.7 nm and 52.2 nm lines (these are weak HeIβ and HeIγ lines present along with HeIα at 58.4 nm), nor is it believed to be due to secondary electron scattering by the residual neutral CO molecules in the system. Another possibility is a breakdown of the Born–Oppenheimer approximation. This is likely to be the case when the vertical Franck–Condon band and the crossing of two potential energy curves overlap. But in CO, the crossing between the $X^2\Sigma^+$ and $A^2\Pi$ curves occurs at 0.148 nm, whereas the Franck–Condon band extends only in the range 0.106–

Table 7.3. Asymmetry Parameters for Various Halomethanes ($h\nu = 21.22$ eV) [678]

Molecule	Molecular Orbital	Vertical Ionization Potential (eV)	β	Molecule	Molecular Orbital	Vertical Ionization Potential (eV)	β
CH_3F	$e(CH_3)$	13.0	0.20	CH_3Cl	$e(Cl)$	11.3	0.85
	$e(CH_3)^a$	13.6	0.15		$a_1(CCl)$	14.4	0.35
	$e(F) + a_1(CF)$	17.1	0.60		$e(CH_3)$	15.4	0.65
CH_2F_2	$b_1(CH_2)$	13.3	0.00	CCl_4	$t_1(Cl_4)$	11.7	0.70
	$a_1(H_2CF_2)$	15.2	0.50		$t_2(Cl_4)$	12.6	0.85
	$b_2(F_2)$	15.4			$e(Cl_4)$	13.4	0.75
	$a_2(F_2)$	15.7	−0.05		$t_2(CCl_4)$	16.6	0.35
	$a_1(F_2) + b_1(F_2) + b_2(CF_2)$	19.0	0.30				
CHF_3	$a_1(CH)$	14.8	0.10	CH_3Br	$e(Br)$	10.5	1.2
	$a_2(F_3)$	15.5	0.05		$e(Br)^b$	10.9	1.15
	$e(F_3)$	16.2	0.45		$a_1(CBr)$	13.5	0.3
	$e(F_3)$	17.2	0.05		$e(CH_3)$	15.1	0.7
CF_4	$t_1(F_4)$	16.2	−0.50	CH_3I	$e(I)$	9.5	1.5
	$t_2(F_4)$	17.5	0.20		$e(I)^b$	10.2	1.46
	$e(F_4)$	18.5	0.15		$a_1(Cl)$	12.5	0.6
					$e(CH_3)$	14.8	0.9

Lone-Pair Band		β (Experiment)	β (Atomic calculation)
CH_3F	$2p\pi$	0.60	0.3
CH_3Cl	$3p\pi$	0.85	1.2
HCl	$3p\pi$	1.4	1.2
CH_3Br	$4p\pi$	1.2	1.4
CH_3I	$5p\pi$	1.5	1.6

^{a} Jahn–Teller splitting.

0.120 nm. It is also unlikely that a very large variation of the electronic transition dipole matrix element takes place such that it would be necessary to account for the variation. The most likely cause could be an electronic autoionization or a shape resonance.

In the electronic autoionization or an internal electronic *Feshbach resonance*, a deeply buried electron is excited such that the total system has an energy higher than the lowest ionization potential. This is followed by an energy transfer to an electron with low ionization energy. As mentioned in Chaper 1, in vibrational autoionization, highly excited Rydberg orbitals cause ionization resulting in a low energy photoelectron. *Shape resonance*, in contrast to Feshbach resonance, involves ionization from the orbital originally excited, but the nature of the interactions of the excited electron with the remaining electrons is such that it results in a delay in the ionization process — the escaping photoelectron needs to overcome a potential energy barrier. In the process, due to coulomb and exchange interactions of the escaping electron with the remaining electrons, a quasibound state exists, and at a photon energy equal to the energy of the state, it affects the photoelectron cross section. This occurs over a few eV of energy. But since the variation in β is observed in about only 0.2 eV range, which is roughly vibrational spacing, shape resonance is unlikely to be the reason. Autoionization resonances, which are typically < 0.1 eV, are a more likely reason for β variation. The observed β can be expressed as

$$\beta_{obs} = \frac{\sigma_{dir}\beta_{dir} + \sigma_{auto}\beta_{auto}}{\sigma_{dir} + \sigma_{auto}} \quad (7.7)$$

where σ_{dir} and σ_{auto} are the total cross sections for the direct and autoionization processes.

When autoionization occurs, vibrational intensities can be written as

$$I \propto F_{if}^2 + qF_{iv}^2 F_{vf}^2 \quad (7.8)$$

where F_{if}^2 is the Franck–Condon factor due to direct photoionization, F_{iv}^2 the factor due to the excited quasibound neutral state, and F_{vf}^2 the factor from the quasibound state to the final state; q is a parameter related to the oscillator strength of the quasiexcited neutral to the ion state and the linewidth of the neutral state. The intensities of the $v' \geqslant 2$ of $X^2\Sigma^+$ and $B^2\Sigma^+$ must be at least partly due to autoionization. If $v' = 2,5$ of $X^2\Sigma^+$ is assumed to be entirely due to autoionization, then by drawing a line between these two points one can obtain a rough energy dependence of β_{auto}. If one further assumes that $v' = 0$ is populated only directly, then energy dependence of β_{dir} can be obtained from the energy dependence of $v' = 0$ by HeIα and HeIβ radiation, which gives β at two different photoelectron energies. With this, along with the observed values of β for $v' = 1$ in the 58.4 nm spectrum, one can obtain for $\sigma_{dir}/\sigma_{auto}$ a value of -0.67, which is not physically meaningful. Therefore an assumption of noninterference between autoionized and directly ionized electron is probably not correct. At present there is inadequate information to determine the extent of autoionization and direct ionization.

There are now also reports in the literature about the study of variation of β of polyatomic organic molecules as a function of photoelectron kinetic energy. A simple way of studying angular distribution at low photoelectron kinetic energies is to use a series of resonance radiation sources such as HeI and NeI, measure β values, and evolve empirical correlations. For example, in a recent investigation of unsaturated hydrocarbons it was observed that with the β's of π orbitals from structurally similar compounds such as cyclohexene, 1,3-cyclohexadiene, 1,4-cyclohexadiene, and norbornadiene, the β/energy variation is about $0.06\,\mathrm{eV}^{-1}$, and for diverse molecules like ethylene, allene, 1,3-butadiene, and benzene, the individual β's vary over a considerable range although the average is zero. For σ orbitals there is a wide variation in the gradients, but for structurally similar compounds the variation is less, the overall average being about $0\,\mathrm{eV}^{-1}$ and thus a lower energy dependence than the π orbitals. The π orbital β values are on the average greater than the σ values. It is observed that in a multi-π electron molecule the nodeless symmetric π-orbital has a lower β value than the antisymmetric ones.

Photoelectron angular distribution measurements have also been used to derive information about the ordering of orbitals. For example, on the basis of knowledge of the orbital orders and β values of the corresponding photoelectron bands of norbornadiene and 1,3-cyclohexadiene, and from β values of 1,4-cyclohexadiene, it has been possible to draw inferences about the orbital arrangement of the latter.

For norbornadiene and 1,3-cyclohexadiene where the homoconjugative through-space effect dominates over the hyperconjugative through-bond effect, it is established that of the two a_1 and b_2 π-orbitals, the antisymmetric b_2 orbital is of higher energy than the symmetric a_1 orbital. In 1,3-cyclohexadiene the hyperconjugative effect is expected to be small, as in 1,3-butadiene where no through-bond effect is possible. In the experimental spectrum, therefore, the first band of highest kinetic energy corresponds to the b_2 orbital, and then a_1, and then the σ orbital. The β's from HeI spectra are:

	$\beta(\pi)b_2$ antisymmetric	$\beta(\pi)a_1$ symmetric	$\beta(\sigma)$
Norbornadiene	0.80	0.55	0.15
1,3-Cyclohexadiene	0.77	0.33	−0.13

In 1,4-cyclohexadiene, on the other hand, the highest kinetic energy band has $\beta = 0.47$, the intermediate one has $\beta = 0.73$, and for the lowest kinetic energy band $\beta = 0.03$. With the ordering of norbornadiene and 1,3-cyclohexadiene unambiguously established, higher β value may be associated with the antisymmetric orbital. This would imply that an orbital with $\beta = 0.47$ for 1,4-cyclohexadiene is due to the a_1 symmetric state, which is of higher energy, and $\beta = 0.73$ is due to the antisymmetric state, which is of lower energy — that is, the orbital order is reversed. The band with $\beta = 0.03$ is assigned to the σ orbital.

We mentioned at the beginning of this section that Eq. (7.1) was derived using dipole approximation on a single-electron model. In the following sections we reconsider this in some detail and present an overview of theoretical studies on the photoelectron angular distribution of gaseous samples before we take up a consideration of angle-resolved photoemission studies of solids.

β FROM THEORY: ONE-ELECTRON SYSTEM

If retardation and relativistic effects are neglected, then certain general results on the angular distribution of photoelectrons for a single-electron atom having a central potential can be obtained by symmetry considerations alone. Let $\psi_{nlm}(\mathbf{r})$ represent the wave function of the initial bound state i of the atom, separable into spherical polar coordinates; let $\psi_f(\mathbf{r})$ be the final state represented by a plane wave $e^{i\mathbf{k}\cdot\mathbf{r}}$ plus an incoming spherical wave; let $\mathbf{u}$ denote a unit

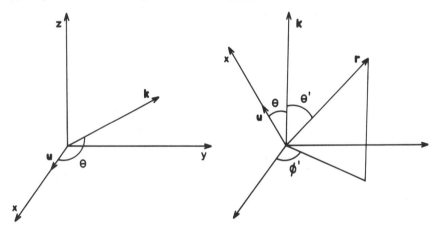

vector in the polarization direction of the photon taken along the x axis, and $\mathbf{k}$ the direction of photoelectron ejection at an angle θ with respect to $\mathbf{u}$. The matrix element of transition, when retardation is neglected, is given by

$$M_i^{\mathbf{u}} \propto \mathbf{u} \cdot \int \psi_f^*(\mathbf{r})\,\mathbf{r}\,\psi_{nlm}(\mathbf{r})\,dv$$

$$\propto \int \psi_f^*(\mathbf{r})\,x\,\psi_{nlm}(\mathbf{r})\,dv$$

(7.9)

M_i^u, hence $d\sigma/d\Omega$, which depends on the square of M_i^u, is not a function of the direction of **u** alone; it depends on the projection of $\mathbf{u} \cdot \mathbf{r}$. Since the integration is carried out along all possible orientations of **r**, the resulting matrix element depends on $\mathbf{u} \cdot \mathbf{k} = k \cos \theta$. The matrix element for the state $\psi_{nlm}(\mathbf{r})$ is also dependent on the quantization direction and the magnetic quantum number m. For a closed shell the magnetic quantum number m can have $2l + 1$ possible values from $-l$ to $+l$. Since each m state is doubly occupied, the cross section from the whole of subshell l is $2(2l + 1)$ times the average $(d\sigma/d\Omega)$. Such an average over all m values is independent of the quantization direction used for the initial state.

Let us now examine the effect of a change in the direction of photoelectron ejection from **k** to $-\mathbf{k}$. If the direction of **u** is kept fixed (i.e., the x axis), then the change in the direction of **k** would require a change in the variable of integration from **r** to $-\mathbf{r}$ and also a reversing of the quantization direction of $\psi_{nlm}(\mathbf{r})$. This causes a change in the sign of the matrix element integral having the form of Eq. (7.9). The cross section, however, depends on the square of the matrix element; it therefore remains unchanged. On the other hand, this change of **k** to the opposite direction $-\mathbf{k}$ causes a change in the sign of $\cos \theta$, $-\theta$ measured from the positive x axis. Since the cross section $(d\sigma/d\Omega)_{av}$ remains unaffected even though $\cos \theta$ changes signs, one concludes that $(d\sigma/d\Omega)_{av}$ is an even function of $\cos \theta$.

In order to derive an explicit expression for $(d\sigma/d\Omega)_{av}$, we use a spherical polar coordinate system (r, θ', ϕ') for **r**, with the direction of **k** as the polar axis (z), which is also the quantization direction for ψ_{nlm}. Then, taking the azimuthal plane that contains **u** as one having $\phi' = 0$, and using $\mathbf{r} = \mathbf{i} r \sin \theta' \cos \phi' + \mathbf{j} r \sin \theta' \sin \phi' + \mathbf{k} r \cos \theta'$ and $\mathbf{u} = \mathbf{i} \sin \theta + \mathbf{k} \cos \theta$, x can be written

$$x = \mathbf{r} \cdot \mathbf{u} = r(\cos \theta \cos \theta' + \sin \theta \sin \theta' \cos \phi') \qquad (7.10)$$

The operator x and hence the matrix element M_i^u therefore depend linearly on $\cos \theta$ and $\sin \theta$. $(d\sigma/d\Omega)_{av}$ is obtained by summing the absolute squares of matrix elements of Eq. (7.9) over all m values. From symmetry argument, coefficients of the cross-term $\cos \theta \sin \theta$ (arising out of the use of Eq. (7.10) in Eq. (7.9) and squaring), which is an odd function, must vanish. This means that the differential cross section will be left with only a $\sin^2 \theta$ and a $\cos^2 \theta$ term:

$$\left(\frac{d\sigma}{d\Omega} \right)_{av} \propto \alpha \sin^2 \theta + \beta' \cos^2 \theta = \alpha + \beta \cos^2 \theta \qquad (7.11)$$

where $\beta' = \beta + \alpha$. We note that Eq. (7.11) has the form of Eq. (7.1).

With the orbital quantum number l' and magnetic quantum number zero, the exact continuum wave function can be written as $\psi_f = \Sigma_{l'} \psi_{f,l'}$; the $\psi_{f,l'}$ is expressed in spherical polar coordinates. The α and β' are evaluated by substituting x from Eq. (7.10) in Eq. (7.9), squaring the dipole matrix element, and summing over all possible m states for a given l. The coefficient of the resulting $\sin^2\theta$ term is α and the coefficient of the $\cos^2\theta$ term is β'; they are given by

$$\alpha = \frac{1}{4} \sum_{m=-1,1} \left| \sum_{l'=l-1,l+1} \int \psi_{f,l'}^*(\mathbf{r}) r \sin\theta' e^{-im\phi'} \psi_{nlm}(\mathbf{r}) dv \right|^2 \quad (7.12)$$

$$\beta' = \beta + \alpha = \left| \sum_{l'=l-1,l+1} \int \psi_{f,l'}^*(\mathbf{r}) r \cos\theta' \psi_{nl0}(\mathbf{r}) dv \right|^2 \quad (7.13)$$

For an initial bound state $s(l=0)$ in any central potential, there are no states ψ_{n0m} with $m=\pm 1$, and therefore the coefficient α in Eqs. (7.11) and (7.12) vanishes. Hence the angular distribution is simply proportional to $\cos^2\theta$. At frequencies ν near the threshold for $2p$ state electrons, one obtains

$$(d\sigma/d\Omega)_{av} \propto 1 + \frac{2(Z-s_2)^2 \text{Ry}}{\nu} \cos^2\theta \quad (7.14)$$

where Z is the nuclear charge, Ry is the Rydberg frequency (3.3×10^{15} sec^{-1}), and $s_2 = 4.15$ is Slater's inner screening constant.

When retardation is taken into account in Born approximation and non-relativistic velocities for the ejected electrons are considered, that is, $(Ze^2/\hbar c)^2 mc^2 \ll h\nu \ll mc^2$, the cross section for the s-state is

$$(d\sigma/d\Omega) \propto \sin^2\theta'' \cos^2\phi'' [1 + 4(v/c) \cos\theta''] \quad (7.15)$$

θ'', ϕ'' are the spherical polar coordinates of the propagation direction $\mathbf{k}$ of the ejected electron, photon propagation direction $\mathbf{k}_\nu$ is taken as the polar (z) axis, polarization direction $\mathbf{u}$ as the x-axis ($\phi''=0$ plane), and $\cos^2\theta = \sin^2\theta'' \cos^2\phi''$. For $2p$ electrons and $Z^2 \text{Ry} \ll h\nu \ll mc^2$, the differential cross section is given by

$$\left(\frac{d\sigma}{d\Omega}\right)_{av} \propto \left[1 + 2\frac{v}{c}\cos\theta''\right] + \sin^2\theta'' \cos^2\phi'' \left[2\frac{(Z-s_2)^2 \text{Ry}}{\nu} + 4\frac{v}{c}\cos\theta''\right] \quad (7.16)$$

Equations (7.1) and (7.2) also express the angular distribution of products in nuclear reactions. In fact, the relationships in Eqs. (7.1), (7.2) are general enough so that whenever an absorption occurs via an electric dipolar process (photoelectrons from atoms and molecules, Auger electrons, electrons that result from two-electron excitation, ions or neutral particles that result from molecular ionization followed by dissociation), the angular distribution is given by Eq. (7.2). Any deviation from this angular distribution indicates participation of processes other than electric dipole.

For β calculation of a one-electron system the initial bound-state wave function can be written as $|\psi_i\rangle = P_{nl}(r)|lm\rangle$. The expression P_{nl} gives the radial behavior of wave function and $|lm\rangle$ represents the angular part; n, l, and m are the principal, azimuthal, and magnetic quantum numbers of the initial state respectively, and r the position coordinate of the electron. The final continuum state of the atom is not a definite angular momentum state, and it is essential that it has the correct asymptotic form of a plane wave plus incoming spherical waves. On this basis, it is possible to write for the continuum state

$$|\psi_f(\mathbf{k})\rangle = 4\pi \sum_{l',m'} (i)^{l'} e^{-i\xi_{l'}} Y^*_{l'm'}(\hat{k}) Y_{l'm'}(\hat{r}) R_{\epsilon l'}(r) \qquad (7.17)$$

where $\hat{k}$ and $\hat{r}$ represent unit vectors along the directions of the ejected electron and the position vector respectively, $Y_{l'm'}$ represents the spherical harmonics, $R_{\epsilon l'}(r) = P_{\epsilon l'}/r$, and $\xi_{l'}$ is the phase shift of the l'th partial scattered wave. Equation (7.17) can be written in a simplified form by representing $a(l', m') = 4\pi(i)^{l'} e^{-i\xi_{l'}} Y^*_{l'm'}(\hat{k})$. The dipole integral $\mathscr{R}_{l'}$ is given by $\mathscr{R}_{l'} = \int_0^\infty P_{nl} r P_{\epsilon l'} dr$. With these expressions if one calculates the photoelectron intensity $I(\theta)$ at an angle θ with respect to the direction of polarization and puts it in the form

$$I(\theta) = \frac{\sigma_{\text{tot}}}{4\pi} [1 + \beta P_2(\cos\theta)] \qquad (7.18)$$

where $P_2(\cos\theta) = \frac{1}{2}(3\cos^2\theta - 1)$, it turns out that for photoejection from the initial state with orbital angular momentum l, β is given by

$$\beta = \{l(l-1)\mathscr{R}^2_{l-1}(\epsilon) + (l+1)(l+2)\mathscr{R}^2_{l+1}(\epsilon)$$
$$- 6l(l+1)\mathscr{R}_{l+1}(\epsilon)\mathscr{R}_{l-1}(\epsilon) \cos[\xi_{l+1}(\epsilon) - \xi_{l+1}(\epsilon)]\}$$
$$\{(2l+1)[l\mathscr{R}^2_{l-1}(\epsilon) + (l+1)\mathscr{R}^2_{l+1}(\epsilon)]\}^{-1} \qquad (7.19)$$

It is easy to check that for $l = 0$ (s electron) $\beta = 2$. For electron with $l \neq 0$ two competing outgoing channels are always present, and the interference between $l - 1$ and $l + 1$ partial waves determines the form of the anisotropy. The interference is sensitive to both the difference $\xi_{l+1} - \xi_{l-1}$ and the relative magnitude of the dipole integrals $\mathscr{R}_{l+1}$ and $\mathscr{R}_{l-1}$.

β FOR MULTIELECTRON ATOMS

For a multielectron atom it may be assumed that both the initial and final states are described satisfactorily by Russell–Saunders coupling and that the total orbital angular momenta and spin angular momenta are well defined quantities in both states. Slater determinants of spin orbitals — antisymmetric with interchange of space and spin coordinates — may be used for both initial and final states. Central field approximation could be used to calculate the radial wave function and phase shifts. Correlation effects can be incorporated by using linear combinations of possible configurations.

In the LS coupling notation, the properly antisymmetrized initial-state wave function is given by

$$\psi_i = \frac{1}{\sqrt{N!}} \sum_P (-1)^P \psi(LM_L SM_s) \qquad (7.20)$$

where P is the permutation operator and $L, M_L, S,$ and M_s respectively stand for the total orbital angular momentum, its z component, total spin angular momentum, and its z component. If the notation $L''M_L''S''M_s''$ is used to describe the final ionic state when the photoelectron ejection takes place, then the final-state wave function ψ_f of an N-electron atom can be expressed as a product of the ionic and free electron wave functions:

$$\psi_f = \frac{1}{\sqrt{N!}} \sum_P \sum_{l',m'} (-1)^P P a(l',m') \psi_{L''M_L''S''M_s''} \phi_{l'm's'm_s'} \qquad (7.21)$$

where $\phi_{l'm's'm_s'}$ represents the spin orbital for the free electron

$$\phi_{l'm's'm_s'} = \delta(\sigma'|m_s') Y_{l'm'}(\hat{r}) R_{el'}(r) \qquad (7.22)$$

σ denotes the spin coordinate of the electron, $Y_{l'm'}(\hat{r})$ and $R_{el'}(r)$ are the angular and radial wave functions. To obtain the differential cross section, a summation over the quantum numbers m_s, M_s'', and M_L'' is needed. The calculation of the matrix element is similar to that in the one-electron case. The dipole transition operator involves summation over all the electrons.

The results of such calculations show that the angular distribution of photoelectrons in an $l \to l'$ electron jump in a multielectron atom whose states are described by LS coupling, is equivalent to the angular distribution for an $l \to l'$ electron jump in a one-electron atom. In the multiconfigurational situation, the different one-electron angular distribution weighted by the square of the configuration-mixing coefficients are to be superimposed. For example, photoionization of the 3s electron from the Mg ground state $1s^2 2s^2 2p^6 3s^2\ ^1S_0$ leads to a $\cos^2\theta$ distribution. If configuration mixing is included, then mixing between configurations such as $1s^2 2s^2 2p^6 3p^2\ ^1S_0$ and $1s^2 2s^2 2p^6 3d^2\ ^1S_0$ is effected and the angular distribution may not any longer be a $\cos^2\theta$ distribution — the degree of departure depends on the nature and extent of configuration mixing. Similar conclusions are obtained if photoelectron transitions between different fine-structure levels of the atom, characterized by the total angular momentum J of the system, are considered. It turns out that the angular distribution from individual fine-structure levels for an $l \to l'$ transition is the same as the $l \to l'$ angular distribution in a one-electron atom — the internal structure of a many-electron atom does not alter the form of the angular distribution.

The above comments are applicable to dipole transition. A multipole of order

l causes a $P_l(\cos\theta)$ dependence of the angular distribution. Further, interference terms between various multipoles occur, and these make the angular distribution dependent also on the direction of light propagation. But as the ratio of the magnitude of the magnetic dipole and electric quadrupole to the electric dipole term is approximately equal to the ratio of the velocity of the photoelectron to the velocity of light, in photoelectron spectroscopy such effects are normally negligible.

Let us now look at the results of a recent theoretical study of angular distribution of photoelectrons from O, N, and C. The atomic photoelectrons originate in a photoionization transition from an initial state having q electrons $(nl)^q$ and designated by the term ^{2S+1}L to a final state $[(nl)^{q-1}\ ^{2S_c+1}L_c]$, $(\epsilon l')\ ^{2S+1}L'$; nl are the principal and azimuthal quantum numbers of the atomic as well as ionic core electrons; l' is the final photoelectron angular momentum; L, L_c, L' the total orbital angular momentum; and S, S_c, S the total spin angular momentum of the neutral atom, residual core ion, and the final state of core ion plus photoelectron respectively. For unpolarized radiation the differential cross section can be written

$$\frac{d\sigma_i(\epsilon)}{d\Omega} = \frac{\sigma_i(\epsilon)}{4\pi}\left[1 - \frac{\beta_i(\epsilon)}{2}P_2(\cos\alpha)\right] \quad (7.23)$$

where α is the angle between the incident beam and the direction of photoelectron ejection and σ_i is the total cross section at photoelectron energy ϵ. As in Chapter 6, σ_i can be expressed as

$$\sigma_i(\epsilon) = \sum_{L'}\sum_{l'=l-1}^{l+1}\sigma_{nl}(LS, L_cS_c, \epsilon l'L') \quad (7.24)$$

σ_{nl} being

$$\sigma_{nl}(LS, L_cS_c, \epsilon l'L') = \frac{4\pi^2\alpha a_0^2}{3g_i}(I+\epsilon)\frac{1}{4l_>^2 - 1}\zeta(LS, L_cS_c, l'L')\gamma|\mathscr{R}_{l'}(\epsilon)|^2 \quad (7.25)$$

Here g_i is the statistical weight of the initial state, a_0 the Bohr radius, α is the fine-structure constant, I is the experimental ionization potential of the ^{2S+1}L state relative to the residual ion core expressed in Rydbergs, $l_>$ is the greater of l and l', and ζ is the relative multiplet strength. Equation (7.25) assumes that the wave functions are satisfactorily approximated by antisymmetrized single particle wave functions P_{nl}^i and P_{nl}^f. The overlap integral γ and the radial dipole matrix element $\mathscr{R}_l(\epsilon)$ are, respectively

$$\gamma = \prod_{\substack{\text{passive}\\ \text{electrons}}}\left|\int_0^\infty P_{nl}^i(r)P_{nl}^f(r)\,dr\right|^2 \quad (7.26)$$

and

$$\mathscr{R}_{l'}(\epsilon) = \int_0^\infty P_{nl}^i(r) r P_{\epsilon l'}^f(r) \, dr \tag{7.27}$$

Continuum wave functions $P_{\epsilon l'}^f$ are normalized per unit energy range;

$$P_{\epsilon l'}^f(r) \xrightarrow[r \to \infty]{} \pi^{-1/2} \epsilon^{-1/4} \sin(\epsilon^{1/2} r + \epsilon^{-1/2} \ln(2\epsilon^{1/2} r) - \tfrac{1}{2} l'\pi + \eta_{l'} + \delta_{l'}) \tag{7.28}$$

here $\eta_{l'}$ is the coulomb phase shift $\eta_{l'} = \arg \Gamma(l' + 1 - i\epsilon^{-1/2})$, and $\delta_{l'}$ is the phase shift with respect to the coulomb waves.

For the LS coupled antisymmetric products of single-particle Hartree–Fock functions, β_i would be a sum over the β's for each of the possible values of angular momentum j_t transferred to the atom by the photon, with proper weightage by the relative cross section:

$$\beta_i = \frac{\sum_{j_t} \beta(j_t) \sigma(j_t)}{\sum_{j_t} \sigma(j_t)} \tag{7.29}$$

If the parity change of the target is $(-1)^{j_t}$, such transitions are said to be parity-favored transitions. $\beta(j_t)_{\text{fav}}$ depends on the magnitude of j_t and photoionization amplitude $\bar{S}_\pm(j_t)$, $S_\pm$ refers to $l' = j_t \pm 1$, and the magnitude depends on coulomb phase shift η and the various radial dipole matrix elements defined above. If, however, the parity change of the target is $-(-1)^{j_t}$, the transition is said to be parity unfavored. For parity unfavored transfers $\beta(j_t)_{\text{unf}} = -1$ (which implies a $\sin^2\theta$ distribution) independently of the dynamics. In the case of photoionization of the oxygen atom, $O(2p^4)^3 P \to {}^2D$, to cite a specific example, β_i reduces to

$$\beta_i = 2 \frac{\mathscr{R}_d^2 - 2 \mathscr{R}_s \mathscr{R}_d \cos(\xi_d - \xi_s)}{2\mathscr{R}_d^2 + \mathscr{R}_s^2} \tag{7.30}$$

where the $\mathscr{R}$'s are the radial dipole matrix elements for d and s waves and $\xi_{l\pm 1} = \delta_{l\pm 1} + \eta_{l\pm 1}$.

The calculated cross sections of photoionization from the O ^3P state using Hartree–Fock length, Hartree–Fock velocity, and Hartree–Slater formulations are shown in Fig. 7.3. The angular distribution is weakly dependent on cross section; the major contribution is from the phase shifts of the continuum waves. Figure 7.4 shows β variation as a function of photoelectron energy.

The variation of β with ϵ happens to be mostly due to the ϵ dependence of $\xi_d - \xi_s$, due to d and s waves. $\xi_d - \xi_s$ consists of a contribution due to non-coulomb shift $\delta_d - \delta_s$, and a coulomb phase shift difference $\eta_d - \eta_s$, given by

$$\eta_d - \eta_s = -\tan^{-1}\frac{1}{2\sqrt{\epsilon}} - \tan^{-1}\frac{1}{\sqrt{\epsilon}} \tag{7.31}$$

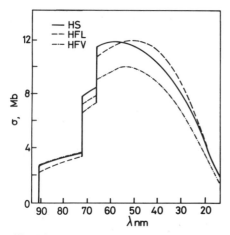

Fig. 7.3. Photoionization cross section of $O(^3P) \to O^+(^4S, {}^2D, {}^2P) + e$ using Hartree-Slater, Hartree-Fock-length, and Hartree-Fock-velocity wave functions [660].

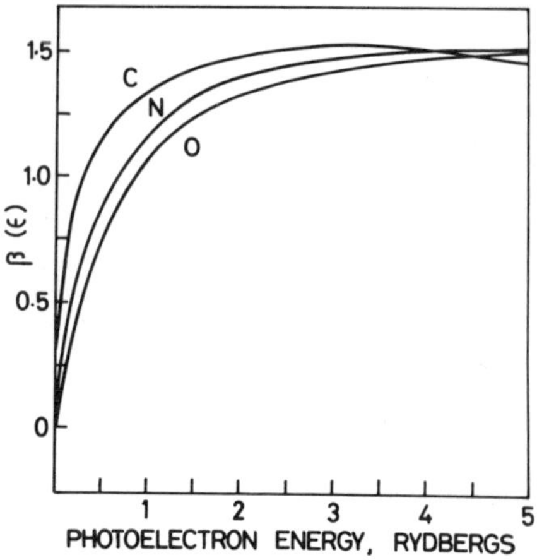

Fig. 7.4. Asymmetry parameter $\beta(\epsilon)$ for photoionization of atomic O, N, and C for photoionization from 2p level using Hartree-Slater wave functions [660].

For $\epsilon = 0$, $\eta_d - \eta_s = -\pi$ and $\delta_d - \delta_s \simeq -\pi$, so the total phase shift difference is $\sim -2\pi$. This makes the numerator of Eq. (7.30) and hence β close to zero at threshold. At energy $\epsilon = 1$ Ry (Rydberg), the only significant change is in the coulomb phase shift difference, which is $\sim -\pi/2$, hence $\xi_d - \xi_s = -3\pi/2$, making $\cos(\xi_d - \xi_s) = 0$ and $\beta \simeq 1$. At higher values of ϵ, both the matrix elements and the phase shifts do not change significantly, hence β does not change much. β values of nitrogen and carbon atom obtained using Hartree–Slater approximation are also shown in Fig. 7.4. These results are quite close to oxygen, since all three have very similar dipole matrix elements and continuum phase shifts.

β IN AUTOIONIZATION

If there are autoionization resonances, the effects of those forces which are ordinarily weak in nonresonant photoionization become pronounced and the photoelectron angular distribution can undergo significant changes as a function of electron energy. We consider here the case of autoionization associated with the photoionization process of Xe $5p^6$.

A photoionization process in which a neutral species A is converted into ion A^+ can be written

$$A(J_0, \pi_0) + \gamma(j_\gamma = 1, \pi_\gamma = -1) \to A^+(J_c, \pi_c) + e[lsj, \pi_e = (-1)^l]$$

(7.32)

where A denotes the target particle with angular momentum J_0 and parity π_0, γ represents the photon with angular momentum 1 and parity -1, and A^+ is the ion with angular momentum J_c and parity π_c. The l, s, and j stand respectively for orbital, spin, and total angular momentum of the photoelectron, and π_e represents its parity. All angular momenta are expressed in units of $\hbar$.

Conservation of total angular momentum J and parity π means that

$$\mathbf{J} = \mathbf{J}_0 + \mathbf{j}_\gamma = \mathbf{J}_c + \mathbf{s} + \mathbf{l} \tag{7.33}$$

and

$$\pi = \pi_0 \pi_\gamma = \pi_c \pi_e = -\pi_0 = \pi_c(-1)^l \tag{7.34}$$

If in the photoionization process $\mathbf{j}_\gamma$ is viewed as the angular momentum input to the system and $\mathbf{l}$ as the angular momentum output, then $\mathbf{j}_\gamma - \mathbf{l}$ may be considered as $\mathbf{j}_t$, the angular momentum transferred to the system. Using Eq. (7.33) one may then write

$$\mathbf{j}_t = \mathbf{j}_\gamma - \mathbf{l} = \mathbf{J}_c + \mathbf{s} - \mathbf{J}_0 \tag{7.35}$$

The various possible values of l in the photoionization process subject to the conservation conditions Eq. (7.33) and Eq. (7.34) constitute the alternative magnitudes of angular momentum transferred in the process.

The angular momentum transfer may be considered a probe of anisotropic electron–target interaction. We may consider the overall process of photoionization as consisting of an initial stage of photoabsorption followed by a second stage of escape of the photoelectron. In the first stage, the photon imparts a momentum of $\mathbf{j}_\gamma$ to the orbital momentum of the electron to yield a photoelectron with $\mathbf{l}'(\mathbf{l}' = \mathbf{j}_\gamma + \mathbf{l}_0$, where $\mathbf{l}_0$ is the initial orbital momentum of the electron), so that the angular momentum transferred in the first stage is $\mathbf{j}_{t'} = \mathbf{j}_\gamma - \mathbf{l}' = -\mathbf{l}_0$, that is, $j_{t'} = l_0$. In the final stage, owing to anisotropic interactions of the escaping electron with the rest of the target, additional angular momentum transfer can take place subject to the constraint of Eq. (7.34). As a result of any additional angular momentum transfer $\mathbf{k}'$, the electronic orbital momentum can change from $\mathbf{l}'$ to $\mathbf{l}$, and for the overall process one can then write $\mathbf{j}_t = \mathbf{j}_\gamma - \mathbf{l} = \mathbf{j}_{t'} - \mathbf{k}'$. There are two major features of autoionization in photoelectron angular distributions: the enhanced importance of alternative angular momentum transfers and the sharp spectral variation of β.

In xenon photoionization

$$\mathrm{Xe}(5p^6\ {}^1S_0\ J_0 = 0, \pi_0 = +1) + \gamma(j_\gamma = 1, \pi_\gamma = -1) \to \mathrm{Xe}^+(5p^5\ {}^2P^0_{3/2,\,1/2})$$
$$+ e(l = 0, 2) \tag{7.36}$$

two ionic states may be formed depending on the photon energy. The ionization potential $I_{3/2}$ for the formation of the ${}^2P^0_{3/2}$ state of the ion doublet ${}^2P^0_{3/2}$, ${}^2P^0_{1/2}$ is 12.127 eV and $I_{1/2}$, for the formation of the higher energy ${}^2P^0_{1/2}$ state, is 13.557 eV. The spectral region between these fine-structure thresholds consists of a Rydberg series of autoionization resonances. The autoionization process, consisting of the escape of an electron initially bound to the $^2P^0_{3/2}$ core, may be written as

$$\left.\begin{array}{l} 5p^5\,(^2P^0_{1/2})\,nd \\ 5p^5\,(^2P^0_{1/2})\,n's \end{array}\right\} \xrightarrow[J=1,\,\pi=-1]{\text{autoionization}} \left\{\begin{array}{l} 5p^5\,(^2P^0_{3/2})\,\epsilon d \\ 5p^5\,(^2P^0_{3/2})\,\epsilon s \end{array}\right.$$

Neutral Rydberg states Final ion and electron (7.37)

where the initial state comprises two series of Rydberg states. The Rydberg electron is either in a series of excited d states with principal quantum number n, or in a series of excited s states with the principal quantum number denoted by n'. The total angular momentum ($J = 1$) and parity ($\pi = -1$) are conserved, and the orbital angular momentum l of the photoelectron can undergo change during the photoionization process.

If photon energies below $I_{1/2}$ are considered, then the photoionization processes given by Eq. (7.36) form ions only in the $^2P^0_{3/2}$ state. The angular momentum and parity quantum numbers are therefore as follows:

Atom	Photon	Ion	Electron	
$J_0 = 0$	$j_\gamma = 1$	$J_c = 3/2$	$l = 0, 2$	(7.38)
$\pi_0 = +1$	$\pi_\gamma = -1$	$\pi_c = -1$	$\pi_l = +1$	

$J_c = 3/2$, because photon energies below $I_{1/2}$ are considered and only the $^2P^0_{3/2}$ are populated. $\pi_l = +1$ is determined by the requirement of parity balance as total $\pi = -1$; $\pi_l = +1$ means that l is even. $l = 0, 2$ are then fixed by the requirement of angular momentum balance, using the total $J = 1$.

From the above values of angular momenta, one obtains

$$j_t = 1 \quad \text{for } l = 0, 2$$
$$j_t = 2 \quad \text{for } l = 2$$

From these values, the angular momentum transferred $\mathbf{j}_t = \mathbf{j}_\gamma - 1$ can be calculated using the angular momentum addition procedure subject to the constraint $\mathbf{j}_t = \mathbf{J}_c + \mathbf{s} - \mathbf{J}_0$.

Since $\pi_0 \pi_c = -1$, odd j_t values ($l = 0, 2$) are parity favored. The $j_t = 2$ case is parity unfavored. The overall β can then be written in terms of the weighed average of these two processes, parity favored and parity unfavored, as

$$\beta = \frac{[\sigma(1)\beta(1) + \sigma(2)\beta(2)]}{\sigma(1) + \sigma(2)} \quad (7.39)$$

Using $\beta_{unf}(2) = -1$ from Eq. (7.29), one may write Eq. (7.39) as

$$\beta = \frac{\sigma(1)\beta(1) - \sigma(2)}{\sigma(1) + \sigma(2)} \quad (7.40)$$

The Rydberg character of the series $(nd, n's)$ of autoionization levels $5p^5(^2P^0_{1/2})nd$, $5p^5(^2P^0_{1/2})n's$ leading to the Xe $^2P^0_{3/2}$ ion can be seen if the spectral region above the Xe $^2P^0_{3/2}$ threshold and below that of Xe $^2P^0_{1/2}$ is mapped. This can be done by using an effective quantum number $\nu_{1/2}$ defined with respect to the upper threshold $I_{1/2}$ of the ion doublet:

$$h\nu = I_{1/2} - \frac{1}{2\nu^2_{1/2}} \quad (7.41)$$

where $h\nu$ is the photon energy, $h\nu$ and $I_{1/2}$ both now in atomic units (1 a.u. = 27.2 eV). Each pair of Rydberg levels $5p^5(^2P^0_{1/2})nd$ and $5p^5(^2P^0_{1/2})n's$ fits within one unit of $\nu_{1/2}$. The resonance profile keeps repeating with period one in $\nu_{1/2}$. The profile width and separation, however, decrease rapidly on the energy scale, from about 0.44 eV ($3.5 \leq \nu_{1/2} \leq 4.5$) just above the $^2P^0_{3/2}$ threshold to 0.005 eV ($20.5 \leq \nu_{1/2} \leq 21.5$) a little below the $^2P^0_{1/2}$ threshold, and to zero ($\nu_{1/2} = \infty$) at the $^2P^0_{1/2}$ threshold.

The results of β calculation for xenon are shown in Fig. 7.5 for one unit range of the effective quantum number $\nu_{1/2}$, from $\nu_{1/2} = 3.5$ to $\nu_{1/2} = 4.5$, from which the spectrum of the whole autoionization can be understood. This particular range of $\nu_{1/2}$ corresponds to $0.196 \leq \epsilon \leq 0.635$ eV, ϵ being the photoelectron

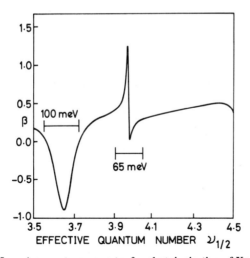

Fig. 7.5. Asymmetry parameter for photoionization of Xe. [652]

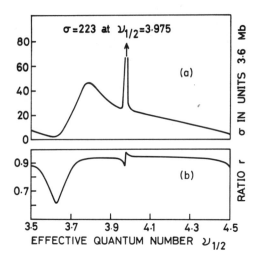

Fig. 7.6. (a) Integrated cross section in Xe photoionization $\sigma = \sigma(1) + \sigma(2)$. (b) Cross section ratio $r = \sigma(1)/\sigma$ [652].

kinetic energy. The integrated cross section is shown in Fig. 7.6. The broad maximum peaking at $\nu_{1/2} \simeq 3.75$ is primarily due to autoionization of $5p^5(^2P^0_{1/2})nd$. The autoionization of $5p^5(^2P^0_{1/2})n's$ is concentrated at the s resonance on the high energy shoulder of the broad peak at $\nu_{1/2} \simeq 3.975$.

The ratio $r = \sigma(1)/[\sigma(1) + \sigma(2)]$, which represents the strength of parity-favored ionization $\sigma(1)$, is also plotted in Fig. 7.6. s waves (i.e., $l = 0$) contribute only to $\sigma(1)$, therefore r also indicates the relative strength of s and d wave ($l = 2$) ionization. With $\nu_{1/2}$ increasing, β shows first a broad dip toward $\beta = -1(j_t = 2)$ and then a rapid oscillation across the s resonance region. The broad minimum in β at $\nu_{1/2} \simeq 3.64$ occurs when σ is a minimum and s wave ionization is negligible and is accompanied by a corresponding minimum of r. With σ rising through a broad d resonance maximum, parity-favored d wave ionization increases, and β increases away from -1. Around the s resonance region, the s and d wave ionization channels couple and produce sharp variations in the relative contributions of s and d wave ionization, resulting in rapid oscillation of β across the s resonance. Just below the s resonance, s wave ionization is depressed and d wave ionization is enhanced, and a sharp rise of β away from the s wave value $\beta = 0$ occurs. Then, s wave ionization rapidly increases to a peak at the s resonance maximum, resulting in β falling to near 0. Thereafter, on the high energy side, s wave ionization drops off, r drops and levels off, and β rises to a smooth plateau. This cycle repeats periodically in $\nu_{1/2}$, with period 1, up to the threshold of the $^2P^0_{1/2}$ level.

A similar β behavior is seen in all of the rare gases, but xenon has the broadest resonances; such resonances in β are to be expected whenever autoionization occurs. There is another autoionization resonance in rare gases at higher photon energies, in the region of $nsnp^6(^2S_{1/2})n'p \to ns^2np^5(^2P_0)[\epsilon s, \epsilon d]$ resonances of the Rydberg series, converging to the $nsnp^6$ $^2S_{1/2}$ level of the ion.

ANISOTROPIC ELECTRON-ION INTERACTION EFFECTS IN OPEN SHELL ATOMS

In certain cases of photoionization, anisotropic interaction effects between photoelectron and residual ion can show up in atomic photoelectron angular distributions. They manifest pronounced differences in the distribution of photoelectron groups corresponding to different LS term levels possible of the residual ion. For closed shell atoms, such effects may not be noticeable as the angular momentum and parity conservation impose severe restrictions, and in the absence of anisotropic interactions, j_t, the angular momentum transferred in the ionization, is restricted to the single value l_0, the photoelectron's initial orbital momentum.

Owing to a dynamical coupling of the orbital motion of the electron to the net orbital motion of the residual ion, the phase shift and the dipole matrix element, which determine the cross section and angular distribution, become dependent on the term level of the residual ion. The dynamical weights of the transition amplitudes are determined by the magnitude of this coupling, which would be different for different values of the orbital momentum of the residual ion. In the photoionization of sulfur, an open shell atom, $S(3p^4\ ^3P) + \gamma \to S^+(3p^3\ L_cS_c) + e(l = 0, 2)$, the effect of this coupling on β has been theoretically calculated. The permissible values of L_cS_c lead to the terms $^4S^0$, $^2D^0$ and $^2P^0$. Ionization to each of these terms can take place through $j_t = l_0 = 1$, both for $l = 0$ and $l = 2$. Further, when $l = 2$, the $^2P^0$ term can also result owing to $j_t = 2$, and both $j_t = 2$ and $j_t = 3$ are possible for the $^2D^0$ term. Because of the differences in the phase shifts that the alternative values of the total L might cause, a simple model where electron–ion interactions are not adequately incorporated, predictions of β may be erroneous. Results of calculations on S show that energy dependence of the asymmetry parameter β for the ^4S, ^2D, ^2P terms differ from one another; the difference is most pronounced near the region of Cooper minimum.

It has also been shown that the photoelectron angular distribution of s electrons in an open shell atom having outer shell configuration ns^2np^q deviates from $\beta = 2$ as is predicted by more approximate theories. Calculation on Cl atom photoionization corresponding to the two possible ionic terms ^1P and ^3P shows that one may expect oscillations in β as a function of energy because of very large effects of anisotropic electron-ion interactions. $\beta(^1P)$ and $\beta(^3P)$

vary differently; deviations of β from 2 are very large near threshold and tend towards 2 from lower values slowly with increase of photoelectron energy.

DIFFERENTIAL CROSS SECTION AND β FOR MOLECULES

We now turn to the derivation of differential photoionization cross section and β expressions employing orthogonalized plane wave (OPW) approximation, an approach that has been used for several molecules. An expression for photoionization cross section can be derived from Fermi's "golden rule rate w." From time-dependent perturbation theory, the following expression may be written for w, the number of transitions to the final state per unit time:

$$w = \frac{2\pi}{\hbar} |\langle k| \mathcal{H}' |m\rangle|^2 \rho(k) \tag{7.42}$$

where $|m\rangle$, $|k\rangle$ denote respectively the initial and final states, $\mathcal{H}'$ represents the perturbation to the total Hamiltonian $\mathcal{H}_0$ of the system, $\rho(k)$ is the density of the final states, that is, $\rho(k)dE_k$ is the number of final states with energies between E_k and $E_k + dE_k$. With $\mathcal{H}' = -(e/mc) \mathbf{A} \cdot \mathbf{P}$, where $\mathbf{A}$ is the vector potential of the electromagnetic field and $\mathbf{P}$ the linear momentum operator, we have, from Eq. (7.42):

$$w = \frac{2\pi e^2 A_0^2}{\hbar m^2 c^2} \left| \mathbf{u} \cdot \left\langle k \left| \sum_n \mathbf{P}_n \right| m \right\rangle \right|^2 \rho(k) \tag{7.43}$$

where outside the matrix element, e, m, c are respectively the electronic charge, electronic mass, and the velocity of light. Here A_0 is the magnutide of the vector potential, $\mathbf{u}$ is the unit vector in the direction of polarization, and $\mathbf{P}_n$ is the linear momentum operator for the nth electron. The energy absorbed by the system is given by $w\hbar\omega$, where $\hbar\omega$ is the energy difference between the initial and final states. This quantity may be equated with the product of the intensity of the radiation expressed by the magnitude of the Poynting vector ($|\mathbf{S}| = \omega^2 A_0^2 / 2\pi c$) and the cross section σ. Thus

$$\sigma \frac{\omega^2 A_0^2}{2\pi c} = w\hbar\omega \tag{7.44}$$

σ, through w in Eq. (7.43), contains information about the photoelectron angular dependence. Dividing σ by 4π steradians we obtain the photoionization cross section per unit solid angle, that is, the differential cross section:

$$\frac{d\sigma}{d\Omega} = \frac{\sigma}{4\pi} \tag{7.45}$$

We thus get, from Eqs. (7.43), (7.44) and (7.45):

$$\frac{d\sigma}{d\Omega} = \frac{\pi e^2}{m^2 c\omega} \left| \mathbf{u} \cdot \left\langle k \left| \sum_n \mathbf{P}_n \right| m \right\rangle \right|^2 \rho(k) \tag{7.46}$$

For simplicity, let us assume an ionization process that involves only an electronic transition. The ground state $|m\rangle$ of the molecule can be approximated by a Slater determinant of doubly occupied orthonormal molecular orbitals ϕ_l. The final state $|k\rangle$ can be taken to be a linear combination of two Slater determinants giving a spin singlet in which one electron has been excited from the molecular orbital ϕ_j to an unbound OPW orbital. The OPW can be described as

$$|\text{OPW}\rangle = N \left[|\text{PW}\rangle - \sum_l^{\text{occ}} \langle l|\text{PW}\rangle |l\rangle \right] \tag{7.47}$$

where N is the normalization constant, given by

$$N = \left[1 - \sum_l^{\text{occ}} \langle l|\text{PW}\rangle\langle \text{PW}|l\rangle \right]^{-1/2} \tag{7.48}$$

and $|\text{PW}\rangle$ represents a plane wave orbital

$$|\text{PW}\rangle = L^{-3/2} e^{i\mathbf{k}\cdot\mathbf{r}} \tag{7.49}$$

that has been normalized in a cubic box of length L. The corresponding density of states is given by

$$\rho(k) = \frac{mkL^3}{2\pi^2\hbar^2} \tag{7.50}$$

where k on the right side of Eq. (7.50) represents the magnitude of the wave vector.

If the shape of the other MO's in the molecule remain the same, the matrix element of Eq. (7.43) on the basis of the final-state orbital defined in Eq. (7.47) can be written

$$\left\langle k \left| \sum_n \mathbf{P}_n \right| m \right\rangle = 2^{1/2} \left[\langle j|\mathbf{P}|\text{PW}\rangle - \sum_l^{\text{occ}} \langle j|\mathbf{P}|l\rangle\langle l|\text{PW}\rangle \right] \tag{7.51}$$

where $|l\rangle$ is one of the doubly occupied MO's and $\langle l|\text{PW}\rangle$ is the overlap integral of $\langle l|$ with the unbound orbital $|\text{PW}\rangle$. In Eq. (7.51), if $|\text{PW}\rangle$ is orthogonal to occupied MO's, then the matrix element reduces to simple $2^{1/2}\langle j|\mathbf{P}|\text{PW}\rangle$ since the second term is zero. Further, since the PW is an eigenfunction of the linear momentum $\mathbf{P} = -i\hbar\nabla$ with an eigenvalue $\hbar k$, Eq. (7.51) becomes

$$\left\langle k \left| \sum_n \mathbf{P}_n \right| m \right\rangle = 2^{1/2}\hbar \left[\mathbf{k}\langle j|\text{PW}\rangle + i \sum_l^{\text{occ}} \langle j|\nabla|l\rangle\langle l|\text{PW}\rangle \right] \tag{7.52}$$

If we now take the scalar product of the left side of Eq. (7.52) with the unit vector u along the direction of polarization, we have

$$\mathbf{u} \cdot \left\langle k \left| \sum_n \mathbf{P}_n \right| m \right\rangle = \mathbf{u} \cdot \mathbf{P}_{km} = 2^{1/2} \hbar \left[k \cos \theta_{ku} \langle j|\text{PW}\rangle + i \sum_l^{\text{occ}} \mathbf{u} \cdot \langle j|\boldsymbol{\nabla}|l\rangle \langle l|\text{PW}\rangle \right] \tag{7.53}$$

where θ_{ku} is the angle between the propagation vector k of the ejected electron and the polarization vector u of photon. Substituting the value of $\mathbf{u} \cdot \mathbf{P}_{km}$ from Eq. (7.53) and the value of $\rho(k)$ from Eq. (7.50) in Eq. (7.46), we have

$$\frac{d\sigma}{d\Omega} = \frac{e^2 k L^3}{2\pi m c \hbar^2 \omega} \left\{ 2^{1/2} \hbar \left[k \cos \theta_{ku} \langle j|\text{PW}\rangle + i \sum_l^{\text{occ}} \mathbf{u} \cdot \langle j|\boldsymbol{\nabla}|l\rangle \langle l|\text{PW}\rangle| \right] \right\}^2 \tag{7.54}$$

If the molecules are randomly oriented, Eq. (7.53) must be averaged over all space defined by Euler angles. This yields

$$\overline{|\mathbf{u} \cdot \mathbf{P}|^2} = 2\hbar^2 (\bar{S}_1 + \bar{S}_2 + \bar{S}_3 + \bar{S}_4 + \bar{S}_5) \tag{7.55}$$

where each $\bar{S}_n$ represents a sum of several terms in the cross section expression. In Eq. (7.55) the first two sums $\bar{S}_1 + \bar{S}_2$ correspond to the plane wave approximation; the terms $\bar{S}_3 + \bar{S}_4 + \bar{S}_5$ are from orthogonalization of the plane wave. The sums $\bar{S}_n$ in Eq. (7.55) can all be expressed in the general form

$$\bar{S}_n = t_n \cos^2\theta_{ku} + u_n + \tfrac{1}{2} v_n (3 \cos^2\theta_{ku} - 1) \tag{7.56}$$

where $u_n = v_n = 0$ for $n = 1, 2$, and 5, and $t_n = 0$ for $n = 3$ and 4.

The angular dependence of differential photoionization cross section can be written from Eq. (7.1) as

$$\frac{d\sigma}{d\Omega} = \frac{\sigma_{\text{tot}}}{4\pi} [1 + \tfrac{1}{2}\beta(3\cos^2\theta_{ku} - 1)] \tag{7.57}$$

This can be simplified to

$$\frac{d\sigma}{d\Omega} = a + b \cos^2\theta_{ku} \tag{7.58}$$

and, in terms of a and b, the asymmetry parameter β can be defined as

$$\beta = \frac{2b}{3a + b} \tag{7.59}$$

From Eq. (7.57), by substituting $\theta_{ku} = 0$ and $\theta_{ku} = \pi/2$ we get for parallel and normal angular distribution respectively:

$$\left(\frac{d\sigma}{d\Omega}\right)_{\parallel} = \frac{\sigma_{\text{tot}}}{4\pi}(1 + \beta) \qquad \text{for } \mathbf{k} \parallel \mathbf{u} \tag{7.60}$$

and
$$\left(\frac{d\sigma}{d\Omega}\right)_\perp = \frac{\sigma_{tot}}{4\pi}[1-\tfrac{1}{2}\beta] \quad \text{for } \mathbf{k}\perp\mathbf{u} \qquad (7.61)$$

In the OPW calculations, the corresponding equations can be obtained by using Eq. (7.56) as follows:

$$\left(\frac{d\sigma}{d\Omega}\right)_\parallel \propto [t_1 + t_2 + t_5 + u_3 + u_4 + v_3 + v_4] \propto (1+\beta) \qquad (7.62)$$

and

$$\left(\frac{d\sigma}{d\Omega}\right)_\perp \propto [u_3 + u_4 - \tfrac{1}{2}(v_3 + v_4)] \propto (1-\tfrac{1}{2}\beta) \qquad (7.63)$$

Thus

and
$$\begin{aligned} a &\propto u_3 + u_4 - \tfrac{1}{2}(v_3 + v_4) \\ b &\propto t_1 + t_2 + t_5 + \tfrac{3}{2}(v_3 + v_4) \end{aligned} \qquad (7.64)$$

Using these values of a and b in Eq. (7.58), we find that the values of β lie between -1 and $+2$. Here $\beta = -1$ represents a pure $\sin^2\theta_{ku}$ distribution, $\beta = 0$ corresponds to a spherical distribution, and $\beta = 2$ signifies a pure $\cos^2\theta_{ku}$ distribution. The total cross section can be calculated from the expression

$$\sigma_{tot} = \frac{4\pi}{3}(3a+b) \qquad (7.65)$$

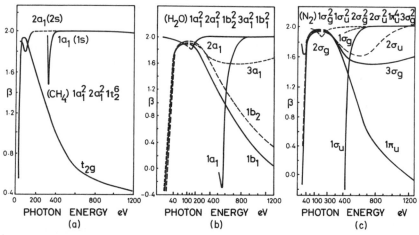

Fig. 7.7. Asymmetry parameter β as a function of photon energy for the occupied orbitals (a) CH_4, (b) H_2O, and (c) N_2 [684].

by using the values of a and b given by Eq. (7.64), using $4e^2 kL^3/mc\omega$ as the proportionality constant.

If the light used is unpolarized, the average over all polarization angles must be taken. For such cases the differential cross section can be written

$$\frac{d\sigma}{d\Omega} = \frac{\sigma_{tot}}{4\pi}[1 - \tfrac{1}{4}\beta(3\cos^2\alpha - 1)] \qquad (7.66)$$

Here α is the angle between the unpolarized photon beam and the ejected photoelectron. This equation is also applicable to OPW calculations. The normal differential cross section, applicable when photoelectrons are collected at right angles to the unpolarized photon beam, $\alpha = \pi/2$, can be written

$$\left[\frac{d\sigma}{d\Omega}\right]_\perp = \frac{\sigma_{tot}}{4\pi}(1 + \tfrac{1}{4}\beta) \qquad (7.67)$$

Values of β for the occupied orbitals of CH_4, H_2O, and N_2 calculated as a function of photon energy using the OPW method are presented in Fig. 7.7.

DIFFERENTIAL CROSS SECTION OF VIBRATIONAL LEVELS

With the availability of energy analyzers having high enough resolution to show detailed vibrational structure, the vibrational transition probabilities associated with photoionization transitions have been measured. Attempts have been made to match these with theoretical Franck–Condon factors to derive information about the nuclear coordinates of ionic species. A limitation of this technique is that it yields only approximate changes in nuclear coordinates, and its accuracy depends on the degree of invariance of the electronic transition moment across the vibrational progression. In this section, we consider a recent work which, for the calculation of photoionization cross sections of diatomic molecules, takes into account variations of the electronic transition moment as a function of both internuclear distance and photoelectron kinetic energy.

As in Eq. (7.46), a general expression for the differential photoionization cross section in the electric dipole approximation for an initial state ψ_0 and a final state ψ_j can be written

$$\frac{d\sigma}{d\Omega} = \frac{\pi e^2}{m^2 c\omega}\left|\mathbf{u}\cdot\left\langle\psi_0\left|\sum_n \mathbf{P}_n\right|\psi_j\right\rangle\right|^2 \rho(\psi_j) \qquad (7.68)$$

Here $\rho(\psi_j)$ is the density of final states and $\mathbf{P}_n$ is the momentum operator $-i\hbar\nabla$; $\sum_n \mathbf{P}_n$ represents summation over all n electrons. The transition moment between the initial $\psi_0(q,Q)$ and the final $\psi_j(q,Q)$ states is given by

$$\mathbf{P}_{0j} = \left\langle \psi_0(q, Q) \middle| \sum_n \mathbf{P}_n \middle| \psi_j(q, Q) \right\rangle \tag{7.69}$$

Here q and Q refer respectively to electronic and nuclear coordinates. The terms arising due to summation of the operator over nuclei vanish due to the orthogonality of the electronic components of $\psi_0(q, Q)$ and $\psi_j(q, Q)$, and are therefore not included. Using the Born–Oppenheimer approximation and writing the total wave function as a product of electronic $\psi_e(q, Q)$, vibrational $\psi_v(Q)$, and rotational $\psi_r(Q)$ wave functions, and further assuming that the rotational wave functions are uncoupled – which permits explicit integration over rotational coordinates – one has from Eq. (7.69):

$$\mathbf{P}_{0j} = \left\langle \psi_{0e}(q, Q) \psi_{0v}(Q) \middle| \sum_n \mathbf{P}_n \middle| \psi_{je}(q, Q) \psi_{jv}(Q) \right\rangle \tag{7.70}$$

For a pure adiabatic electronic transition at fixed-ground-state nuclear positions Q_0 and with fixed photoelectron energy ϵ_0, if one recognizes that $\mathbf{P}_n$ operates only on electronic coordinates, the differential cross section of a vibrational band can be written as

$$\begin{aligned}
\frac{d\sigma_v}{d\Omega} &= \frac{e^2}{2m^2 vc} \left| \mathbf{u} \cdot \left\langle \psi_{0e}(q, Q_0) \middle| \sum_n \mathbf{P}_n \middle| \psi_{je}(q, Q_0) \right\rangle \right|^2 |\langle \psi_{0v}(Q) | \psi_{jv}(Q) \rangle|^2 \rho(\psi_j) \\
&= \frac{d\sigma}{d\Omega}(Q_0, \epsilon_0) F_v
\end{aligned} \tag{7.71}$$

where F_v denotes the square of the vibrational overlap integral $|\langle \psi_{0v}(Q) | \psi_{jv}(Q) \rangle|^2$. The initial state $\psi_{0e}(q, Q_0)$ may be approximated by a Slater determinant of doubly occupied orthonormal MO's ϕ_l, and the final state $\psi_{je}(q, Q_0)$ can be approximated by a linear combination of two Slater determinants giving a spin singlet in which an electron has been excited from the MO ϕ_j to an unbound Schmidt OPW orbital $|PW(k)\rangle = L^{-3/2} e^{i\mathbf{k}\cdot\mathbf{r}}$, k being the wave vector of the emitted electron. Calculations show that $d\sigma(Q_0, \epsilon_0)/d\Omega$ provides an approximation to the total integrated intensity of a photoelectron transition. The intensities of various vibrational bands are then determined by the relative magnitude of the Franck–Condon factors for the individual bands.

In order to explore the dependence of the electronic transition moment on Q, the Q can be retained as such in Eq. (7.69) and integration can be performed over q and Q. This is difficult to work out, however, because for most molecules there is a lack of knowledge on the nature of wave function variation with Q. As an alternative method, the dependence of the transition moment on Q can be considered in terms of the r-centroid. The r-centroid is defined by $\bar{r}_{v0vj} = \langle \psi_{0v} | r | \psi_{jv} \rangle / \langle \psi_{0v} | \psi_{jv} \rangle$, which may be interpreted as the weighed average or the expectation value of r experienced by a molecule in both states of a transition $\psi_{jv} \leftarrow \psi_{0v}$. Using the r-centroid, the transition probabilities can be evaluated at the r-centroid coordinates. Equation (7.71) can be written as $d\sigma_v/d\Omega = (d\sigma(\bar{r}_{v0vj}, \epsilon_0)/d\Omega)F_v$. Furthermore, for a given energy of the incident photon, various vibrational bands will have different photoelectron energies. The combined effect of electron kinetic energy and variation of nuclear coordinate can be determined by calculating the cross section as a function of both $\bar{r}_{v0vj}$ and ϵ_k, that is, by calculating $d\sigma_v/d\Omega = [d\sigma(\bar{r}_{v0vj}, \epsilon_k)/d\Omega] F_v$.

Calculations have been carried out by assuming that the bound orbitals are identical in the initial and final states and that the unbound orbital is a plane wave orthogonalized to all occupied bound orbitals. INDO molecular wave functions have been used because they require less computation time. They have the disadvantage, however, of not being orthogonalized to the 1s core orbitals and thus not useful for calculations involving high-energy photons like X-rays. For UV photoionization, this problem is negligible. Besides the incident photon energy, other essential inputs to the calculation are ionization energies and appropriate molecular coordinates of the state concerned. For the calculation of Franck–Condon overlaps and r-centroids the vibrational wave functions used are the Morse oscillators, which are given by

$$\psi_v = (\gamma/N_v)^{1/2} \exp\left(\frac{-Y}{2}\right) Y^{\alpha/2} L_v^\alpha(Y) \qquad (7.72)$$

where

$$N_v = \sum_{s=0}^{v} \Gamma \frac{s+\alpha}{s!} \qquad (7.73a)$$

$$\alpha = (\omega_e/\omega_e x_e) - 2v - 1 \qquad (7.73b)$$

$$\gamma = 1.2177 \times 10^7 (4\mu\omega_e x_e)^{1/2} \qquad (7.73c)$$

$$Y = \frac{\omega_e}{\omega_e x_e} \exp[-\gamma(r - r_e)] \qquad (7.73d)$$

and the generalized Laguerre polynomials are given by

$$L_v^\alpha(Y) = \sum_{s=0}^{v} \frac{(-1)^{v-s}\Gamma(\alpha+1+v)Y^{v-s}}{s!(v-s)!\Gamma(\alpha+v+1-s)} \qquad (7.73e)$$

In the above set of conditions, r_e is the equilibrium internuclear distance, ω_e is the harmonic vibrational frequency, $\omega_e x_e$ the anharmonicity constant of the given electronic state, and μ is the reduced mass of the species. Three different types of calculation have been performed: (1) for $d\sigma/d\Omega(Q_0, \epsilon_k)$ with Q fixed and using appropriate ϵ_k for each vibrational band, (2) ϵ_k kept fixed and

$d\sigma(\bar{r}_{v0vj}, \epsilon_0)/d\Omega$ calculated for a wide range of r values, and (3) $d\sigma_v/d\Omega = (d\sigma(\bar{r}_{v0vj}, \epsilon_k)/d\Omega)F_v$ evaluated for each vibrational transistion $v_j \leftarrow v_0$ with $\bar{r}_{v0vj}$ and ϵ_k used corresponding to that vibrational transition and INDO wave functions determined at $r = \bar{r}_{v0vj}$. All bound orbitals are assumed identical in the initial and final states, and for the unbound orbital OPW approximation is used. Such calculations have been carried out for N_2 and H_2, and experimental and theoretical results have been compared.

A plot of the ratio of the experimental vibrational band intensity to the Franck–Condon factors is shown in Fig. 7.8. If the experimental transition moment remains constant and the intensity changes are merely reflections of changes in Franck–Condon factors, a line with ratio = 1 is to be expected. All the plots, however, show positive slopes and indicate that the transition moment increases across the progression. For the N_2 $^2\Sigma_g^+$, $^2\Pi_u$ and $^2\Sigma_u^+$ bands, an inspection of the calculated $d\sigma(Q_0, \epsilon_k)/d\Omega$ values — that is, the cross section at a fixed internuclear distance Q_0 and variable photoelectron energy ϵ_k (the data are not shown here) — indicates that the cross section increases across the envelope of the $^2\Sigma_g^+ \leftarrow {}^1\Sigma_g^+$ transition and decreases for the $^2\Pi_u \leftarrow {}^1\Sigma_g^+$ and $^2\Sigma_u^+ \leftarrow {}^1\Sigma_g^+$ transitions. This apparently anomalous trend suggests that in the cross section versus ϵ_k curves, the ϵ_k's of these bands are on the opposite sides of the maximum. For the same reason, in photoionization of H_2 to $^2\Sigma_g^+$ using He II excitation, an increase in the value of $d\sigma(Q_0, \epsilon_k)/d\Omega$ across the progression is observed, whereas with He I excitation a decrease is noted.

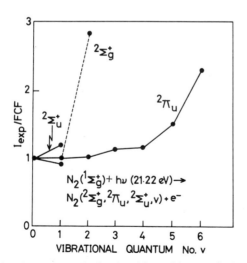

Fig. 7.8. The ratio of experimental vibrational intensities to calculated Franck–Condon factors for N_2 [591].

Table 7.4. Calculated Differential Photoionization Cross Sections $d\sigma(\bar{r}_{v0vj}, \epsilon_k)/d\Omega$ and Total Transition Probabilities $(d\sigma(\bar{r}_{v0vj}, \epsilon_k)/d\Omega)F_v$, $(d\sigma(Q_0, \epsilon_k)/d\Omega)F_v$, and $(d\sigma(\bar{r}_{v0vj}, \epsilon_0)/d\Omega)F_v$ for N_2 and H_2 [591][a]

Transition	v	$\bar{r}_{v0vj}$ (nm)	$(d\sigma/d\Omega)(\bar{r}_{v0vj}, \epsilon_k)$	Franck–Condon Factor (F_v)	Experimental Vibrational Intensity	$(\bar{r}_{v0vj}, \epsilon_k)$	$(d\sigma/d\Omega)F_v$ (Q_0, ϵ_k)	$(\bar{r}_{v0vj}, \epsilon_0)$
$N_2\ ^2\Sigma_g^+$	0	0.1112	1.474	0.9023	100.0	100.0	100.0	100.0
(He I)	1	0.1013	1.554	0.0906	6.94 ± 0.66	10.6	10.6	10.3
	2	0.0925	1.620	0.0065	0.32 ± 0.21	0.85	0.75	0.73
$N_2\ ^2\Pi_u$	0	0.1139	8.492	0.2447	87.1 ± 3.1	76.1	81.6	76.1
(He I)	1	0.1114	8.784	0.3107	100.0 ± 2.8	100.0	100.0	100.0
	2	0.1090	9.046	0.2253	76.3 ± 1.5	74.5	70.0	72.7
	3	0.1068	9.302	0.1236	43.5 ± 2.9	42.2	37.3	40.0
	4	0.1047	9.546	0.0574	19.1 ± 2.4	19.9	16.8	18.7
$N_2\ ^2\Sigma_u^+$	0	0.1091	1.682	0.8912	100.0	100.0	100.0	100.0
(He I)	1	0.1186	1.384	0.1070	9.84 ± 2.51	9.9	9.6	9.5
$H_2\ ^2\Sigma_g^+$	0	0.0895	2.181	0.0921	40 ± 3	42.4	45.7	42.4
(He II)	1	0.0847	2.305	0.1750	83 ± 3	84.9	88.1	86.3
	2	0.0803	2.422	0.1959	100	100.0	100.0	100.0
	3	0.0763	2.530	0.1670	96 ± 3	89.0	83.4	87.8
	4	0.0726	2.631	0.1274	81 ± 4	70.6	66.7	69.2
	5	0.0691	2.727	0.0873	65 ± 4	50.2	46.2	48.5
	6	0.0659	2.815	0.0566	47 ± 4	33.6	30.3	32.2
$H_2\ ^2\Sigma_g^+$	0	0.0895	10.21	0.0921	46.3	47.3	48.3	47.3
(He I)	1	0.0847	10.20	0.1750	86.0	89.8	90.6	88.5
	2	0.0803	10.15	0.1959	100.0	100.0	100.0	100.0
	3	0.0763	10.06	0.1670	88.4	84.5	83.9	85.9
	4	0.0726	9.948	0.1274	70.0	63.7	62.9	65.9
	5	0.0691	9.811	0.0873	57.9	43.1	42.3	45.5
	6	0.0659	9.651	0.0566	47.2	27.5	26.9	29.6

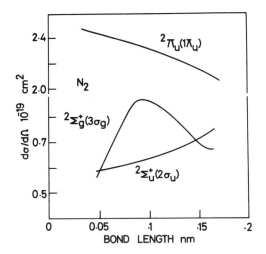

Fig. 7.9. Calculated differential cross sections for N_2 as a function of bond length. The orbital ionized is shown in parentheses [591].

A plot of $(d\sigma(r, \epsilon_0)/d\Omega)$ versus r is shown in Fig. 7.9. The behaviors of the three nitrogen curves are different; primarily this reflects the nature of variation of the INDO wave function with r. It is found that the differential cross section $d\sigma(r, \epsilon_0)/d\Omega$ of the molecular orbitals follows the coefficient of the atomic orbital that has the highest cross section. In the low kinetic energy region, the $2s$ atomic orbital cross section is higher than that of the $2p$, and thus the differential cross section tracks the $2s$ atomic orbital coefficient. In the $2\sigma_u$ molecular orbital, the curve reflects the increase in the value of $2s$ coefficients. In the $3\sigma_g$ molecular orbital, the $2s$ coefficients increase, go through a maximum, and then decrease. In the $1\pi_u$ molecular orbital, with increasing r, the $2p_x$ and $2p_y$ characters of the molecular orbital begin to separate, leading to pure $2p_x$ and $2p_y$ orbitals, and the cross section approaches that of a nitrogen $2p$ atomic orbital. For H_2, the character of molecular orbital does not depend on the bond length; therefore, the variation of $d\sigma(r, \epsilon_0)/d\Omega$ is entirely due to the variation in r. This r dependence appears in the argument of the spherical Bessel functions which are used to expand the plane wave. For H_2, $d\sigma(r, \epsilon_0)/d\Omega$ for $^2\Sigma_g^+(1\sigma_g)$ shows a substantial change from 0.3×10^{-18} cm^2 to 0.16×10^{-18} cm^2 for a bond length change from 0.04 to 0.12 nm. This shows that the cross section may have a strong dependence on r although the molecular orbital wave functions remain unchanged.

The electronic cross section $d\sigma(\bar{r}_{v0vj}, \epsilon_k)/d\Omega$ presented in Table 7.4 shows an increase across all of the vibrational envelopes. Exceptions are the N_2 $^2\Sigma_u^+$ and H_2 $^2\Sigma_g^+$ bands with $h\nu = 21.22$ eV, which probably reflect the inadequacies of

OPW approximation for photoelectrons of low kinetic energy. Photoionization of H_2 by $h\nu = 40.82$ eV shows the largest variation in cross section, with about 40% increase across the band envelope. Other than these two exceptions, intensities predicted by Franck–Condon factors for high vibrational quantum number are lower than vibrational band intensities, and the total transition probability $d\sigma(\bar{r}_{v0vj}, \epsilon_k)/d\Omega)F_v$ predictions of intensities are definitely better than from the Franck–Condon factors alone.

ANGULAR DISTRIBUTION DATA AND MOLECULAR ORBITAL PARAMETERS

Drastic simplifications are necessary for the calculation of cross sections and β values in the photoionization of molecules, and one has to resort to approximations. Formulas have been derived for differential cross sections of molecules in which the cross section has been factored into an angular part and a radial part. The angular part is calculated by use of angular momentum algebra. For the radial part, explicit expressions for each partial wave component of both the initial molecular orbital and the photoelectron are required. We now look at one method that does this by using a single-center partial wave expansion of the molecular orbitals. The theory provides a procedure to invert experimental angular distribution data to determine the parameters of the initial molecular orbital.

In a photoionization process involving the ejection of an electron with momentum $\hbar \mathbf{k}$ and the change of electron orbital angular momentum by one unit ($l' = l \pm 1$, where l is the electron angular momentum in the bound state and l' that in the free state), $I(\theta)$ the angular distribution of the photoelectron and β the asymmetry parameter, in the united atom limit, have the following properties.

Ionization from an s orbital:

$$ns(l = 0) \rightarrow kp(l' = 1), I(\theta) \sim \cos^2 \theta, \beta = 2.0 \qquad (7.74)$$

Ionization from a p orbital:

$$np(l = 1) \rightarrow ks(l' = 0), I_1(\theta) \sim 1, \beta = 0.0 \qquad (7.75)$$

$$np(l = 1) \rightarrow kd(l' = 2), I_2(\theta) \sim (1 + 3\cos^2 \theta), \beta = 1.0 \qquad (7.76)$$

$I(\theta) = I_1(\theta) + I_2(\theta) +$ an interference term between the $s(l' = 0)$ and $d(l' = 2)$ partial waves proportional to $\cos(\xi_0 - \xi_2)$, where ξ_0 and ξ_2 are the phase shifts for the $l = 0$ and $l = 2$ partial waves

(7.77)

Ionization from a d orbital:

$$nd\,(l = 2) \to kp\,(l' = 1), \quad I_1(\theta) \sim (3 + \cos^2 \theta), \quad \beta = 0.2 \tag{7.78}$$

$$nd\,(l = 2) \to kf\,(l' = 3), \quad I_2(\theta) \sim (1 + 2\cos^2 \theta), \quad \beta = 0.8 \tag{7.79}$$

$I(\theta) = I_1(\theta) + I_2(\theta) +$ an interference term between the $p(l' = 1)$ and $f(l' = 3)$ partial waves proportional to $\cos(\xi_1 - \xi_3)$, where ξ_1, ξ_3 are the phase shifts for the $l' = 1$ and $l' = 3$ partial waves (7.80)

Here θ is the angle between the unit vector of polarization of the radiation and the direction of the ejected electron.

In terms of Born theory, cross sections of orbitals depend on both k and an effective charge parameter ζ. The dependence is characteristic of the orbital from which the electron is lost, and if $k \gg \zeta$, one obtains the following functional dependence:

$$\sigma_s \sim \zeta_s^5 k^{-7}, \quad \sigma_p \sim \zeta_p^7 k^{-9}, \quad \sigma_d \sim \zeta_d^9 k^{-11} \tag{7.81}$$

The magnitude of k is known from photoelectron kinetic energy ϵ ($\epsilon = \hbar^2 k^2/2m$). Therefore, determination of the cross section in the atomic limit as a function of electron energy provides information on the nature of dependence on k and hence on the symmetry of the orbital. When the symmetry is determined, measurement of the cross section at 54.73° yields the effective charge parameter ζ, since k dependence is already known.

For diatomic molecules, one may express the molecular orbitals, in the single-center partial wave expansion, in a parametric form:

$$\psi \sim s(\zeta_s) + \lambda_1 p(\zeta_p) + \ldots, \quad m = 0 \;(\sigma\text{-type}) \tag{7.82}$$

$$\psi \sim p(\zeta_p) + \lambda_1 d(\zeta_d) + \ldots, \quad m = \pm 1 \;(\pi\text{-type}) \tag{7.83}$$

$$\psi \sim s(\zeta_s) + \lambda_1 d(\zeta_d) + \ldots, \quad m = 0 \;(\sigma_g\text{-type}) \tag{7.84}$$

$$\psi \sim p(\zeta_p) + \lambda_1 f(\zeta_f) + \ldots, \quad m = 0 \;(\sigma_u\text{-type}) \tag{7.85}$$

Here ζ_j is the effective charge parameter appropriate to the orbitals, and λ_j is a mixing parameter. The labels s, p, d, and so on, designate orbitals having angular momentum quantum numbers $l = 0, 1, 2$, respectively, and m is the azimuthal quantum number. In the molecular case the dominant angular dependence will be that of the major contributor to the orbital, with minor departures due to the presence of other components. The extent of these departures depends on the behavior of individual components. From this, the λ_1 contributions can be determined. Just as measurements of β can provide knowledge about the mixing parameter, measurements of cross section as a function of energy can yield information about the effective charge parameters. Furthermore, since the effective charge parameter in diatomic molecules changes with the internuclear distance – for example, for $H_2^+ \sigma_g$ molecular orbital $1 \leq \zeta_j \leq 2$ ($\zeta_j = 1$ in the separated atom limit and $\zeta_j = 2$ in the united atom limit) – a study of σ and β as a function of the initial vibrational state of the molecule can provide information on the effective charge parameter as a function of the internuclear distance.

The sequence of steps to evolve orbital information from angular distribution data would be as follows. For a given system, a molecular orbital is first assumed and a set of λ_j's obtained from β values as a function of energy. Once the λ_j's are known, the cross section – determined by measurements of intensities at 54.73° – may be used to find ζ_j's. This may be accomplished by fitting the cross section data in an equation of a suitable form. For example a $\zeta_j^{m-2} k^{-m}$ type of dependence should give for Eq. (7.82) a cross section dependence of the form

$$\sigma \sim \zeta_s^5 k^{-7} + \lambda_1 \zeta_s^{5/2} \zeta_p^{7/2} k^{-8} + \lambda_1^2 \zeta_p^7 k^{-9} \qquad (7.86)$$

Finally, λ_j's and ζ_j's as a function of internuclear distance can be obtained by measurements as a function of vibrational states. The inversion of data to obtain orbital properties from β values is possible if the ejected electron wave function can be obtained from theory. For this purpose, the initial molecular orbital wave function must be taken in the form of single-center partial wave expansion as indicated above, and for the photoelectron continuum wave function, consecutive use of the distorted-wave Born and Coulomb–Born approximations can be made.

Applying this to tetrahedral CH_4 and noting that ionization can occur from molecular orbitals of a_1 or t_2 symmetry, the symmetry orbitals of the T_d point group can be written:

$$\psi_{a_1} \sim s_C + \lambda(s_1 + s_2 + s_3 + s_4) \qquad (7.87)$$

$$\psi_{t_2,x} \sim p_{x,C} + \lambda(s_1 - s_2 + s_3 - s_4) \qquad (7.88)$$

$$\psi_{t_2,y} \sim p_{y,C} + \lambda(s_1 + s_2 - s_3 - s_4) \qquad (7.89)$$

$$\psi_{t_2,z} \sim p_{z,C} + \lambda(s_1 - s_2 - s_3 + s_4) \qquad (7.90)$$

where $s_C, p_{x,C}, p_{y,C}, p_{z,C}$ are the orbtials for the $n = 2$ shell of the carbon atom and s_j's are $1s$ orbitals centered on protons. These have the following forms:

$$s_C \sim S_{1s,2s} e^{-\zeta_1 r} - r e^{-\zeta_2 r} \qquad (7.91)$$

$$p_{x,C} \sim r e^{-\zeta_2 r} \sin\theta \cos\phi \qquad (7.92)$$

$$p_{y,C} \sim r e^{-\zeta_2 r} \sin\theta \sin\phi \qquad (7.93)$$

$$p_{z,C} \sim r e^{-\zeta_2 r} \cos\theta \tag{7.94}$$

$$s_j \sim e^{-r_j} \tag{7.95}$$

$S_{1s,2s}$ is a coefficient chosen to orthogonalize s_C to C $1s$, ζ_1 and ζ_2 are effective charge parameters chosen to be $\zeta_1 = 5.75$ and $\zeta_2 = 1.63$. The λ is taken as 0.6, which is compatible with a larger electronegativity of carbon. The differential cross sections for photoionization from a_1 and t_2 orbitals are shown in Figs. 7.10 and 7.11. For $l_1 = 2$ and $l_1 = 0$, the results are not different because of the very small permanent quadrupole moment for molecules of T_d symmetry. (Here l_1 represents the orbital angular momentum quantum number for the direction of the momentum $\hbar k$ of the ejected electron measured in a molecule-fixed frame.) Only at $l_1 = 4$ is significant deviation from $\beta = 2$ noticed. Around $l_1 = 6$ the results converge at $\beta = 1.84$. For ionization from each triply degenerate t_2 orbital, the cross section has a major contribution from the $l_1 = 1$ or p radial component of the molecular orbital. The β value of 1.55 is obtained because of constructive interference in the dominant p orbital contribution.

At a high ratio of photoelectron velocity to the velocity of the bound electron, the motion of the ejected electron approaches the limit of motion in a spherically symmetric, screened coulomb potential. The screened coulomb potential can be replaced by a coulomb potential of strength Z, the latter taken as an adjustable parameter. Since at high energies the photoelectric conversion takes place with most efficiency near the nuclei, this is a good approximation with Z equal to the strength of the full nuclear charge, and then

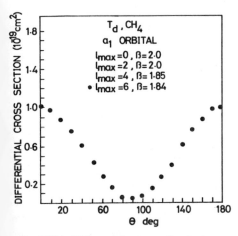

Fig. 7.10. Differential cross section for ionization from the a_1 orbital of CH$_4$ ($\bullet$ $l_{max} = 6$) [682].

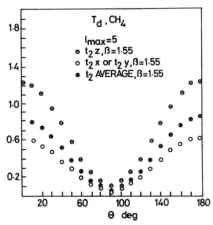

Fig. 7.11. Differential cross section for ionization from the t_2 orbitals of CH$_4$ [682].

correction can be applied for screening effects to first order in the Coulomb–Born series. With this formulation, β can be obtained in analytic form and one can study the effects of Z, k, the velocity of the photoelectron, and the charge and anisotropy parameters of the jth component of the molecular orbital, expanded about the center of mass. As a further improvement, one can have a numerical or Coulomb–Born solution of the partial wave equations of the ejected electron in an appropriate screened coulomb potential, thus eliminating Z as an adjustable parameter.

MEAN FREE PATH ESTIMATES FROM ANGLE-RESOLVED PHOTOEMISSION FROM SOLIDS

Whereas the angular distribution of photoelectrons from gaseous samples yields β values that can be correlated with the orbital from which the photoelectron is ejected, photoelectron angular distribution from solid samples, in one application, can be used to determine inelastic mean free paths λ in adsorbates and in the substrate. For determination of λ, information on the angular dependence and a value of mean thickness is needed. The presence of any islands on the sample surface causes large modification of the angular dependence.

If the solid sample is rotated about an axis perpendicular to both the direction of the photon flux and the direction of observation of the photoelectrons, then the intensity of a photoelectron line from a homogeneous adsorbate of thickness d is

$$I_a(\epsilon, \theta) = C(\epsilon)\lambda_a(\epsilon)\left[1 - \exp\left(-\frac{d}{\lambda_a(\epsilon)\cos\theta}\right)\right] \quad (7.96)$$

where θ is the angle of observation with respect to the normal to the surface. The intensity of a line from the substrate at energy ϵ', on the other hand, is

$$I_s(\epsilon', \theta) = C'(\epsilon')\lambda_s(\epsilon')\exp\left(-\frac{d}{\lambda_a(\epsilon')\cos\theta}\right) \quad (7.97)$$

The C and C' contain the respective differential cross sections that introduce the energy dependence, and these can be eliminated from Eq. (7.96) and Eq. (7.97) by measurements on reference samples of the corresponding materials. For this purpose, one reference sample is taken where for the adsorbate layer $d \gg \lambda_a$. This results in a contribution of unity from the term in the square brackets of Eq. (7.96) for I_a^{ref}. Another reference sample of the substrate with clean surface is taken ($d = 0$), which makes the exponential term in Eq. (7.97) equal to unity for I_s^{ref}. Therefore, the relative intensities can be written

$$\frac{I_a}{I_a^{\text{ref}}}(\epsilon,\theta) = 1 - \exp\left(-\frac{d}{\lambda_a(\epsilon)\cos\theta}\right) \tag{7.98}$$

and

$$\frac{I_s}{I_s^{\text{ref}}}(\epsilon',\theta) = \exp\left(-\frac{d}{\lambda_a(\epsilon')\cos\theta}\right) \tag{7.99}$$

and are independent of experimental parameters if the measurements are carried out under identical conditions. It is obvious from Eqs. (7.98) and (7.99) that measurements of I_a/I_a^{ref} or I_s/I_s^{ref} at fixed angle yield λ, provided d is known. A mean value of d can be obtained, for example, by the method of the frequency variation of a quartz crystal.

Measurements of the angular distribution and the application of either Eq. (7.96) or Eq. (7.97) provide d/λ_a from the shape alone, and the normalization procedure mentioned above is unnecessary. When adjacent photoelectron lines are taken, $\epsilon \simeq \epsilon'$ and the energy dependence of λ_a may be neglected. Under such conditions, one obtains from Eqs. (7.96) and (7.97):

$$\frac{I_s}{I_a}(\epsilon,\theta) = \frac{C'\lambda_s}{C\lambda_a}\left[\exp\left(\frac{d}{\lambda_a\cos\theta}\right) - 1\right]^{-1} \tag{7.100}$$

In the limiting case when $d \ll \lambda_a \cos\theta$, Eq. (7.100) can be simplified to give

$$\frac{I_s}{I_a} = \frac{C'\lambda_s}{Cd}\cos\theta \tag{7.101}$$

Even in the presence of a second layer of another substance on the surface, these two relationships remain valid.

If the adsorbate is not homogeneous but consists of islands of thickness $\bar{d}/\alpha$, where $\bar{d}$ represents the mean value of the thickness and α is the fraction of the surface covered ($\bar{d}/\alpha = d$), and further if $\bar{d}$ is small compared with the linear extension of an island, then the relative intensities of Eqs. (7.97) and (7.98) can be expressed as

$$\frac{I_a}{I_a^{\text{ref}}} = \alpha\left[1 - \exp\left(-\frac{d}{\alpha\lambda_a\cos\theta}\right)\right] \tag{7.102}$$

and

$$\frac{I_s}{I_s^{\text{ref}}} = (1-\alpha) + \alpha\exp\left(-\frac{d}{\alpha\lambda_a\cos\theta}\right) \tag{7.103}$$

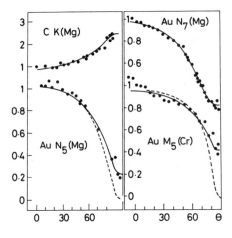

Fig. 7.12. Photoelectron angular distribution, normalized at 0°, from a gold sample with carbon contamination. Symbols in parentheses show the type of x-ray source. Solid lines represent best fits with Eqs. (7.102) and (7.103). The dashed lines correspond to the distribution without islands. [731].

Both λ_a and α can be determined simultaneously from relative intensity measurements. However, if measurements are also made at $\theta = 0$, it is possible to determine λ_a and α with only one sample.

Some of the results obtained in an experiment with a gold reference sample having a surface carbon layer are shown in Fig. 7.12, with intensities normalized at $\theta = 0$. The lines drawn through the experimental points were obtained from best fits of Eqs. (7.102) and (7.103) (solid lines) and Eq. (7.97) (dashed lines). The situation at larger angles is modified owing to the influence of islanding. Similar results have been obtained for carbon contamination on a nickel reference sample. The $\lambda/\bar{d}$, in the range 400–4560 eV, has an energy dependence of $\epsilon^{0.46}$. The model describes the distribution from the surface and from the

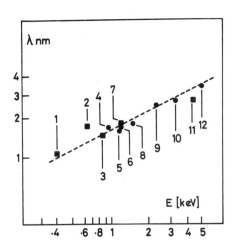

Fig. 7.13. Inelastic mean free path of electrons in gold as a function of kinetic energy. The numbers refer to various photoelectron lines. Sample thickness is 1.1 nm. The numbers 1, 4, 6, and 7 represent Mg 1253.6 eV excitation; 2, 5, and 8, Al 1486.6 eV excitation; 3 is a Ni Auger line; 9, Ti 4510.9 eV excitation; and 10, 11, and 12 are for Cr 5414.7 eV excitation. ■ Ni, ● Au [731].

bulk equally well, and to the extent that the model has been tested it has been found that it is independent of the nature of the substrate.

Results for the electron inelastic mean free path in gold at different energies are shown in Fig. 7.13. The values have been obtained from Eqs. (7.98) and (7.99) at $\theta = 0$ and assuming that the gold layer is homogeneous. The figure shows an $\epsilon^{0.5}$ energy dependence (dashed line). However, if the layer is homogeneous, the angular distribution should have a cosine shape as an upper limit, in terms of Eqs. (7.100) and (7.101). If it is assumed that islands exist, a better agreement is obtained with Eqs. (7.102) and (7.103). The resulting values are: $\alpha = 0.57$ and $\lambda_{Au}(1 \text{ keV}) = 0.9$ nm. This value of λ is approximately a factor of 2 smaller than the value obtained by assuming a homogeneous gold layer. The best fit for energy dependence is obtained with $\epsilon^{0.80}$.

SURFACE SENSITIVITY ENHANCEMENT AT GRAZING ELECTRON EXIT ANGLES AND GRAZING X-RAY INCIDENCE ANGLES

The mean depth of no-loss photoemission from a solid, which gives a measure of the distance most of the electrons are able to travel in the solid and still emerge through the surface, varies as a function of θ, where θ denotes the direction of photoelectron emergence from the surface. If λ_e is assumed to be a direction-independent property of the material, then this mean depth at an angle θ is given by $\lambda_e \sin \theta$. Thus, if θ decreases from 90° to 5° the mean depth decreases by a factor of 6, hence a corresponding enhancement of emission should result. A decrease of mean depth at low angles of emission also means that under that condition most of the emitted electrons are from shallower depths, that is, the surface region is accentuated. In an elemental substrate with a surface coating of oxide, as the photoemission angle is reduced one may expect an enhancement of photoemission intensity from the surface layers,

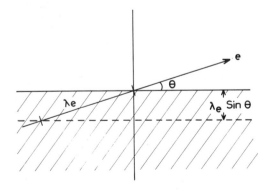

the effect becoming particularly significant at grazing angles of emergence of photoelectrons. This technique has often been used for increasing surface sensitivity. In a way, it provides a parallel method to the etching-free depth profiling technique mentioned in Chapter 5, which employs the electron energy dependence of escape depths.

At low angles of emergence, surface plasmon losses manifest more prominently. A decrease in the angle of emergence decreases the mean depth of emission, thereby increasing the probability of exciting surface plasmons. From the intensity of surface plasmon peaks caused by photoelectrons from adsorbates, one can thus trace the extent to which penetration of the adsorbate has taken place. At the surface the coordination number of the surface atoms is less; therefore, for solid sample species like transition metals the d band widths should change. Such d-band narrowing has been observed in Cu, when measurements are made at low emission angles, that is, when photoemission from the surface region is emphasized.

Another interesting enhancement effect is observed when very low X-ray incidence angles are used. At angles $\lesssim 1°$ the mean X-ray penetration depth (10^2–10^4 nm for $\phi \gg 1°$, where ϕ is the angle of incidence with respect to the surface) decreases markedly and attains values on the order of electron attenuation length λ_e. At low incidence angles X-rays cannot penetrate the substrate much due to increase of refraction; thereby near-surface photoemission is accentuated. The onset of enhancement as a function of ϕ is sharp, as has been observed in an Si $2p$ (oxide)/Si $2p$ (element) ratio with varying oxide layer thicknesses on silicon. Surface roughness shadows are a critical factor in such studies, and proper surface preparation is essential. Also, the measurement and interpretation of low ϕ data is less simple than measurements at low θ values.

SYMMETRY SELECTION RULES AND SINGLE CRYSTAL EFFECTS IN ANGLE-RESOLVED PHOTOEMISSION

In a solid that is exposed to a radiation of energy $h\nu$, the magnitude of a one-electron optical transition cross section depends on the matrix element $\langle \psi_{\mathbf{k}}^f | \mathbf{A} \cdot \nabla | \psi_{\mathbf{k}} \rangle$, where $\psi_{\mathbf{k}}$ and $\psi_{\mathbf{k}}^f$ present respectively the initial and final state Bloch functions of the crystal, and $\mathbf{A}$ is the vector potential corresponding to the radiation field. From the translational symmetry properties of Bloch functions it follows that the matrix element is nonzero only when the following relationship holds:

$$\mathbf{k}^f = \mathbf{k} + \mathbf{g} \qquad (7.104)$$

where $\mathbf{k}^f$ and $\mathbf{k}$ are respectively the wave vectors of the final and initial states of the electron, and **g** is a reciprocal lattice vector. In Eq. (7.104) we have assumed that $\mathbf{k}^f$ undergoes negligible change as the photoelectron escapes from the surface (i.e. $\mathbf{k}^f = \mathbf{K}$ where **K** is the wave vector outside the solid surface when surface perturbation is taken into account). Transitions that obey selection rules such as Eq. (7.104) are called *direct* (or **k** conserving) transitions and those that do not, *nondirect* transitions.

A direct transition may be viewed as a vertical transition in the reduced zone scheme. A nondirect transition, where phonons participate, may be considered to consist of a vertical transition by the absorption of a photon followed by a transition to an appropriate upper band minimum by the emission or absorption of a phonon of requisite energy. Because of phonon absorption, **k** changes and Eq. (7.104) is no longer valid. The direct transitions are dominant at low temperature. At high temperatures, phonon-assisted nondirect transitions become important. The latter may sometimes obscure direct transition effects in photoelectron spectra of most materials even at room temperatures. Nondirect processes also become important at high photon intensities. Thus, both temperature and photon energy determine the ratio between the direct and nondirect transitions.

When photon energy is high, such as in XPS experiments, the photon wave vector $\mathbf{k}_\nu$ also needs to be included in the selection-rules Eq. (7.104). Then one has

$$\mathbf{k}^f = \mathbf{k} + \mathbf{g} + \mathbf{k}_\nu \qquad (7.105)$$

At the photon energy of Al K_α radiation, 1486.6 eV, typical values for the magnitudes of the wave vectors for valence electron emission are

$$|\mathbf{k}^f| \simeq \frac{2\pi}{\lambda_e} \simeq 197\,\text{nm}^{-1} \qquad |\mathbf{k}| \lesssim 20\,\text{nm}^{-1}, \qquad |\mathbf{k}_\nu| = \frac{2\pi}{\lambda} \simeq 7\,\text{nm}^{-1},$$

$$(7.106)$$

where λ is the photon wavelength and λ_e the de Broglie wavelength associated with the excited photoelectron.

The photoelectron kinetic energy ϵ in terms of the magnitude k^f of the final state wave vector $\mathbf{k}^f$ can be written $\epsilon = \hbar^2 k^{f2}/2m$. In an angle-resolved photoemission experiment, it is useful to resolve the wave vector $\mathbf{k}^f$ into two components, $\mathbf{k}^f_\parallel$ and $\mathbf{k}^f_\perp$, that is, parallel and perpendicular with respect to the photoemitting surface. This can be written

$$\mathbf{k}^f = \mathbf{k}^f_\parallel + \mathbf{k}^f_\perp \qquad (7.107)$$

From Eq. (7.104) $\mathbf{k}^f_\parallel$ can be expressed as

$$\mathbf{k}^f_\parallel = \mathbf{k}_\parallel + \mathbf{g}_\parallel \qquad (7.108)$$

where $g_\parallel$ is the parallel component of the reciprocal lattice vector $\mathbf{g}$. Equation (7.108) means that the wave vector parallel to the surface is conserved. The perpendicular component $\mathbf{k}_\perp^f$ of the wave vector, $\mathbf{k}_\perp^f = \mathbf{k}_\perp + \mathbf{g}_\perp$, however, is indeterminate. Whereas the photoelectron ϵ is

$$\epsilon = \frac{\hbar^2}{2m}(k_\parallel^{f2} + k_\perp^{f2}) = E_f(\mathbf{k}) - E_{\text{vac}} \qquad (7.109)$$

(E_{vac} is the vacuum level), in the *three-step model* involving (1) electron excitation inside the solid, (2) its transport to the surface, and (3) escape through the surface, the constraint of electron escape only requires that the perpendicular component of energy is positive, that is:

$$\frac{\hbar^2}{2m} k_\perp^{f2} > 0 \qquad (7.110)$$

Since the right side of Eq. (7.110) can have any positive value, this means that $\mathbf{k}_\perp$ is not uniquely defined. Because of this uncertainty in the normal momentum of electrons in the solid, it is not usually possible to derive the full energy-band structure of the solid by measurement of angle-resolved photoelectron energy distribution curves. The uncertainty in $\mathbf{k}_\perp$ also causes momentum broadening, which affects structure in the photoemission energy distribution.

In XPS valence electron spectra, anisotropies in photoemission intensities have been observed in many single crystals, such as in Au, Ag, Cu, Pt, and in the layer compounds MoS_2, $GaSe_2$ and $SnSe_2$. Two models have been proposed, the *direct transition model* and the *plane wave matrix element model*. In the direct transition model, Eq. (7.104) is used rigorously. The magnitude of $\mathbf{k}^f$ is determined from the photoelectron kinetic energy $\epsilon = \hbar^2 k^{f2}/2m$. Since the energy width of a valence band is small, at XPS energies the range of $\mathbf{k}^f$ is narrow over a valence band; for example, in Au with $a = 0.408$ nm, $\mathbf{k}^f$ is in the range $12.84\,(2\pi/a) \leqslant \mathbf{k}^f \leqslant 12.88\,(2\pi/a)$, with $(2\pi/a)$ the reduced-zone radius. The direction of $\mathbf{k}^f$ is experimentally determined with respect to crystal orientation. To this, a correction $-\mathbf{k}_\nu$ is applied because of the photon. This $(\mathbf{k}^f - \mathbf{k}_\nu)$ determines uniquely the set of $\mathbf{k}$ values in the reduced zone from which allowed transitions can occur. If it is assumed that the magnitude of the matrix element is the same for all $\mathbf{k} \rightarrow \mathbf{k}^f$ transitions, then an angle-resolved photoelectron spectrum can be predicted which would be proportional to the density of electronic states in the allowed $\mathbf{k}$ region. This theory satisfactorily explains several detailed angle-resolved photoelectron spectra; a notable one is that of Cu in the photon energy range 40–200 eV. In the case of Au, a few discrepancies are noted for certain emission directions. Theory predicts large changes, whereas the experiment shows otherwise — this could be due to some reduced zone averaging or phonon smearing. At lower temperatures, this direct transition model is expected to be a better approximation.

The plane wave matrix element model starts with the k-conservation model of Eq. (7.104) and additionally assumes that final-state complexities enter the photoemission process in a way that causes all k values in the reduced zone to contribute to emission in any given direction. It is also assumed that the angle-dependence in XPS valence spectra arises out of the characteristic nature and magnitude of directional matrix elements summed over all occupied initial states. A plane wave final state $\psi_{\mathbf{k}}^f(\mathbf{r}) = \exp(i\mathbf{k}^f \cdot \mathbf{r})$ and an LCAO initial state of the form

$$\psi_{\mathbf{k}}(\mathbf{r}) = \sum_{\mathbf{R}_i} \exp(i\mathbf{k} \cdot \mathbf{R}_i) \left[\sum_{\mu} C_{\mu \mathbf{k}} X_{\mu}(\mathbf{r} - \mathbf{R}_i) \right] \quad (7.111)$$

are used to calculate the matrix elements. Here $\mathbf{R}_i$ denotes the position of an atomic center in the lattice, $X_{\mu}(\mathbf{r} - \mathbf{R}_i)$ represents an atomic orbital centered at $\mathbf{R}_i$, $X_{\mu}(\mathbf{r}) = R_{\mu}(r) Y_{\mu}(\theta, \phi)$, and $C_{\mu \mathbf{k}}$ is an expansion coefficient. Examination of Eq. (7.111) shows that the initial state consists of a linear combination of Fourier transforms of various atomic orbitals. The angular dependence of these Fourier transforms in the $\mathbf{k}^f$ space and that of the atomic functions in the real space are the same. Therefore, one may write $X_{\mu}(\mathbf{k}^f) = f_{\mu}(k^f) Y_{\mu}(\theta_{\mathbf{k}^f}, \phi_{\mathbf{k}^f})$. The angles $\theta_{\mathbf{k}^f}, \phi_{\mathbf{k}^f}$ indicate the direction of $\mathbf{k}^f$, and $f_{\mu}(k^f)$ is a radial integral function only of the magnitude k^f. For a polarization direction $\mathbf{u}$, the matrix element has the form

$$|\langle \psi_{\mathbf{k}}^f | \mathbf{A} \cdot \mathbf{P} | \psi_{\mathbf{k}} \rangle|^2 \propto (\mathbf{u} \cdot \mathbf{k}^f)^2 | \sum_{\mu} C_{\mu \mathbf{k}} X_{\mu}(\mathbf{k}^f)|^2 \quad (7.112)$$

Dependence of the matrix element on the square of the second factor, which contains the spherical harmonics [i.e., $Y(\theta_{\mathbf{k}^f}, \phi_{\mathbf{k}^f})$ in $X_{\mu}(\mathbf{k}^f)$] indicates that orbital symmetry properties thus directly influence angle-resolved photoemission intensity. For example, contributions from the orbital $d_{x^2-y^2}$ may be expected to be maximum along the $\pm x$ and $\pm y$ directions. The results show good agreement in the valence spectra of Au, Cu, MoS_2, and $GaSe_2$. For predicting room temperature XPS results, though the validity of the free electron plane wave final states is questionable, this is still the preferred model.

CHEMISORPTION, SURFACE STATES, AND ANGLE-RESOLVED PHOTOEMISSION

In photoexcitation, the momentum k associated with an electron state is normally conserved. It has been observed that in Si and Ge a large fraction of electrons created in direct transitions retain their component of momentum parallel to the crystal surface $\mathbf{k}_{\parallel}$.

It is possible to determine a two-dimensional energy band diagram (E versus

$k_\parallel$) of surfaces by angle-resolved photoemission measurements. One proceeds as follows. A photoelectron spectrum from the sample is obtained by placing the detector at an angle θ with respect to the surface normal and orienting the sample along a certain crystal direction.

For a photoemission peak of kinetic energy ϵ, $k_\parallel$ and the initial-state energy E are evaluated as follows:

$$k_\parallel = \left(\frac{2m\epsilon}{\hbar^2}\right)^{1/2} \sin\theta \qquad E = h\nu - \epsilon - \xi \qquad (7.113)$$

where ξ represents ionization potential (to vacuum level) from the level with respect to which E is expressed (such as valence band maximum). Thus, one set of $\epsilon(\theta)$ data yields a pair of $k_\parallel, E$ values, that is, one point on the $E, k_\parallel$ diagram. By varying the angle θ, and from corresponding shifts in energy of peaks, the entire range of $E, k_\parallel$ variation is obtained. This can be repeated for various crystallographic directions. Such techniques have been applied for deriving energy band diagrams of layer compounds such as $TaSe_2$ and TaS_2.

The technique has also been applied to studies on saturation adsorption of Cl on cleaved Si(111). In the angle-resolved valence electron spectrum induced by ultraviolet radiation of photon energy 20 eV, the EDC along the ΓK direction shows the Cl-induced peaks in the range -10 to -3 eV relative to the valence band maximum E_V. There are two likely sites for Cl atom adsorption, one in which a Cl atom rests on the top of a surface Si atom, and the other in which it is in the space between three surface Si atoms. The former is a onefold covalent site involving the Si dangling orbital and the Cl p_z orbital; the latter is an ionic site of threefold symmetry and involves somewhat preferentially the $p_{x,y}$ orbitals in bonding.

There are two prominent peaks in the angle-resolved spectrum, one at -7.4 eV, and the other in the range of -6 to -4 eV. The former emission maximizes at $\theta = 0°$, therefore it indicates a p_z symmetry. The latter, which maximizes at higher θ values (30–50°), indicates $p_{x,y}$ orbitals, but overall the covalent site predominates. Two-dimensional band energy diagram for ΓKM and $\Gamma M\Gamma$ azimuth are shown in Fig. 7.14. The projection of the bulk Si band structure into the surface Brillouin zone is shown by the shaded area, and the experimentally determined Cl induced bands are shown by the lines b, π, σ, and s. Corresponding theoretical results of pseudopotential calculation using Cl atoms in the covalent site are also shown in the figure. Except that the $\pi-\sigma$ gap is relatively smaller and the s level is found to be of higher energy in the experiment, there is good agreement. The discrepancies could be due to the use of an overly strong Cl pseudopotential.

If this is taken as an indication of the essential correctness of the pseudopotential model, then the charge densities predicted by the model are instruc-

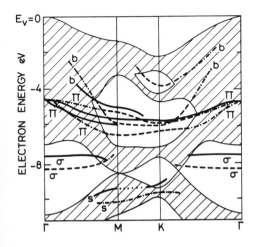

Fig. 7.14. Comparison of theoretically calculated and experimentally measured surface energy band structures for Cl adsorbed on a Si(111) surface. The continuous and dotted lines represent experimental values; the dashed and dot-dashed lines represent calculated values. The shaded area is a two-dimensional projection of the bulk Si band structure [710].

tive. Chlorine atom charge distribution corresponding to the peak $\sigma(-7.4\,\text{eV})$ shows that it is strongly p_z-like, the π peak (-5 to $-6\,\text{eV}$) is clearly $p_{x,y}$-like, the b states involve backbonding or transverse orbitals between the first and second atomic layers of Si, and s states are similar to σ but involve Si s rather than Si p_z. One could thus unravel a detailed picture of the chemisorptive bond.

Such $E(k_\parallel)$ analysis has also been applied to study surface states, such as in the (110) face of GaAs and the (111) face of Si.

ANGLE-RESOLVED PHOTOEMISSION FROM ORIENTED MOLECULES ON SINGLE CRYSTALS

In cases where the adsorbed molecules do not dissociate and strong adsorbate–substrate interactions are absent, the photoelectron spectrum of an adsorbate-covered surface is determined solely by the adsorbate molecule itself independent of the substrate. In a gas phase spectrum, averaging over all possible molecular orientation takes place, but in a solid, the adsorbate spectrum may be due exclusively to a preferred orientation of the adsorbate molecules on the substrate. In that case, the photoelectron angular distribution from the surface will be a characteristic of the orientation of the molecule. It will also depend on the polarization direction of the radiation used, the latter affecting the $\mathbf{A}\cdot\mathbf{P}$ term of the dipole matrix element, and on photoelectron energy, as the final-state wave function would change accordingly. Recently there have been calculations on such oriented molecules (e.g., CO, N_2) with specific orientation and polarization direction. Figure 7.15 gives polar plots of the

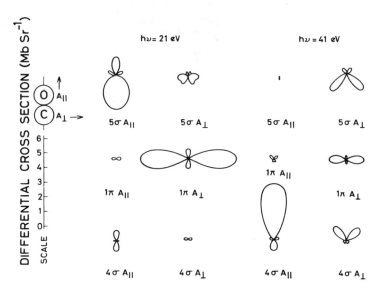

Fig. 7.15. Calculated angular distributions of photoelectrons from principal orbitals of an oriented CO molecule. A differential cross section scale is shown. The results are at photon energies of 21 eV and 41 eV; $A_\parallel$ and $A_\perp$ are polarization directions parallel and perpendicular to the C–O axis [711, 712].

results of differential cross section calculations on 5σ, 1π, and 4σ orbitals of CO at photon energies of 21 eV and 41 eV and polarization directions parallel and perpendicular to the molecular axis. Such results can be used to infer the orientation of adsorbed molecules on a surface where the only role of the surface is to orient the adsorbate molecules.

Early investigation of the adsorption of CO on Ru and of CO on a (100) surface of Ni showed that the experimental observations are consistent with an undissociated CO molecule oriented perpendicular to the surface. In CO adsorption on an Ni(100) surface the 4σ orbital appears at -11 eV relative to the Fermi level E_F, and the 1π and 5σ levels appear as a composite peak at about -8 eV, both identified from earlier data on angle integrated spectra of gaseous CO and nickel carbonyls. In comparison with gaseous CO, in an adsorbed CO spectrum the 5σ peak, which represents the C lone-pair orbital, moves to lower energies, indicating that the C end is more involved in bonding with the substrate. The 4σ orbital is primarily around the O end of the CO molecule and is least perturbed by the surface. Angle-resolved spectral measurements of the 4σ peak at the photon energy of 35 eV confirm that the CO molecule is vertically oriented within 5° on the Ni surface.

PARITY CONSIDERATIONS AND ANGULAR DISTRIBUTION

Wave function parity can also be used for investigating models of the electronic structure of surfaces. For this purpose one needs to carry out an angle-resolved experiment in which photoelectrons propagating along the surface normal or in a plane of mirror-symmetry normal to the sample surface are isolated and detected. In such a mirror plane, the initial-state and final-state wave functions are of either even or odd parity depending on whether or not the wave function possesses reflection symmetry. The mirror plane emission does not contain any odd-parity final states. This can be seen by noting that the photoelectron reaching the detector located away from the surface is in a plane-wave state with its momentum vector lying in the mirror plane. The wave fronts of the plane wave are orthogonal to the mirror plane, hence the parity of the final state is even. The surface component of the experimentally observed momentum $p_\parallel$ in the mirror plane is equal to $k_\parallel$, and the final state is invariant under the reflection operation.

The matrix element of transition is defined as

$$M = \langle f | \mathbf{A} \cdot \mathbf{P} | i \rangle \tag{7.114}$$

M has a nonzero value only if the integrand has a component that remains invariant under crystal symmetry operations. In other words, it is nonzero if, among the final-state wave function, the operator $\mathbf{A} \cdot \mathbf{P}$, and the initial-state wave function, either all are of even parity or one is even and the remaining two are odd. The momentum vector lying in the mirror plane being of even parity, the operator $\mathbf{P}$ is also of even parity; parity of the operator $\mathbf{A} \cdot \mathbf{P}$ is therefore the parity of $\mathbf{A}$. The operator $\mathbf{A} \cdot \mathbf{P}$ is thus of even parity if the polarization vector $\mathbf{A}$ is parallel to the mirror plane $(A_\parallel)$ and is of odd parity if $\mathbf{A}$ is perpendicular to the mirror plane $(A_\perp)$. Therefore, in cases where spin–orbit effects are not important, it can definitely be said that for $A_\perp$ a nonzero matrix element is obtained only when the initial-state wave function $|i\rangle$ is odd, or for $A_\parallel$ only when the initial-state wave function $|i\rangle$ is even.

This polarization-dependent angle-resolved photoelectron spectroscopy method has been applied to the analysis of angle-resolved data from the W(001) surface with a saturation coverage of H. In this case, there are two normal planes, one containing the [100] azimuth and the other containing the [110] azimuth. With synchrotron radiation, for the H-induced level with an initial-state band energy of $-6.5\,\mathrm{eV}$ (below Fermi level), the (100) mirror plane emission shows no increase in emission with $A_\perp$ but an increase in emission with $A_\parallel$; therefore, the initial-state parity is even. The $-4.3\,\mathrm{eV}$ band in (110) mirror plane emission has high intensity in the $A_\perp$ mode and not in the $A_\parallel$ mode; therefore the initial state here has odd parity. This assignment might appear discordant, as H $1s$ orbital is even. But since there are two H atoms for each surface W atom, both symmetric and antisymmetric combinations of the H atoms are necessary. It has been possible to conclude that the principal component of the W–H chemisorptive bond is a coupling between the antisymmetric H $1s$ combination and W $5d_{x^2-y^2}$ orbitals. This has parity properties in agreement with experiment.

CORE LEVEL PHOTOEMISSION AND PHOTOELECTRON DIFFRACTION

In experimental measurements of angle-resolved energy-integrated valence and core spectral intensities, extensive fine structure is observed, such as in single crystal studies of NaCl and Au. These structures are sharp, having FWHM values of only 5–10°, and intense, having high peak-height-to-background ratios, such as 2:1. The structures originate because of electron diffraction from various sets of planes in the crystal and are related to low-energy electron diffraction bands obtained using electron kinetic energies $\epsilon_{kin} \gtrsim 300$ eV. These results also resemble high-energy 0.01–1 MeV electron channeling phenomena observed from radioactive nuclei embedded in single crystals. Electron diffraction and electron channeling phenomena indicate that one may expect enhanced photoelectron emission parallel to each set of planes having Miller indices hkl within the first order Bragg angle θ_{hkl} ($\lambda_e = 2d_{hkl} \sin \theta_{hkl}$, where λ_e is the electron de Broglie wavelength and d_{hkl} is the interplanar spacing).

These bands, known as *Kikuchi bands*, drop off sharply at the edge of $\pm \theta_{hkl}$ limit. The overall photoelectron intensity distribution contains a superposition of such Kikuchi bands for the various low-index planes within the crystal. Such band measurements can yield information on near-surface atomic order of crystal orientation. Recent experimental results show that whenever anisotropy is observed in core level photoemission of an atom adsorbed on a single crystal surface, it is related to nearest neighbor atomic geometry. This indicates that such measurements can be usefully employed for information on the bonding geometries of adsorbate atoms or molecules.

Surface structure of adsorbate species can also be determined from angle-resolved core level photoemission studies. If an atom is located at the origin of a coordinate system and a photoelectron is emitted from the core level of the atom, then at a position **r**, the wave function of the photoelectron can be expressed as

$$\psi(\mathbf{r}) = \psi^\circ(\mathbf{r}) + \psi'(\mathbf{r}) \tag{7.115}$$

Here ψ° represents the direct wave, directly propagating towards the detector, and ψ' stands for the indirect waves, singly or multiply scattered by other atoms before arriving at the detector. Interference of ψ' waves with each other and with ψ° leads to emission anisotropies. This photoelectron diffraction phenomena is similar to LEED or Extended X-ray Absorption Fine Structure (EXAFS) but has a distinct advantage – unlike LEED, the adsorbed layer does not have to be periodic, and one can selectively study photoemission from core levels specific to surface atoms.

The best condition for such experiments is to have photoelectrons in the 30–200 eV range, energy appropriate for low-energy electron diffraction. Synchrotron radiation may therefore be considered essential. The use of such a photoelectron energy range has the advantage that available knowledge of LEED studies can be utilized. It would also ensure strong backscattering, scattering by the substrate atoms lying beneath the adsorbate atoms. For a study of azimuthal anisotropy of the scattering, measurements are needed for a large range of polar emission angle θ (i.e., the angle with respect to surface normal) and for a number of emergent kinetic energies.

We describe here the results of a recent experiment, one of the first of its kind, in which most of the above requirements are satisfied. The experiment involves the (001) surface of Ni with Na or Te atoms adsorbed on it in the $c(2 \times 2)$ configuration. The surface arrangement is checked by medium energy (4.5 keV) grazing-incidence electron diffraction. Synchrotron radiation in the 60–120 eV range is used. In these experiments, the photoemission angle θ is kept fixed and the photoemitted electrons are monitored while the sample is rotated about its surface normal. The plane polarized radiation strikes the surface at $-45°$ to the surface normal. This measurement procedure eliminates those anisotropies of core emission that are associated with free atoms and

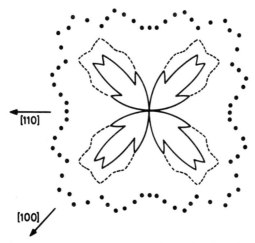

Fig. 7.16. Comparison between theory and experiment for the azimuthal dependence of emission from the Na $2p$ levels for Ni (001)–Na c (2 ×2) at a photon energy of 80 eV and a polar emission angle of 30° [716].

isolates those associated with photoelectron diffraction effects. Figure 7.16 shows the result of azimuthal dependence of emission from $2p$ levels of Na for Ni(001)–Na $c(2 \times 2)$ at a photon energy of 80 eV and $\theta = 30°$. The full circles are the results of multiple scattering calculations. The data points have been subjected to a fourfold symmetrization. The solid line represents the same data after common minimum values have been subtracted. The dashed line represents mirror-symmetrized experimental data. Note that not only are the main lobes along the direction [100] azimuth predicted, but the detailed fine structure on these lobes are also reproduced. Such diffraction structure varies, often dramatically, with photoelectron energy as well as with polar emission angles, providing a wide range of checks for confirmation of an adsorbate structure. Such angle-resolved photoemission studies, along with comparison of results with theoretical predictions, constitute a powerful method for determining adsorbate structure by means of photoelectron spectroscopy.

REFERENCES AND NOTES

This final section contains references to the original sources from which this book has been drawn, references to supplementary research papers, and other data that might have been digressions had they been incorporated into the text.

Since the field of photoelectron spectroscopy encompasses several disciplines, a terminology that is commonplace to one is often quite alien to another. With this consideration in mind, notes have been added. Many of them are elementary, yet they may be of some help to some readers. Although the notes are presented in sequence, they are sufficiently independent for browsing. It was found impossible to compile a list of terms for which a reader might need assistance; some readers may have to take recourse to reading materials suggested in the references.

The figures in the book have been drawn from data presented in the references cited. Figures from these sources were either redrawn for uniformity in appearance or combined with other drawings into a single diagram for a compact view. Some have been augmented with new data. Although the principal features of the redrawn figures are accurate, for critical measurements the original diagrams should be consulted. Since this book is intended to be used as a general reader, references to the original sources are not explicitly provided in the body of the text, but as most of the text material has accompanying figures, the text references can be inferred from the figure references. For passages of text that have no figures or tables, sources can be found in the notes under the appropriate heading. An effort has been made to limit the references to a reasonable number while providing a representative picture of the diversity of the field.

The following publishers kindly gave permission to use material from some of their publications: The Academic Press, Almqvist & Wiksells Boktryckeri Ab, American Chemical Society, American Institute of Physics, The American Physical Society, Butterworth & Co. Ltd., The Chemical Society, Elsevier Scientific Publishing Co., Europhysics journal group, Groupement pour L'Avancement des Méthodes Spectroscopiques et Physico-Chemiques D'Analyse, Heyden & Son Ltd., North-Holland Publishing Co., The Pergamon Press Ltd., Plenum Publishing Corporation, The Royal Society of Chemistry (Faraday Division), The Royal Society, Société Française de Microscopie Electronique, Springer-Verlag, Taylor & Francis Ltd., and Wiley-Interscience.

BOOKS AND CONFERENCE REPORTS

[1] K. Siegbahn, C. Nordling, A. Fahlman, R. Nordberg, K. Hamrin, J. Hedman, G. Johansson, T. Bergmark, S. E. Karlsson, I. Lindgren, and B. J. Lindberg, *ESCA: Atomic, Molecular and Solid State Structure Studied by Means of Electron Spectroscopy*, Nova Acta Regiae Soc. Sci. Upsaliensis, Ser. IV, Vol. 20, 1967.

[2] G. V. Marr, *Photoionization Processes in Gases*, Academic, London, 1967.

[3] J. A. R. Samson, *Techniques of Vacuum Ultraviolet Spectroscopy*, Wiley, New York, 1967.

[4] K. Siegbahn, C. Nordling, G. Johansson, J. Hedman, P. F. Heden, K. Hamrin, U. Gelius, T. Bergmark, L. O. Werme, R. Manne, and Y. Baer, *ESCA Applied to Free Molecules*, North Holland, Amsterdam, 1969.

[5] D. W. Turner, A. D. Baker, C. Baker, and C. R. Brundle, *High Resolution Molecular Photoelectron Spectroscopy*, Wiley, New York, 1970.

[6] D. A. Shirley, Ed., *Electron Spectroscopy* (Proceedings of the International Conference at Asilomar, September 7–10, 1971), North Holland, Amsterdam, 1972.

[7] A. D. Baker and D. Betteridge, *Photoelectron Spectroscopy*, Pergamon, Oxford, 1972.

[8] *Photoelectron Spectroscopy, Faraday Discussions of the Chemical Society* (at Sussex, September 14–16, 1972) 54, 1972.

[9] K. D. Sevier, *Low Energy Electron Spectrometry*, Wiley-Interscience, New York, 1972.

[10] W. Dekeyser, L. Fiermans, G. Vanderklen, and J. Vennik, Eds., *Electron Emission Spectroscopy* (Proceedings of the NATO Summer Institute at the University of Gent, Belgium, August 28–September 7, 1972), D. Reidel, Dordrecht, 1973.

[11] J. H. D. Eland, *Photoelectron Spectroscopy*, Butterworths, London, 1974.

[12] J. B. Pendry, *Low Energy Electron Diffraction*, Academic, London, 1974.

[13] R. Caudano and J. Verbist, Eds., *Electron Spectroscopy* (Proceedings of the International Conference at Namur, April 16–19, 1974), Elsevier, Amsterdam, 1974.

[14] T. A. Carlson, *Photoelectron and Auger Spectroscopy*, Plenum, New York, 1975.

[15] *Electron Spectroscopy of Solids and Surfaces, Faraday Discussions of the Chemical Society* (at Vancouver, July 15–17, 1975) 60, 1975.

[16] D. Chattarji, *The Theory of Auger Transitions*, Academic, London, 1976.

[17] J. W. Rabalais, *Principles of Ultraviolet Photoelectron Spectroscopy*, Wiley-Interscience, New York, 1977.

[18] D. Briggs, Ed., *Handbook of X-Ray and Ultraviolet Photoelectron Spectroscopy*, Heyden, London, 1977.

[19] H. Ibach, Ed., *Electron Spectroscopy for Surface Analysis*, Springer-Verlag, Berlin, 1977.

[20] J. Hedman and K. Siegbahn, Eds., *Electron Spectroscopy* (Proceedings of the International Conference on Electron Spectroscopy at Uppsala, May 9–12, 1977), *Physica Scripta* 16 (1977), 169–465.

[21] C. R. Brundle and A. D. Baker, Eds., *Electron Spectroscopy: Theory, Techniques and Applications*, 4 vols, Academic, London, 1977–1980.

[22] M. Cardona and L. Ley, Eds., *Photoemission in Solids*, Vol. 1 [principles] 1978; Vol. 2 [case studies] 1979, Springer-Verlag, Berlin.

[23] L. Fiermans, J. Vennik, and W. Dekeyser, Eds., *Electron and Ion Spectroscopy in Solids*, Plenum, New York, 1978.
[24] R. E. Ballard, *Photoelectron Spectroscopy and Molecular Orbital Theory*, Hilger, Bristol, 1978.
[25] J. Berkowitz, *Photoabsorption, Photoionization, and Photoelectron Spectroscopy*, Academic, New York, 1979.
[26] R. C. G. Leckey and J. Liesegang, Eds., *Australian Conference on Electron Spectroscopy* (Bundoora, Australia, 1978), *J. Electron Spectrosc.* **15,** 1979.
[27] V. V. Nemoshkalenko and V. G. Aleshin, *Electron Spectroscopy of Crystals*, Plenum, New York, 1979.
[28] K. Kimura, S. Katsumata, Y. Achiba, T. Yamazaki, and S. Iwata, *Handbook of He I Photoelectron Spectra of Fundamental Organic Compounds*, Japan Scientific Societies Press, Tokyo, and Halsted, New York, 1981.

REVIEW ARTICLES

This list is limited to reviews published since 1973. References to reviews published before 1973 are available in some of the articles, for example, [43]. See also [84].

[29] W. C. Price, "Photoelectron Spectroscopy and the Electronic Structure of Matter," in W. C. Price, S. S. Chissick, and T. Ravensdale, Eds., *Wave Mechanics, The First Fifty Years*, Butterworths, London, 1973, pp. 315–331 (11 references).
[30] J. L. Bahr, "Photoelectron Spectroscopy," *Contemp. Phys.* **14** (1973), 329–355 (37 references).
[31] W. E. Swartz Jr., *Anal. Chem.* **45** (1973), 789A–800A (36 references).
[32] J. C. Riviere, "Auger Electron Spectroscopy," *Contemp. Phys.* **14** (1973), 513–539 (13 references).
[33] D. A. Shirley, "ESCA," *Adv. Chem. Phys.* **23** (1973), 85–159 (109 references).
[34] W. Steckelmacher, "Energy Analyzers for Charged Particle Beam," *J. Phys. E* **6** (1973), 1061–1071 (170 references).
[35] K. Siegbahn, "Electron Spectroscopy – An Outlook," *J. Electron Spectrosc.* **5** (1974), 3–97 (136 references).
[36] U. Gelius, "Recent Progress in ESCA Studies of Gases," *J. Electron Spectrosc.* **5** (1974), 985–1057 (110 references).
[37] D. C. Frost, "Progress in Ultraviolet Photoelectron Spectroscopy," *J. Electron Spectrosc.* **5** (1974), 99–132 (60 references).
[38] D. Betteridge and M. A. Williams, "Electron Spectroscopy: Ultraviolet Photoexcitation," *Anal. Chem.* **46** (1974), 125R–133R (295 references).
[39] D. M. Hercules and J. C. Carver, "Electron Spectroscopy: X-Ray and Electron Excitation," *Anal. Chem.* **46** (1974), 133R–150R (545 references).
[40] W. C. Price, "Photoelectron Spectroscopy," *Adv. At. Mol. Phys.* **10** (1974), 131–171 (54 references).
[41] C. K. Jorgensen, "Photoelectron Spectra Showing Relaxation Effects in the Continuum

and Electrostatic and Chemical Influences of the Surrounding Atoms," *Adv. Quantum Chem.* **8** (1974), 137–182 (108 references).

[42] B. Wannberg, U. Gelius, and K. Siegbahn, "Design Principles in Electron Spectroscopy," *J. Phys. E.* **7** (1974), 149–159 (44 references).

[43] S. Leach, "Photoelectron Spectroscopy: Principles and Instrumentation," in N. Damany, J. Romand, and B. Vodar, Eds., *Some Aspects of Vacuum Ultraviolet Radiation Physics.* Pergamon, Oxford, 1974, pp. 192–238 (209 references).

[44] J. Berkowitz, "Photoelectron and Photoion Spectroscopy of Molecules," in E. Koch, R. Haensel, and C. Kunz, Eds., *Vacuum Ultraviolet Radiation Physics* (Proceedings of the 4th International Conference on Vacuum Ultraviolet Radiation Physics, Hamburg, July 22–26, 1974), Pergamon-Vieweg, 1974, pp. 107–136 (53 references).

[45] H. Raether, "Recent Development in Electron Energy Loss Spectroscopy," in [44], pp. 591–606 (40 references).

[46] C. Kunz, "Perspectives of Synchrotron Radiation" (Report on a Panel Discussion), in [44], pp. 753–771 (48 references).

[47] C. S. Fadley, R. J. Baird, W. Siekhaus, T. Novakov, and S. A. L. Bergstrom, "Surface Analysis and Angular Distribution in X-Ray Photoelectron Spectroscopy," *J. Electron Spectrosc.* **4** (1974), 93–137 (48 references).

[48] M. O. Krause, "Electron Spectrometry," in B. Craseman, Ed., *Atomic Inner Shell Processes,* Academic, London, 1975, Vol. 2, pp. 33–81 (165 references).

[49] A. F. Orchard, "Photoelectron Spectroscopy and Allied Techniques: General Introduction," in P. Day, Ed., *Electronic States of Inorganic Compounds: New Experimental Techniques,* D. Reidel, Dordrecht, 1975, pp. 267–304 (65 references).

[50] S. T. Manson, "Atomic Photoelectron Spectroscopy," *Adv. Electronics & Electron Phys.* Part I, **41** (1976), 73–111 (153 references), Part II, **44** (1977), 1–32 (82 references).

[51] D. M. Hercules. "Electron Spectroscopy: X-Ray and Electron Excitation," *Anal. Chem.* **48** (1976), 294R–313R (842 references).

[52] A. D. Baker, M. A. Brisk, and D. C. Liotta, Electron Spectroscopy: Ultraviolet Excitation," *Anal. Chem.* **48** (1976), 281R–294R (293 references).

[53] D. Berenyi, "Recent Applications of Electron Spectroscopy," *Adv. Electronics & Electron Phys.* **42** (1976), 55–111 (257 references).

[54] W. E. Spicer, I. Lindau, J. N. Miller, D. T. Ling, P. Pianetta, P. W. Chye, and C. M. Garner, "Studies of Surface Electronic Structure and Surface Chemistry using Synchrotron Radiation," in [20], pp. 388–397 (41 references).

[55] E. J. Arsolado, "Uses of Synchrotron Radiation," *Contemp. Phys.* **18** (1977), 527–546 (34 references).

[56] K. Wittel and S. P. McGlynn, "The Orbital Concept in Molecular Spectroscopy," *Chem. Rev.* **77** (1977), 745–771 (88 references).

[57] J. E. Collin, "Autoionization in Atomic and Molecular Physics," *Endeavour* (new series) **1** (1977), 122–128 (37 references).

[58] J. E. Castle, "The Use of X-Ray Photoelectron Spectroscopy in Corrosion Science," *Surf. Sci.* **68** (1977), 583–602 (75 references).

[59] W. L. Jolly, "Inorganic Applications of X-Ray Photoelectron Spectroscopy," *Topics Curr. Chem.* **71** (1977), 149–182 (129 references).

[60] R. L. Martin and D. A. Shirley, "Many-Electron Theory of Photoemission," in [21], Vol. 1, pp. 75–117 (82 references).

[61] E. Heilbronner and J. P. Maier, "Some Aspects of Organic Photoelectron Spectroscopy," in [21], Vol. 1, pp. 205–292 (229 references).

[62] M. Thompson, P. A. Hewitt, and P. S. Wooliscroft, "An Introduction to the Chemical Applications of Vacuum Ultraviolet Photoelectron Spectroscopy," in [18], pp. 341–379 (124 references).

[63] C. S. Fadley, "Basic Concepts of X-Ray Photoelectron Spectroscopy," in [21], Vol. 2, pp. 1–156 (420 references).

[64] A. D. Baker, M. A. Brisk, and D. C. Liotta, "Electron Spectroscopy: Ultraviolet and X-Ray Excitation," *Anal. Chem.* **50** (1978), 328R–346R (432 References).

[65] R. H. Williams, "Electron Spectroscopy of Surfaces," *Contemp. Phys.* **19** (1978), 389–414 (37 references).

[66] H. W. Werner, "Introduction to Secondary Ion Mass Spectrometry (SIMS)," in [23], pp. 324–441 (187 references).

[67] S. P. McGlynn, H. Aldrich, and D. Dougherty, "The Impact of Photoelectron Spectroscopy on Biology," *J. Mol. Struc.* **45** (1978), 119–138 (25 references).

[68] J. H. D. Eland, "Molecular Photoelectron Spectroscopy," *J. Phys. E* **11** (1978), 969–977 (91 references).

[69] C. J. Powell and P. E. Larson, "Quantitative Surface Analysis by X-Ray Photoelectron Spectroscopy," *Appl. Surf. Sci.* **1** (1978), 186–201 (58 references).

[70] F. P. Larkins, "Australian Conference on Electron Spectroscopy: Conference Summary," *J. Electron Spectrosc.* **15** (1979), 323–328.

[71] T. Bergmark, "Autoionization in Electron Spectroscopy," Univ. Uppsala Report No. UUIP-820 (28 references).

[72] A. F. Carley and R. W. Joyner, "The Application of Deconvolution Methods in Electron Spectroscopy: A Review," *J. Electron Spectrosc.* **16** (1979), 1–23 (69 references).

[73] J. A. R. Samson, "Vacuum Ultraviolet Photoelectron Spectroscopy of Atoms and Molecules," *J. Electron Spectrosc.* **15** (1979), 257–267 (42 references).

[74] I. Lindau and W. E. Spicer, "Photoelectron Spectroscopy in the Energy Region 30–800 eV using Synchrotron Radiation," *J. Electron Spectrosc.* **15** (1979), 295–306 (14 references).

[75] C. N. R. Rao, P. K. Basu, and M. S. Hegde, "Systematic Organic UV Photoelectron Spectroscopy," *Appl. Spectrosc. Rev.* **15** (1979), 1–193 (536 references).

[76] S. N. Karchaudhari and K. L. Cheng, "Recent Study of Solid Surfaces by Photoelectron Spectroscopy," *Appl. Spectrosc. Rev.* **16** (1980), 187–297 (765 references).

[77] A. D. Baker, M. A. Brisk, and D. C. Liotta, "Electron Spectroscopy: Ultraviolet and X-Ray Excitation," *Anal. Chem.* **52** (1980), 161R–174R (349 references).

[78] R. H. Williams, G. P. Srivastava, and I. T. McGovern, "Photoelectron Spectroscopy of Solids and Their Surfaces," *Rep. Prog. Phys.* **43** (1980), 1357–1414 (200 references).

[79] S. Trajmar, "Electron Impact Spectroscopy," *Acc. Chem. Res.* **13** (1980), 14–20 (59 references).

[80] P. M. Johnson, "Molecular Multiphoton Ionization Spectroscopy," *Acc. Chem. Res.* **13** (1980), 20–26 (36 references).

[81] A. G. Akimov and L. P. Kazanskii, "Electron Spectroscopy and the Study of

Chemical Reactions on the Surface of Metals," *Russian Chemical Reviews* **50** (1981), 1–13 (76 references).

[82] J. G. Jenkin, "The Development of Angle-Resolved Photoelectron Spectroscopy: 1900–1960," *J. Electron Spectrosc.* **23** (1981), 187–273 (232 references).

[83] N. H. Turner and R. J. Colton, "Surface Analysis: X-ray Photoelectron Spectroscopy, Auger Electron Spectroscopy, and Secondary Ion Mass Spectrometry," *Anal. Chem.* **54** (1982), 293R–322R (935 references).

The article below provides a good review of photoelectron spectroscopic work done before 1971. A fairly comprehensive list of recent UPS and XPS work is available in [77] and [83].

[84] S. D. Worley, "Photoelectron Spectroscopy in Chemistry," *Chem. Rev.* **71** (1971), 295–314 (187 references).

CHAPTER 1

1 nm (nanometer) = 10 Å. 1 eV = 8067 cm^{-1} = 1.602 × 10^{-12} erg/molecule = 23060 cal/mole.

General-purpose book on molecular structure

[85] P. J. Wheatley, *The Determination of Molecular Structure*, Oxford University Press, Oxford, 1968.

Optical spectroscopy

[86] C. N. Banwell, *Fundamentals of Molecular Spectroscopy*, McGraw-Hill, New York, 1966.

[87] G. Herzberg, *Atomic Spectra and Atomic Structure*, Dover, New York, 1944.

[88] G. Herzberg, *Spectra of Diatomic Molecules*, D. Van Nostrand, New York, 1950.

[89] G. Herzberg, *Infrared and Raman Spectra of Polyatomic Molecules*, D. Van Nostrand, New York, 1964.

[90] G. Herzberg, *Electronic Spectra and Electronic Structure of Polyatomic Molecules*, D. Van Nostrand, New York, 1966.

Nuclear magnetic resonance

[91] J. D. Roberts, *Nuclear Magnetic Resonance*, McGraw-Hill, New York, 1959.

[92] J. A. Pople, W. G. Schneider, and H. J. Bernstein, *High Resolution Nuclear Magnetic Resonance*, McGraw-Hill, New York, 1959.

Diffraction methods

[93] J. B. Cohen, *Diffraction Methods in Materials Science*, Macmillan, New York, 1966.

[94] Chemical Society, *Molecular Structure by Diffraction Methods*, Vol. I, 1973.

Mössbauer spectroscopy

[95] G. M. Bancroft, *Mössbauer Spectroscopy*, McGraw-Hill, New York, 1973.

Electron impact phenomena and mass spectrometry

[96] F. H. Field and J. L. Franklin, *Electron Impact Phenomena and the Properties of Gaseous Ions*, Academic, New York, 1970.
[97] C. A. McDowell, Ed., *Mass Spectrometry*, McGraw-Hill, New York, 1963.
[98] F. W. McLafferty, Ed., *Mass Spectrometry of Organic Ions*, Academic, New York, 1963.

Several examples of multiple ionization by photon impact are now known, such as one involving double ionization of CO_2 reported in the following article.

[99] J. A. R. Samson, P. C. Kemeny, and G. N. Haddad, *Chem. Phys. Lett.* 51 (1977), 75.

For high-resolution electron impact ionization studies and determination of ionization efficiency curves, a retarding potential difference method is sometimes used, wherein ionization studies are carried out by increasing the electron energy in small steps. The technique was initiated by the group cited in [100] below. Among the books that describe it are [96] and [97].

[100] R. E. Fox, W. M. Hickam, D. J. Grove, and T. Kjeldaas Jr., *Rev. Sci. Instrum.* 26 (1955), 1101.

Linewidth of monochromatic radiation: A plot of the radiation intensity versus wavelength gives an energy distribution profile of the monochromatic radiation; the width of the line radiation is often specified as the full width at half maximum (FWHM).

In electron impact ionization, the interaction time of the electron with neutral species may be considerably shorter than the rearrangement time of the orbitals. At 10 eV range, this interaction time is of the order of 10^{-16} sec. This is much shorter than the 10^{-14} sec a 10 eV photon spends on the basis of 50 cycles of the radiation field. With an orbital reorganization time of the order of 10^{-15} sec, a 10 eV photoionization therefore may be expected to be an adiabatic process. Since the interaction time in electron impact ionization is short, the molecular electron makes a vertical transition in the Franck–Condon sense, and it constitutes a partially diabatic process [43].

In Fig. 1.1, since the diagram is based only on electronic energy, $E_{int} = 0$ means that the ion M^+ is in its electronic ground state. When the neutral species is a diatomic or a polyatomic molecule, however, it can well happen that the ion formed is in its electronic ground state but in some excited vibrational or rotational state.

Very early experiments in x-ray photoelectron spectroscopy

[101] H. Robinson and W. F. Rawlinson, *Phil. Mag.* **28** (1914) 277.

The first observations that core photoelectron peak intensities could be used for quantitative analysis

[102] R. G. Steinhardt and E. J. Serfass, *Anal. Chem.* **25** (1953), 697.

[103] R. G. Steinhardt, F. A. D. Granados, and G. I. Post, *Anal. Chem.* **27** (1955), 1046.

Description of the first angle-resolved photoelectron spectroscopy experiments

[104] J. G. Jenkin, J. Liesegang, R. C. G. Leckey, and J. D. Riley, *J. Electron Spectrosc.* **15** (1979), 307.

Some of the early work in UPS are described in the following articles.

[105] F. I. Vilesov, B. L. Kurbatov, and A. N. Terenin, *Dokl. Akad. Nauk SSSR* **138** (1961), 1329 [Trans. *Soviet Phys.-Doklady* **6** (1961), 490].

[106] B. L. Kurbatov, F. I. Vilesov, and A. N. Terenin, *Dokl. Akad. Nauk SSSR* **140** (1961), 797 [Trans. *Soviet Phys.-Doklady* **6** (1961), 883].

[107] D. W. Turner and M. I. Al-Joboury, *J. Chem. Phys.* **37** (1962), 3007.

Some of the early work in XPS are described in [1] and [4]. For a general review of the development of XPS, see the following articles.

[108] J. G. Jenkin, R. C. G. Leckey, and J. Liesegang, *J. Electron Spectrosc.* **12** (1977), 1.

[109] J. G. Jenkin, J. D. Riley, J. Liesegang, and R. C. G. Leckey, *J. Electron Spectrosc.* **14** (1978), 477.

Comments on the early work in UPS

[110] W. C. Price, *J. Electron Spectrosc.* **13** (1978), 153.

Although isolated XPS and UPS experiments were reported earlier, the present concerted activity in the field began from the time of K. Siegbahn's XPS experiments in Sweden [1981 Nobel lecture in physics: *Rev. Mod. Phys.* **54** (1982), 709−728] and D. W. Turner's UPS experiments in England.

For Auger transitions, [9] and [14] give detailed lists of various Auger transition energies. The expression for energy of Auger transition is due to the following source.

[111] M. F. Chung and L. H. Jenkins, *Surf. Sci.* **22** (1970), 479.

The natural width ΔE of an energy level (state) is determined by the uncertainty principle relationship $\Delta E \Delta t \sim h$, where Δt represents the lifetime of the state and h is the Planck's constant.

Binding energies derived from photoelectron spectra are invariant regardless of the photon energy used. For example, if the photon energy is 21.22 eV and the electron kinetic energies of two spectral peaks are 6.8 eV and 12.5 eV, the respective binding energies are $21.22 - 6.8 = 14.42$ eV and $21.22 - 12.5 = 8.72$ eV. If a photon energy of 40.8 eV is used, then the corresponding photoelectron peaks appear at 26.38 and 32.08 eV of electron kinetic energy; the respective binding energies therefore are again 14.42 eV and 8.72 eV.

HBr^+ spectrum

[112] J. Delwiche, P. Natalis, J. Momigny, and J. E. Collin, *J. Electron Spectrosc.* **1** (1972/1973), 219.

Ion absorption spectrscopy experiments using flash discharge

[113] G. Herzberg, Proceedings International Conference on Spectroscopy, January 9–18, 1967, Department of Atomic Energy, Bombay, Vol. 1, p. 158.

For nomenclature of atomic orbitals and spectroscopic terms, see [87]. For molecular orbitals and term values thereof, see [86], [88], [89], and [90].

The Franck–Condon principle states that electronic transitions are so much faster than vibrational motion of nuclei that during electronic transitions the nuclear coordinates practically do not change. If internuclear distance is plotted on the x-coordinate and electronic energy on the y-coordinate, an electronic transition can be represented by a vertical line. This is what is meant by vertical transition, and it relates to the vertical ionization potential when the upper state belongs to the molecular ion.

Franck–Condon factors represent the relative transition probabilities to populate the vibrational levels of higher electronic states. They can be calculated from overlap integrals of the vibronic wave functions of the upper and the lower states.

For methods of determining vibrational constants from spectral lines, see, for example, [88] and [89].

Cross sections (of cm^2 dimension) are quantities by which atomic and molecular processes can be quantitatively described. The differential cross section $d\sigma/d\Omega$ per unit solid angle gives angle-dependent properties of the cross section. Integration of the differential cross section over 4π steradians yields the total cross section σ. For further details, see Chapters 6 and 7.

Koopmans's theorem

[114] T. Koopmans, *Physica* **1** (1933), 104 (in German).
[115] G. G. Hall, *Faraday Disc. Chem. Soc.* **54** (1972), 7.

An example of a non-Koopmans's-theorem effect in HeI UPS of *p*-quinodimethane derivative

[116] T. Koenig and S. Southworth, *J. Amer. Chem. Soc.* **99** (1977), 2807.

For Hartree–Fock and Hartree–Fock–Slater wave functions, electron correlation, and configuration mixing, see the notes to Chapter 3.

The atomic relaxation energy values reported are from the following article.

[117] U. Gelius and K. Siegbahn, *Faraday Disc. Chem. Soc.* **54** (1972), 257.

Rydberg states originate when the molecular potential effectively resembles that of atoms, and the energy levels are given approximately by Rydberg-type formulas. For a discussion of Rydberg states, see the following book.

[118] A. B. F. Duncan, *Rydberg Series in Atoms and Molecules,* Academic, New York, 1971.

For autoionization, see [43], [57], [71], chapter 7 of [2], and the following articles.

[119] P. Natalis, J. Delwiche, and J. E. Collin, *Chem. Phys. Lett.* **13** (1972), 491.
[120] P. Natalis, J. Delwiche, and J. E. Collin, *Faraday Disc. Chem. Soc.* **54** (1972), 98.

Transient species have also been studied using photoelectron spectroscopy. Short-lived radicals or free atoms can be studied if they can be transferred sufficiently rapidly to a photoelectron spectrometer. Spectra have been obtained for atomic H, O, N, the halogens, several vibrationally excited molecular species, CS, NF_2, ClO_2, $(CF_3)_2 NO$, $SO_3 F$, and others.

UPS of the CS free radical

[121] N. Jonathan, A. Morris, M. Okuda, K. J. Ross, and D. J. Smith, *Faraday Disc. Chem. Soc.* **54** (1972), 48.

UPS of free radicals NF_2, ClO_2, $(CF_3)_2 NO$ and $SO_3 F$

[122] A. B. Cornford, D. C. Frost, F. G. Herring, and C. A. McDowell, *Faraday Disc. Chem Soc.* **54** (1972), 56.

In biochemical applications, the systems that have been studied include metal ion binding to proteins and amino acids. For example, complexes of Zr and Hf with glycine; Cu and Co complexes with histidine, lysine, tyrosine, and apoerythrocuprein; Zn and Cd complexes with cystine, cysteine, and thioacetic acid. Ferredoxin and chromatium have also been studied. XPS has also been used for determinations of grain protein. Sources are cited in [39].

In biological structure determinations, several nucleosides and t-RNA have been studied.

[123] L. D. Hulett and T. A. Carlson, *Clinical Chem.* **16** (1970), 677.

For a recent list of UPS–XPS of biomolecules, see [67] and [77].

Study of phthalocyanines by UPS

[124] J. Berkowitz, *J. Chem. Phys.* **70** (1979), 2819.

UPS studies of psychotropic drugs

[125] L. N. Domelsmith and K. N. Houk, *Int. J. Quantum Chem.* **5** (1978), 257.

Teichoic acids are unusual phosphate polymers that constitute cell walls and intracellular components of certain gram-positive bacteria. They are of two types, one containing the chains of D-glycerol and the other containing the chains of D-ribitol. Both types are joined by phosphate diester linkages.

D-alanine residues linked to the glycerol phosphate backbone

D-alanine ester linkage at C–2 or C–3 ribitol; monosaccharide units present are 2-acetamido-2-deoxy-D-glucose

Phthalocyanine   Salazopyrin

Compton effect: The scattering of photons by matter, in which the photons lose part of their energy. The scattered photons, as a result, have lower frequency. The Compton effect also takes place in photoelectron spectroscopy experiments, but at energies less than about 5 keV the Compton scattering cross section is insignificant in comparison with photoionization cross section.

Review of Compton scattering and electron momentum

[126] M. J. Cooper, *Contemp. Phys.* **18** (1977), 489.

Review of spin-polarized electron emission experiments (thus far, experiments have been conducted only with UV radiation)

[127] M. Campagna, D. T. Pierce, F. Meier, L. Sattler, and H. C. Siegmann, *Adv. Electron. and Electron Phys.* **41** (1976), 113.

Search for polarized photoelectrons from molecules

[128] U. Heinzman, J. Kessler, and E. Kuhlmann, *J. Chem. Phys.* **68** (1978), 4753.

Spin polarization of electrons ejected from unpolarized atoms by unpolarized and linearly polarized light

[129] N. A. Cherepkov, *J. Phys. B.* **11** (1978), L435.

Only single photon transitions are considered in this book. Multiphoton transitions are important when radiation intensities are particularly high, such as in laser radiations. For discussion on multiphoton transitions, see, for example, ref. [80].

For threshold photoelectron spectroscopy, see notes to Chapter 2.

For further reading on Auger spectroscopy, see

[14], chapter 6, p. 279; [16], chapter 7; [18], p. 249; [23], chapter 6, p. 230; and [32].

CHAPTER 2

For a discussion of resolution, sensitivity, and general design constraints of electron energy analyzers, see [34], [42], and [43].

Vacuum ultraviolet emission from plasmas

[130] C. Breton and J. Schwob, in N. Damany, J. Romand, and B. Vodar, Eds., *Some Aspects of Vacuum Ultraviolet Radiation Physics*, Pergamon, Oxford, 1974, p. 241.

Electron-impact excitation in He plasmas and resonance lamp properties

[131] H. W. Drawin, F. Emard, and K. Katsonis, *Z. Naturforsch.* **28a** (1973), 1422.
[132] J. A. R. Samson, *Rev. Sci. Instrum.* **40** (1969), 1174, has examined He I 58.4 nm radiation emitted by four line sources. Line broadening becomes noticeable at 0.2 torr, and at 0.8 torr abundant self-absorption is observed. The results are also presented in [43].

Étendue

[133] M. C. Tobin, *Laser Raman Spectroscopy*, Wiley-Interscience, New York, 1971, p. 35.

Cylindrical grid analyzer

[134] M. I. Al-Joboury and D. W. Turner, *J. Chem. Soc.* (1963), 5141.

Spherical grid analyzer

[135] D. C. Frost, C. A. McDowell, and D. A. Vroom, *Phys. Rev. Lett.* **15** (1965), 612.
[136] D. A. Huchital and J. D. Rigden, in [6], p. 79.
[137] P. Staib, *J. Phys. E.* **5** (1972), 484.

Double filter retarding potential analyzer

[138] J. D. Lee, *Rev. Sci. Instrum.* **43** (1972), 1291.

Differential retarding field analyzer

[139] I. Lindau, J. C. Helmer and J. Uebbing, *Rev. Sci. Instrum.* **44** (1973), 265.
[140] H. Hotop and G. Hubler, *J. Electron Spectrosc.* **11** (1977), 101.

Cylindrical 127° analyzer

[141] A. L. Hughes and V. Rojansky, *Phys. Rev.* **34** (1929), 284.
[142] D. W. Turner, *Proc. Roy. Soc. London* **A307** (1968), 15.

Hemispherical analyzer, besides [1]

[143] B. Wannberg, G. Engdahl, and A. Skollermo, *J. Electron Spectrosc.* **9** (1976), 111.
[144] M. E. Gellender and A. D. Baker, *J. Electron Spectrosc.* **4** (1974), 249.

Photoelectron spectrometer with monochromatized X-rays (also partly reported in [35] and [117].)

[145] U. Gelius, E. Basilier, S. Svensson, T. Bergmark, and K. Siegbahn, *J. Electron Spectrosc.* **2** (1973), 405.
[146] H. Fellner-Feldegg, U. Gelius, B. Wannberg, A. G. Nilsson, E. Basilier and K. Siegbahn, *J. Electron Spectrosc.* **5** (1974), 643.

Cylindrical mirror analyzer

[147] E. Blauth, *Z. Phys.* **147** (1957), 228.
[148] W. Mehlhorn, *Z. Phys.* **160** (1960), 247.
[149] H. Z. Sar-El, *Rev. Sci. Instrum.* **38** (1967), 1210.
[150] K. Maeda and T. Ihara, *Rev. Sci. Instrum.* **42** (1971), 1480.
[151] S. Aksela, *Rev. Sci. Instrum.* **42** (1971), 810.

Cylindrical mirror electron spectrometer, for studies of gases and metal vapors

[152] J. Vayrynen and S. Aksela, *J. Electron Spectrosc.* **16** (1979), 423.

Focusing properties are discussed in [150] and [151] for l_S and l_I which are respectively the distance of the source slit and the image slit from the inner cylinder perpendicular to the axis. The example of Fig. 2.11 is for the special case where $l_S = l_I = r_1$. The apparatus of Fig. 2.11 also employs preretardation of electrons.

Bessel box analyzer

[153] J. D. Allen, Jr., J. P. Wolfe, and G. K. Schweitzer, *Int. J. Mass Spectrom. Ion Phys.* **8** (1972), 81.

[154] J. D. Allen, Jr., J. D. Durham, G. K. Schweitzer, and W. E. Deeds, *J. Electron Spectrosc.* **8** (1976), 395.

Polarized vacuum ultraviolet radiation and x-radiation

[155] J. A. R. Samson, *Nucl. Instrum. Methods* **152** (1978), 225.

Polarization by gratings used at grazing incidence

[156] E. T. Arakawa, M. W. Williams, and J. A. R. Samson, *Appl. Opt.* **17** (1978), 2502.

Angular distribution measurements

[157] D. L. Ames, J. P. Maier, F. Watt, and D. W. Turner, *Faraday Disc. Chem. Soc.* **54** (1972), 277.

Angular distribution measurements with polarized He I radiation

[158] W. H. Hancock and J. A. R. Samson, *J. Electron Spectrosc.* **9** (1976), 211.

Instrumentation on angular distribution of photoelectrons from solid surfaces

[159] W. McMahon and L. Heroux, *Appl. Opt.* **13** (1974), 438.

Miniature plane mirror analyzer and angular distribution measurements

[160] P. E. Best, *Rev. Sci. Instrum.* **46** (1975), 1517.
[161] T. S. Green and G. A. Proca, *Rev. Sci. Instrum.* **41** (1970), 1409.
[162] N. V. Smith, P. K. Larsen, and M. M. Traum, *Rev. Sci. Instrum..* **48** (1977), 454.

Most of the above-mentioned energy analyzers have been used in conjunction with photoelectron spectrometers. In [34], a general review of energy analyzers, a classified reference list is provided for retarding fields at collector, electrostatic filter lens, inverse retarding field analyzers, retarding potential difference method, fountain analyzers, cylindrical radial field analyzers, cylindrical sector field analyzers, cylindrical mirror analyzers, inverse second power field, electrostatic lens as energy analyzer, quadrupole filter lens, electrostatic prism spectrometer, wien filter, trochoidal analyzers, time-of-flight analyzers, and analyzers that use deceleration—acceleration systems in conjunction with deflection energy analyzers.

Books on electron optics

[163] O. Klemperer, *Electron Optics,* Cambridge University Press, Cambridge, 1953.
[164] B. Paszkowski, *Electron Optics,* Iliffe, London, 1968.
[165] P. Grivet, *Electron Optics,* Pergamon, Oxford, 1972.

Photoelectron spectrometer for liquids

[166] H. Siegbahn and K. Siegbahn, *J. Electron Spectrosc.* **2** (1973), 319.

UPS of liquids

[167] H. Aulich, L. Nemec, L. Chia, and P. Delahay, *J. Electron Spectrosc.* **8** (1976), 271.
[168] L. Nemec, L. Chia, and P. Delahay, *J. Electron Spectrosc.* **9** (1976), 241.

XPS of liquids

[169] K. Siegbahn, *Pure Appl. Chem.* **48** (1976), 77.
[170] L. Nemec, H. J. Gaehrs, L. Chia, and P. Delahay, *J. Chem. Phys.* **66** (1977), 4450.
[171] S. C. Avanzino and W. L. Jolly, *J. Amer. Chem. Soc.* **100** (1978), 2228.

Photoelectron spectrometer for frozen solutions

[172] K. Burger and E. Fluck, *Inorg. and Nucl. Chem. Lett.* **10** (1974), 171.
[173] K. Burger, F. Tschismarov, and H. Ebel, *J. Electron Spectrosc.* **10** (1977), 461.
[174] G. Miksche, H. Miksche, H. Murauer, and K.Persy, *J. Electron Spectrosc.* **10** (1977), 423 (in German).
[175] H. Kuroda, T. Ohta, and Y. Sato, *J. Electron Spectrosc.* **15** (1979), 21.

Photoelectron spectroscopy of water and inorganic cations and anions in aqueous solutions

[176] I. Watanabe, J. B. Flanagan, and P. Delahay, *J. Chem. Phys.* **73** (1980), 2057.
[177] K. V. Burg and P. Delahay, *Chem. Phys. Lett.* **78** (1981), 287.
[178] P. Delahay, *Acc. Chem. Res.* **15** (1982), 40.

UPS in a strong magnetic field

[179] G. Beamson, S. J. Pearce, and D. W. Turner, *Physica Scripta* **16** (1977), 186.
[180] G. Beamson, S. J. Pearce, and D. W. Turner, *Chem. Phys. Lett.* **56** (1978), 5.
[181] G. Beamson, H. Q. Porter, and D. W. Turner, *J. Phys. E.* **13** (1980), 64.

Ion cyclotron resonance spectrometry

[182] J. M. S. Henis, in J. L. Franklin, Ed., *Ion—Molecule Reactions,* Plenum, New York, 1972, Vol. 2, p. 395.

Photoelectron spectromicroscopy uses a new principle of electrostatic field-free imaging. It employs a strong axially symmetric divergent magnetic field to generate an enlarged image of a photoemissive object; the process preserves the original total energy distribution. At a field strength of 16 Tesla with 5 eV electrons, a lateral resolution of 50 nm may be expected (5 nm at 0.05 eV electrons) in comparison with a limiting value of 30 nm in scanning Auger microscopy when a cold cathode field emitting source is used. Depending on the photon energy used, the technique possesses the ability to produce either a molecular or an atomic map of the surface, with a factor of 1000 less power dissipation than Auger electron spectroscopy and correspondingly less chance of radiation damage.

[183] G. Beamson, H. Q. Porter, and D. W. Turner, *Nature* **290** (1981), 556.

Molecular beam studies of high-temperature molecules

[184] A. W. Potts, T. A. Williams, and W. C. Price, *Proc. Roy. Soc.* **A341** (1974), 147.
[185] A. W. Potts and W. C. Price, *Physica Scripta* **16** (1977), 191.
[186] J. Berkowitz, in [21], Vol. 1, p. 355.

Fig. 2.22 is from the following source.

[187] J. E. Pollard, D. J. Trevor, Y. T. Lee, and D. A. Shirley, Lawrence Berkeley Laboratory Report LBL-11250 (1981) *Rev. Sci. Instrum.* **52** (1982), 1837.

Operational details of the molecular beam apparatus were kindly provided by the authors in a private communication, 1981.

Supersonic nozzle beams

[188] J. B. Anderson, R. P. Andres, and J. B. Fenn, in J. Ross, Ed., *Advances in Chemical Physics*, Vol. 10, 1966, p. 275.

Quadrupole mass spectrometry

[189] P. H. Dawson, Ed., *Quadrupole Mass Spectrometry and its Applications*, Elsevier, Amsterdam, 1976.

Helium resonance lamps

[190] R. T. Poole, J. Liesegang, R. C. G. Leckey, and J. G. Jenkin, *J. Electron Spectrosc.* **5** (1974), 773.
[191] F. Burger and J. P. Maier, *J. Electron Spectrosc.* **5** (1974), 783.

Vacuum ultraviolet source of the line radiation of the rare gas ions

[192] F. Burger and J. P. Maier, *J. Phys. E.* **8** (1975), 420.
[193] S.–T. Lee, R. A. Rosenberg, E. Matthias, and D. A. Shirley, *J. Electron Spectrosc.* **10** (1977), 203.

High intensity discharge lamp for monochromatized angle-resolved UPS

[194] N. J. Shevchik, *J. Electron Spectrosc.* **14** (1978), 411.

Charge particle oscillator and hollow cathode He II lamps

[195] L. L. Coatsworth, G. M. Bancroft, D. K. Creber, R. J. D. Lazier, and P. W. M. Jacobs, *J. Electron Spectrosc.* **13** (1978), 395.

Vacuum ultraviolet resonance line radiation source from rare gas atoms and ions

[196] G. M. Lancaster, J. A. Taylor, A. Ignatiev, and J. W. Rabalais, *J. Electron Spectrosc.* **14** (1978), 143.

He II resonance lamp for UPS

[197] S. Katsumata, K. Nomoto, K. Ohmori, Y. Kirihata, T. Yamazaki, Y. Achiba, and K. Kimura, *J. Electron Spectrosc.* **16** (1979), 485.

He II radiation source

[198] F. Burger and J. P. Maier, *J. Electron Spectrosc.* **16** (1979), 471.

SiK_α radiation

[199] J. E. Castle, L. B. Hazell, and R. D. Whitehead, *J. Electron Spectrosc.* **9** (1976), 247.

YM_ζ radiation source

[200] R. Nilsson, R. Nyholm, A. Berndtsson, J. Hedman, and C. Nordling, *J. Electron Spectrosc.* **9** (1976), 337.

Np radiation source

[201] M. O. Krause, C. W. Nestor Jr., and J. H. Oliver, *Phys. Rev. A* **15** (1977), 2335.

FK_α radiation source

[202] A. Berndtsson, R. Nyholm, R. Nilsson, J. Hedman, and C. Nordling, *J. Electron Spectrosc.* **13** (1978), 131.

Characteristic Lines from X-ray Sources and Low-Energy Resonance Sources

X-rays	Energy (eV)	X-rays	Energy (eV)	Line	λ(nm)	Line	λ(nm)
Cu $K_{\alpha 1}$	8047.8	Mg $K_{\alpha 1}$	1253.7	He I	58.4334	Ar I	106.6659
Cr $K_{\alpha 1}$	5414.7	Mg $K_{\alpha 2}$	1253.4	He II	30.3783	Kr I	116.4867
Ti $K_{\alpha 1}$	4510.9	Na $K_{\alpha 1,2}$	1041.0	Ne I	73.5895	Kr I	123.5838
Al $K_{\alpha 1}$	1486.7	Zr M_ζ	151.4	Ne I	74.3718	H(Ly$_\alpha$)	121.5700
Al $K_{\alpha 2}$	1486.3	Y M_ζ	132.3	Ar I	104.8219	Xe	129.5586

Photoelectron spectroscopy with synchrotron radiation

[203] T. Sagawa, R. Kato, S. Sato, M. Watanabe, T. Ishii, I. Nagakura, S. Kono, and S. Suzuki, *J. Electron Spectrosc.* 5 (1974), 551.
[204] K. Thimm, *J. Electron Spectrosc.* 5 (1974), 755.
[205] P. Pianetta and I. Lindau, *J. Electron Spectrosc.* 11 (1977), 13.
[206] I. Lindau and H. Winick, *J. Vac. Sci. Technol.* 15 (1978), 977.
[207] P. Dhez, P. Jaegle, F. J. Wuilleumier, E. Kallne, V. Schmidt, M. Berland, and A. Carillon, *Nucl. Instrum. Methods* 152 (1978), 85.

Synchrotron Radiation Sources[a, b]

Storage Rings

PETRA (Hamburg, Germany) 18, 18, 192, 67.4; 15, 80, 192, 39.0; 10, 50, 192, 11.6
CESR (Cornell University, USA) 8, 100, 32.5, 35; 4, 50, 32.5, 4.4
VEPP-4 (Novosibirsk, USSR) 7, 10, 16.5, 46.1; 4.5, 10, 16.5, 12.2
DORIS (Hamburg, Germany) 5, 50, 12.1, 22.9; 2.5, 300, 12.1, 2.9
SPEAR (Stanford University, USA) 4.0, 100, 12.7, 11.1; 2.5, 300, 12.7, 2.7
VEPP-3 (Novosibirsk, USSR) 2.25, 100, 6.15, 4.2
DCI (Orsay, France) 1.8, 500, 4.0, 3.63
ADONE (Frascati, Italy) 1.5, 60, 5.0, 1.5
VEPP-2M (Novosibirsk, USSR) 0.67, 100, 1.22, 0.54
ACO (Orsay France) 0.54, 100, 1.1, 0.32
SOR Ring (Tokyo, Japan) 0.40, 250, 1.1, 0.13
SURF II (Washington DC, USA) 0.25, 25, 0.84, 0.041
TANTALUS I (University of Wisconsin, USA) 0.24, 200, 0.64, 0.048
PIB (Braunschweig, Germany) 0.14, 150, 0.46, 0.013
N-100 (Karkhov, USSR) 0.10, 25, 0.50, 0.004
PEP (Stanford University, USA, 1979) 18, 10, 165.5, 78; 15, 55, 165.5, 45.2; 10, 35, 165.5, 13.4
PHOTON FACTORY (Tsukuba, Japan, 1982) 2.5, 500, 8.33, 4.16
NSLS (Brookhaven National Laboratory, USA, 1981) 2.5, 500, 8.17, 4.2
SRS (Daresbury, UK, 1980) 2.0, 500, 5.55, 3.2
ALADDIN (University of Wisconsin, USA, 1980) 1.0, 500, 2.08, 1.07
BESSY (Berlin, Germany, 1982) 0.80, 500, 1.83, 0.62
NSLS (Brookhaven National Laboratory, USA, 1980) 0.70, 500, 1.90, 0.40

Synchrotrons

DESY (Hamburg, Germany) 7.5, 10–30, 31.7, 29.5
ARUS (Erevan, USSR) 4.5, 1.5, 24.6, 8.22
BONN I (Bonn, Germany) 2.5, 30, 7.6, 4.6
SIRIUS (Tomsk, USSR) 1.36, 15, 4.23, 1.32
INS-ES (Tokyo, Japan) 1.3, 30, 4.0, 1.22
PAKHRA (Moscow, USSR) 1.3, 300, 4.0, 1.22
LUSY (Lund, Sweden) 1.2, 40, 3.6, 1.06
FIAN C-60 (Moscow, USSR) 0.68, 10, 1.6, 0.44
BONN II (Bonn, Germany) 0.5, 30, 1.7, 0.16

[a] This list was obtained from Stanford Electronics Laboratories, Stanford University.
[b] The numbers in each group stand respectively for E(GeV), I(mA), R(meters), ϵ_c(KeV).

[208] F. Wuilleumier, M. Y. Adam, P. Dhez, N. Sandner, V. Schmidt, and W. Mehlhorn, *Jpn. J Appl. Phys.* **17** (1978), Supplement 17–2, 44.
[209] M. Y. Adam, F. Wuilleumier, S. Krummacher, N. Sandner, V. Schmidt, and W. Mehlhorn, *J. Electron Spectrosc.* **15** (1979), 211.

Synchrotron radiation properties are also discussed in [46].

Calibration of photoelectron spectrometers

[210] J. L. Gardner and J. A. R. Samson, *J. Electron Spectrosc.* **8** (1976), 469.

Absolute calibration of an X-ray photoelectron spectrometer

[211] M. F. Ebel, *J. Electron Spectrosc.* **8** (1976), 213.

Experimental determination of analyzer brightness

[212] M. Vulli and K. Starke, *J. Phys. E* **10** (1977), 158.

Experimental determination of the transmission factor of a photoelectron spectrometer

[213] B. Barbaray, J. P. Contour, and G. Mouvier, *Analusis* **5** (1977), 413 (in French).

Measurement of absolute intensity of radiation in XUV

[214] T. Sasaki, T. Oda, and H. Sugawara, *Appl. Opt.* **16** (1977), 3115.

XPS calibration and deconvolution

[215] M. F. Ebel and N. Gurker, *Proceedings 7th International Vacuum Conference 3rd International Conference on Solid Surfaces,* Vienna, 1977.

Energy calibration of electron spectrometers

[216] K. Richter and B. Peplinski, *J. Electron Spectrosc.* **13** (1978), 69.

Relativistic effects in calibration of electrostatic electron energy analyzers: toroidal analyzers, parallel plate analyzers, and cylindrical mirror analyzers

[217] O. Keski-Rahkonen and M. O. Krause, *J. Electron Spectrosc.* **13** (1978), 107.
[218] O. Keski-Rahkonen, *J. Electron Spectrosc.* **13** (1978), 113.
[219] S. Evans, in [18], p. 121.

The discussion on calibration is based primarily on [210] and [219]. For Fermi level and work function, see the notes to Chapter 5.

Threshold photoelectron spectroscopy (TPES) is a technique that involves the detection of threshold photoelectrons as a function of incident photon energy. Threshold photoelectrons are those photoelectrons that are formed when the incident photon energy just exceeds the ionization potential of the sample molecule; these photoelectrons therefore have nominally zero kinetic energy. The method requires a radiation source of variable photon energy. At a given photon energy, threshold photoelectrons produced in the ionization zone are accelerated and then analyzed with an electron energy analyzer. The analysis energy of the analyzer is chosen as equal to the acceleration energy and is kept fixed. It has certain advantages with regard to electron collection efficiency, energy resolution, constancy of electron transmission function, and detection of autoionization states that produce zero-energy electrons. Further, if the photoions produced along with the threshold photoelectrons are also detected, reactions of such energy-selected ions can be studied.

[220] D. Villarejo, R. R. Herm, and M. G. Inghram, *J. Chem. Phys.* **46** (1967), 4995.

[221] R. Stockbauer and M. G. Inghram, *J. Chem. Phys.* **54** (1971), 2242.

[222] R. Spohr, P. M. Guyon, W. A. Chupka, and J. Berkowitz, *Rev. Sci. Instrum.* **42** (1971), 1872.

[223] T. Hsieh, J. P. Gilman, M. J. Weiss, G. G. Meisels, and P. M. Hierl, *Int. J. Mass Spectrom. Ion Phys.* **36** (1980), 317.

In *threshold photoelectron spectroscopy by electron attachment* (TPSA), some electron trapping molecules such as SF_6 or $CFCl_3$ are present in the photoionization chamber along with the sample molecules. The threshold photoelectrons attach to the trapping molecules, and the products (in this case SF_6^-, or Cl^- by dissociative attachment of $CFCl_3$) are detected mass spectrometrically. A plot of the negative ion yield as a function of incident photon energy provides the threshold photoelectron spectrum. Advantages of this method are (1) the resolution in final ion states may approach 3–5 meV (FWHM), convolution of the photon bandpass and the electron attachment width, (2) no electron optics are required, (3) a photoionization mass spectrometer can be easily converted into a TPSA apparatus. The threshold spectrum of Xe shows additional peaks below the $^2P_{3/2}$ ionization threshold from collisional ionization and capture of high Rydberg state electrons by SF_6 or $CFCl_3$. In CO, due to autoionization transistions, striking differences in vibrational intensities are observed in TPSA and PES spectra. Ar, Xe, N_2 CO, C_2H_2, SF_6, and $CFCl_3$ have been studied by the TPSA technique. For the last two gases, lineshapes and cross sections for the thermal electron attachment process have been measured. Ionizations to the Xe $^2P_{1/2}$ level was used as the source of monoenergetic, low-energy electrons.

[224] J. M. Ajello and A. Chutjian, *J. Chem. Phys.* **65** (1976), 5524.

[225] J. M. Ajello, A. Chutjian, and R. Winchell, *J. Electron Spectrosc.* **19** (1980), 197.

[226] A. Chutjian, *Phys. Rev. Lett.* **46** (1981), 1511; *erratum* **48** (1982), 289.

[227] A. Chutjian, *J. Phys. Chem.* **86** (1982), 3518.

Microcomputer automation of the sample probe of an XPS spectrometer

[228] J. W. Fischer, R. M. Downey, L. R. Schrawyer, and R. G. Meisenheimer, *J. Electron Spectrosc.* **16** (1979), 475.

Study of photoions by photoion–photoelectron coincidence

[229] T. Baer, *J. Electron Spectrosc.* **15** (1979), 225.

Electron coincidence spectrometer with position sensitive detectors

[230] S. T. Hood and E. Weigold, *J. Electron Spectrosc.* **15**, (1979), 237.

Photoelectron–photoion coincidence spectrometer with double field time-of-flight mass analysis

[231] Y. Niwa, T. Nishimura, H. Nozoye, and T. Tsuchiya, *Int. J. Mass Spectrom. Ion Phys.* **30** (1979), 63.

A 'teaching' photoelectron spectrometer

[232] W. C. Price and R. Ibrahim, *J. Phys. E.* **11** (1978), 618.

Reference [232] describes the construction of a simple photoelectron spectrometer that has been used in a teaching laboratory for five years with little maintenance. It consists of a helium discharge source, a slotted grid retarding potential analyzer, and an inexpensive transistorized electrometer. In sample spectra (resolution 25 meV FWHM obtained on Ar^+ doublet) vibrational structure of H_2^+, D_2^+, N_2^+, and O_2^+ are well resolved.

Iterative deconvolution of XPS generated by an achromatic source

[233] A. F. Carley and R. W. Joyner, *J. Electron Spectrosc.* **13** (1978), 411.

Inversion of self-convolution

[234] V. Martinez, *J. Electron Spectrosc.* **17** (1979), 33.

Deconvolution is the procedure by which spurious contributions from spectra can be eliminated, provided the nature of such extraneous contributions are known. In photoelectron spectroscopy, such undesirable contributions can arise from spectral distribution of the incident X-rays, imperfections in the analyzer, and the detection systems. An efficient deconvolution can lead to higher resolution.

Deconvolution using the Fourier transform technique

[235] N. Beatham and A. F. Orchard, *J. Electron Spectrosc.* **9** (1976), 129.

[236] D. C. Champeney, *Fourier Transforms and Their Physical Applications*, Academic, London, 1973.

The basis of the deconvolution procedure using the Fourier transform technique is as follows [235].

If $f(x)$ represents the intrinsic or natural line shape and $g(x)$ is the line shape actually observed (x denoting the photoelectron kinetic energy), $g(x)$ is formed out of convolution of $f(x)$ with $k(x)$, the latter being the extraneous broadening and having components $k_1(x)$, $k_2(x)$, and so on. This can be expressed as

$$g(x) = f(x)(0)\,k(x),\ k(x) = k_1(x)(0)\,k_2(x)$$

(0) representing the convolution operation.

For the two functions $f_1(x)$ and $f_2(x)$, the operation (0) explicitly means

$$f_1(x)(0)f_2(x) = \int_{-\infty}^{\infty} f_1(x')f_2(x-x')\,dx'$$

The Fourier transforms of $g(x)$ and $k(x)$ can be written

$$G(s) = \int_{-\infty}^{\infty} g(x)\exp(2\pi isx)\,dx,\quad K(s) = \int_{-\infty}^{\infty} k(x)\exp(2\pi isx)\,dx$$

This yields the convolution theorem:

$$F(s) = \frac{G(s)}{K(s)} = \int_{-\infty}^{\infty} f(x)\exp(2\pi isx)\,dx$$

The function $f(x)$ can be recovered by the inverse Fourier transform:

$$f(x) = \int_{-\infty}^{\infty} F(s)\exp(-2\pi isx)\,ds$$

Although photoionization chambers are operated at low pressures, an increase in the pressure sometimes leads to interesting phenomena. For example, anomalous peaks appear at high pressures due to temporary negative ion states formed by electron–molecule collisional attachments. When the negative ions reemit the electrons, the initial energy of the reemitted electrons is reduced by the number of vibrational quanta of the neutral ground state in which the neutral is formed. Thus, with an increase in pressure, the main photoelectron peak degrades into a series of low-energy bounds having intervals corresponding to the vibrational states of the ground state molecule. A situation like this arises in HeI ionization of N_2 to the $^2\Sigma_u^+$ state at 18.76 eV. Also, if the pressure is high and two species A and B are present, sometimes the photoelectrons due to species A can cause electron absorption spectrum of species B, which gives a vibrational pattern from the ground state of the negative ion B^-. See [40].

[237] D. G. Streets, A. W. Potts, and W. C. Price, *Int. J. Mass Spectrom. Ion Phys.* **10** (1972/1973), 123.

CHAPTER 3

Free-atom core binding energies from XPS: Zn, Cd, Na, K, Rb, Cs, and Mg

[238] M. S. Banna, D. C. Frost, C. A. McDowell, and B. Wallbank, *J. Chem. Phys.* **68** (1978), 696.

[239] M. S. Banna, B. Wallbank, D. C. Frost, C. A. McDowell, and J. S. H. Q. Perera, *J. Chem. Phys.* **68** (1978), 5459.

XPS of atomic Na

[240] R. L. Martin, E. R. Davidson, M. S. Banna, B. Wallbank, D. C. Frost, and C. A. McDowell, *J. Chem. Phys.* **68** (1978), 5006.

XPS of Mg, Zn, Sr, Cd, Hg, and Pb

[241] W. Mehlhorn, B. Breuckmann, and D. Hausamann, *Physica Scripta*, **16** (1977), 177.

Core binding energies of metals with $Z = 20$, 22–29, 31, 38–42, 44–52, 56, 72–89, 90, and 92

[242] J. C. Fuggle and N. Martensson, *J. Electron Spectrosc.* **21** (1980), 275.

A theoretical study of $1s$ spectrum of atomic Li

[243] F. P. Larkins, P. D. Adeney, and K. G. Dyall, *J. Electron Spectrosc.* **22** (1981), 141.

Binding energy shifts presented in Fig. 3.1 are from the following source:

[244] D. T. Clark, in [18], p. 211.

Electrostatic calculation of XPS chemical shifts

[245] Z. B. Maksic, K. Kovacevic, and H. Metiu, *Croatica Chemica Acta* **46** (1974), 1.

XPS chemical shifts and Hammett substituent constants in substituted benzene derivatives

[246] B. Lindberg, S. Svensson, P. A. Malmquist, E. Basilier, U. Gelius, and K. Siegbahn, *Chem Phys. Lett.* **40** (1976), 175.

Internal standards for measuring chemical shifts in coordination compounds

[247] R. Larsson and B. Folkesson, *Physica Scripta* **16** (1977), 357.

Ab initio study of chemical shifts in some chlorine-containing systems

[248] D. B. Adams, *J. Electron Spectrosc.* **10** (1977), 247.

Single orbital relaxations accompanying core ionization in CH_4, NH_3, H_2O, HF, and Ne

Characteristic Elemental Binding Energies (eV) of the Deepest-lying Core Level Having Energy < 1486.7 eV (Al K_α Line)[a,b]

H	13.6	$K1$	Cr	697.5 (+ 0.6) $L1$	Ag	718.5 (+ 1.2) $M1$	Yb	479.7 (+ 1.4) $N1$
He	24.59		Mn	769.2 (+ 0.4)	Cd	770.9 (+ 1.2)	Lu	506.4 (+ 0.8)
Li	54.0 (+ 1.0)[a]		Fe	848.5 (+ 0.4)	In	825.6 (+ 1.6)	Hf	538.0
Be	111.1 (+ 0.6)		Co	926.4 (+ 0.4)	Sn	883.5 (+ 0.8)	Ta	562.0 (+ 2.0)
B	188.0		Ni	1009.9 (+ 1.0)	Sb	943.0 (+ 2.0)	W	591.0 (+ 2.0)
C	284.0 (+ 0.7)		Cu	1096.1 (+ 3.2)	Te	1005.0 (+ 2.0)	Re	624.2 (+ 1.6)
N	409.8 (+ 0.2)		Zn	1200.4 (+ 0.6)	I	1072.0	Os	655.0
O	542.9 (+ 0.4)		Ga	1298.3 (+ 1.4)	Xe	1148.2 (+ 1.0)	Ir	689.8 (+ 1.6)
F	696.66 (+ 0.1)		Ge	1413.9 (+ 1.4)	Cs	1210.0 (+ 2.0)	Pt	724.0 (+ 2.0)
Ne	870.1 (+ 0.2)		As	1358.4 (+ 1.4) $L2$	Ba	1292.0 (+ 2.0)	Au	758.0 (+ 2.0)
Na	1070.5 (+ 1.3)		Se	1473.6 (+ 1.4)	La	1361.0 (+ 2.0)	Hg	798.0 (+ 4.0)
Mg	1302.9 (+ 0.2)		Br	257.0	$M1$	Ce	1435.0 (+ 2.0)	Tl	844.0 (+ 4.0)
Al	117.3 (+ 1.1) $L1$		Kr	292.6 (+ 0.4)	Pr	1342.0	$M2$	Pb	892.0 (+ 2.0)
Si	148.8 (+ 1.5)		Rb	325.7 (+ 2.0)	Nd	1408.0		Bi	938.0 (+ 2.0)
P	189.0		Sr	357.7 (+ 0.6)	Pm	1476.0		Po	995.0
S	230.2 (+ 1.4)		Y	391.2 (+ 1.6)	Sm	1422.0	$M3$	At	1042.0
Cl	270.0		Zr	431.0	Eu	1484.0		Rn	1097.0
Ar	326.2 (+ 0.2)		Nb	469.0	Gd	1221.1 (+ 1.6) $M4$	Fr	1153.0
K	378.3 (+ 0.6)		Mo	504.1 (+ 1.6)	Tb	1274.8 (+ 4.8)	Ra	1208.0
Ca	437.9 (+ 1.4)		Tc	544.0	Dy	1335.0	Ac	1269.0
Sc	497.7 (+ 0.6)		Ru	585.0	Ho	1395.0	Th	1330.0
Ti	561.1 (+ 0.6)		Rh	626.8 (+ 1.2)	Er	1456.0	Pa	1387.0
V	626.8 ((+ 0.8)		Pd	671.1 (+ 1.2)	Tm	1471.0	$M5$	U	1438.0 (+ 2.0)

[a] The binding energies are with respect to (1) the vacuum level for gases, (2) the Fermi level for metals, and (3) the top of the valence band for semiconductors. $K = 1, L = 2, M = 3, N = 4; 1 = s_{1/2}, 2 = p_{1/2}, 3 = p_{3/2}, 4 = d_{3/2},$ and $5 = d_{5/2}$.
[b] The number in parentheses, when added to the binding energy, provides the range of binding energy values available for the level; this includes results of various measurements and also the range of errors in the experiments. For references to original work, see [22], [14], and [1].

[249] D. T. Clark, B. J. Cromarty, and A. Sgamellotti, *J. Electron Spectrosc.* **13** (1978), 85.

Correlation of inner shell binding energies with charge transfer band shifts

[250] H. Klingelhöfer and G. Lehmann, *J. Electron Spectrosc.* **16** (1979), 259.

Chemical shifts in Al–Si compounds

[251] J. E. Castle, L. B. Hazell, and R. H. West, *J. Electron Spectrosc.* **16** (1979), 97.

Role of relaxation energy in chemical shifts

[252] F. O. Ellison and M. G. White, *J. Chem. Ed.* **53** (1976), 430.

Reference [252] describes a simplified procedure for calculating relaxation energies in molecules. The hydrogen fluoride case is treated in detail.

The two relationships that form the basis of the method are as follows. (1) The Slater–Zener–Snyder (SZS) atomic energy expression, which for an atom A is

$$\epsilon^A = -(13.605 \text{ eV}) \sum_n E_n^A (N_n^A, S_n^A) \qquad (3-1)$$

$$E_n^A(N_n^A, S_n^A) = \frac{N_n^A}{n^2}(Z^A - S_n^A)^2 - 2\frac{N_n^A}{n^2}(Z^A - S_n^\circ)(Z^A - S_n^A)$$

where ϵ is the total electronic energy of the atom A, n is the principal quantum number, N_n is the electronic population of the nth quantum state, Z the nuclear charge, S_n the shielding parameters for atomic orbitals, and S_n° the corresponding optimum shielding parameters, which can be calculated from Slater's rules. (2) The change of energy due to a population change. For a diatomic molecule AB this can be written

$$d\epsilon = \frac{\partial \epsilon^A}{\partial N^B} dN^A + \frac{\partial \epsilon^B}{\partial N^B} dN^B \qquad (3-2)$$

and this can be used for minimization of the energy as a function of the population. Under this minimization condition $d\epsilon = 0$. Since the total charge is conserved, it follows that $dN^A = -dN^B$, which implies that $(\partial \epsilon^A/\partial N^A) = (\partial \epsilon^B/\partial N^B)$. Since $(\partial \epsilon/\partial N)$ is equal to the electronegativity in Mulliken's scale, one may conclude that charge flow in the molecule will take place until an equalization of electronegativity (EOE) is established, which, however, may cause partial charges to develop on atoms A and B. It is an iterative application of SZS and EOE equations, used to determine the charge distribution that gives the minimum energy and hence the value of the energy at the minimum. One may start with an assumed electron population in various quantum states of A and B, then calculate from Eq. (3–1), minimize with respect to the population, as in Eq. (3–2), then determine the resulting charge distribution, which will be an improvement over the initially assumed values. Iteration continues until the resulting energy and charge values in two consecutive iteration cycles differ by less than a prescribed value. Such calculations can be made for the initial molecular state, then for the Koopmans's state with loss of an electron but leaving the rest of the system unperturbed, and then for the final relaxed state of the ion. The energy difference between the latter two cases gives the relaxation energy.

Charge potential model and C 1s group shifts

[253] U. Gelius, P. F. Heden, J. Hedman, B. J. Lindberg, R. Manne, R. Nordberg, C. Nordling, and K. Siegbahn, *Physica Scripta* 2 (1970), 70.

[254] B. J. Lindberg, K. Hamrin, G. Johansson, U. Gelius, A. Fahlman, C. Nordling, and K. Siegbahn, *Physica Scripta* 1 (1970), 286.

Following is the list of molecules referred to in Figs. 3.2 and 3.4, arranged according to the type of hybridization. The binding energy refers to the atom in italic type.

1a ($H_3CCH_2NH_2$), 2a (H_3CCH_2OH), 3a (H_3CCH_2Cl),

4a ($H_3CCH_2OCH_2CH_3$), 5a ($H_3CCH_2OOCCH_3$), 6a ($H_3CCH_2OOCCl_3$)

7a ($H_3CCH_2OOCCF_3$), 8a ($H_3CCH\begin{smallmatrix}CH_3\\OCH(CH_3)_2\end{smallmatrix}$),

9a ($H_3CCH\begin{smallmatrix}OCH(CH_3)\\OCH(CH_3)\end{smallmatrix}O$), 10a ($H_3CCOH_3$), 11a ($H_3CCOCF_3$),

12a ($H_3CCOONa$), 13a (H_3CCOOH), 5b ($H_3CCOOCH_2CH_3$),

14 (H_3COH), 15a ($H_3COCH(OCH_3)_2$), 16a ($H_3COCOOCH_3$),

17 ($H_2C\begin{smallmatrix}CH_2-CH_2\\CH_2-CH_2\end{smallmatrix}CH_2$), 1b ($H_2C\begin{smallmatrix}CH_3\\NH_2\end{smallmatrix}$),

18 ($H_2C\begin{smallmatrix}NCH_2\\CH_2\\NCH_2-N\\CH_2\\NCH_2\end{smallmatrix}$), 2b ($H_2C\begin{smallmatrix}CH_3\\OH\end{smallmatrix}$), 4b ($H_2C\begin{smallmatrix}CH_3\\OCH_2CH_3\end{smallmatrix}$),

19a (H₂C(COOH)(OCH₂COOH)), 5c (H₂C(CH₃)(OOCCH₃)), 7b (H₂C(CH₃)(OOCCF₃)),

6b (H₂C(CH₃)(OOCCl)), 3b (H₂C(CH₃)(Cl)), 20 (H₂C(Br)(Br)),

21 (H₂C(Cl)(Cl)), 8b (HC(CH₃)(CH₃)(OCH(CH₃)₂)), 22a (HC(CH(OH)COOH)(COOK)(OH)),

23 (HC(CH(OH)CH(OH))(OH)(CH(OH)CH(OH))HCOH, 9b (HC(CH₃)(O—CH(CH₃))(O—CH(CH₃))O),

15b (HC(OCH₃)(OCH₃)(OCH₃)), 24 (HC(Cl)(Cl)(Cl)), 25 (HC(F)(F)(F)),

26a (Cl₃CCOONa), 27a (Cl₂FCCClF₂), 28 (Cl₃CCCl),

27b (ClF₂CCCl₂F), 11b (F₃CCOCH₃), 29a (F₃CCOONa),

7c (F₃CCOOCH₂CH₃), 30 (F₃CF), 31 (O=C(H)(H)),

10b (O=C(CH₃)(CH₃)), 11c (O=C(CH₃)(CH₃)), 32 (O=C(CO—CO)(CO—CO)C=O),

33 (O=C(H)(ONH₄)), 12b (O=C(CH₃)(ONa)), 34 (O=C(CH₂NH₂)(ONa)),

22c (O=C(CH(OH)CH(OH)COOH)(OK)), 35 (O=C(COONa)(ONa)),

26b (O=C(CCl₃)(ONa)), 29a (O=C(CF₃)(ONa)), 36 (O=C(NH₂)(NH₂)),

13b (O=C(CH$_3$)(OH)), 22d (O=C(CH(OH)CH(OH)COOK)(OH)),

19b (O=C(CH$_2$OCH$_2$COOH)(OH)), 37 (O=C(COOH)(OH)), 5d (O=C(CH$_3$)(OCH$_2$CH$_3$)),

7d (O=C(CF$_3$)(OCH$_2$CH$_3$)), 38 (O=C(ONa)(ONa)), 39 (O=C(ONa)(OH)),

16b (O=C(OCH$_3$)(OCH$_3$)), 6c (O=C(OCH$_2$CH$_3$)(Cl)), 40 (O=C(F)(F)),

41 (HC=CH–CH=CH cycle), 42a (HC=C(OH)–C(OH)=CH cycle, with COH), 43 (HC–N=CH–NH–N cycle),

42b (HOC=CH–C(OH)=CH cycle, with CH), 44 (FC–CF=CF–CF=CF cycle),

45 (S=C=S), 46 (O=C=O), 47 (K$^+$C$^-$≡N).

Valence potential model

[255] M. E. Schwartz, J. D. Switalski, and R. E. Stronski, in [6], p. 605.

Following is the list of molecules referred to in Fig. 3.3. When the C atoms in the molecule are not all equivalent, the binding energy shift refers to the atom in italic type.

1 (C$_2$H$_2$), 2 (C$_2$H$_4$), 3 (CH$_4$), 4 (C$_2$H$_6$), 5 (C$_6$H$_6$),

6 (CH$_3$CH$_2$OH), 7 (C$_6$H$_5$F, C$_\gamma$), 8 (C$_6$H$_5$F, C$_\beta$), 9 (CH$_3$COOH),

10 (CH$_3$CHO), 11 (CO), 12 (CH$_3$OH), 13 (CH$_3$*C*H$_2$OH),

14 (*C*H$_3$CN), 15 (CH$_3$Cl), 16 (CH$_3$F), 17 (HCN),

18 (C$_6$H$_5$F, C$_\alpha$), 19 (CH$_3$CN, CH$_3$*C*N), 20 (CH$_3$*C*HO),

21 (HCHO), 22 (CH$_2$Cl$_2$), 23 (CH$_3$*C*OOH), 24 (HCOOH),

25 (CS$_2$), 26 (CH$_2$F$_2$), 27 (CO$_2$), 28 (CHCl$_3$), 29 (CHF$_3$),

30 (CCl$_4$), 31 (CF$_4$).

Effective charges in tetrachlorometallate and hexachlorometallate ions from XPS

[256] R. Larsson and B. Folkesson, *Chemica Scripta* **11** (1977), 5.

Correlation of thermodynamic data and XPS shifts, equivalent cores

[257] W. L. Jolly in [6], p. 629.
[258] P. Finn and W. L. Jolly, *J. Amer. Chem. Soc.* **94** (1972), 1540.
[259] D. B. Adams, *J. Electron Spectrosc.* **9** (1976), 251.
[260] W. L. Jolly and C. Gin, *Int. J. Mass Spectrom. Ion Phys.* **25** (1977), 27.

Comments on correlations of XPS shifts with other physical and chemical data

[261] D. A. Shirley, *J. Electron Spectrosc.* **5** (1974), 135.

XPS–NMR correlations

[262] B. J. Lindberg, *J. Electron Spectrosc.* **5** (1974), 149.

Halomethanes

[263] U. Gelius, G. Johansson, H. Siegbahn, C. J. Allan, D. A. Allison, J. Allison, and K. Siegbahn, *J. Electron Spectrosc.* **1** (1972/1973), 285.

Monosubstituted methanes, NMR data

[264] H. Suhr, *Anwendungen der kernmagnetischen Resonanz in der organischen Chemie*, Springer Verlag, Berlin, 1965.

Fluoromethanes, using C $1s$ and F $1s$

[265] D. Zeroka, *Chem. Phys. Lett.* **14** (1972), 471.

Organometal phenyl sulfides

[266] S. Pignataro, L. Lunazzi, C. A. Boicelli, R. Di Marino, A. Ricci, A. Magnini, and R. Danieli, *Tetrahedron Lett.* (1972), 5341.

p-Fluorophenyl sulfur compounds, NMR

[267] J. W. Emsley and I. Phillips, in *Progress in NMR spectroscopy*, J. W. Emsley, J. Feeney, and L. H. Sutcliffe, Eds., Pergamon, Oxford, 1971.

Dithiole derivatives

[268] B. J. Lindberg, R. Pinel, and Y. Mollier, *Tetrahedron* **30** (1974), 2537.
[269] B. J. Lindberg, S. Hogberg, G. Malmsten, J. E. Bergmark, O. Nilsson, S. E. Karlsson, A. Fahlman, U. Gelius, R. Pinel, M. Stavaux, Y. Mollier, and N. Lozach, *Chemica Scripta* **1** (1971), 183.

Mössbauer—XPS correlations

[270] U. Gelius, *Physica Scripta* **9** (1974), 133.
[271] M. Barber, P. Swift, D. Cunningham, and M. J. Frazer, *Chem. Commun.* (1970), 1338.
[272] I. Adams, J. M. Thomas, G. M. Bancroft, K. D. Butler, and M. Barber, *Chem. Commun.* (1972), 751.
[273] J. Blomquist, *J. Electron Spectrosc.* **14** (1978), 73.

C 1s binding energy shifts in substituted benzene derivatives

[274] B. Lindberg, S. Svensson, P. A. Malmqvist, E. Basilier, U. Gelius, and K. Siegbahn, Uppsala Univ. Report UUIP-910, December 1975.

The Hammett equation, based on linear free energy relationship (LFER), correlates the equilibrium constants of a reaction R involving a benzene derivative M having a side chain with a reactive center and that of MS (M with a meta or para substitutent S on benzene ring) with the equilibrium constants of benzoic acid and substituted benzoic acid having the same substituent S. It is expressed as

$$\log \frac{K}{K_0} = \rho_R \log \frac{K'}{K_0'} = \rho_R \sigma_S$$

where K, K_0, K', and K_0' respectively stand for the equilibrium constants of M, MS, benzoic acid, and substituted benzoic acid. ρ_R is a characteristic constant of the reaction independent of the substituent, and σ_S is characteristic of the substituent and is independent of the reaction.

Binding energy shifts of vanadium complexes

[275] R. Larsson, B. Folkesson, and G. Schon, *Chemica Scripta* **3** (1973), 88.

XPS of biguanide complexes

[276] W. E. Swartz Jr. and R. A. Alfonso, *J. Electron Spectrosc.* **4** (1974), 351.

The high resolution XPS of NH_4NO_3 using monochromatized X-rays [36] shows that the N 1s width is 1.46 eV in NH_4^+ and 1.18 eV in NO_3^- as would be expected with vibrational broadening of core lines. The 1.8 eV value observed for both NH_4^+ and NO_3^- in [276], therefore, might be having significant instrumental contribution. If such contribution, however, is present also in the N 1s width of the complexes then the conclusions derived in [276] may still be expected to be valid.

XPS studies of axial and equatorial CO groups in $Fe(CO)_5$

[277] S. C. Avanzino, W. L. Jolly, P. A. Malmquist, and K. Siegbahn, *Inorg. Chem.* **17** (1978), 489.

χ^2 and F tests

[278] H. A. Laitinen, *Chemical Analysis,* McGraw-Hill, New York, 1960, p. 552.

XPS of trimethylenemethane iron tricarbonyl and butadiene iron tricarbonyl

[279] J. W. Koepke, W. L. Jolly, G. M. Bancroft, P. A. Malmquist, and K. Siegbahn, *Inorg. Chem.* **16** (1977), 2659.

XPS studies on platinum blues

[280] J. K. Barton, S. A. Best, S. J. Lippard, and R. A. Walton, *J. Amer. Chem. Soc.* **100** (1978), 3785.

XPS of porphyrins

[281] J. P. Macquet, M. M. Millard, and T. Theophanides, *J. Amer. Chem. Soc.* **100** (1978), 4741.

XPS of Ni-phosphine complexes

[282] A. Furlani, G. Polzonetti, M. V. Russo, and G. Mattogno, *Inorg. Chim. Acta* **26** (1978), L39.

For the shake-up spectrum of Ne, see [4], [35], and [36].

Role of electron spin in shake-up spectra

[283] T. X. Carroll and T. D. Thomas, *J. Electron Spectrosc.* **10** (1977), 215.

Shake-up satellite in the XPS of metal hexacarbonyls

[284] G. M. Bancroft, B. D. Boyd, and D. K. Creber, *Inorg. Chem.* **17** (1978), 1008.

Shake-off in Ne and the interaction of the continuum states

[285] D. Chattarji, W. Mehlhorn, and V. Schmidt, *J. Electron Spectrosc.* **13** (1978), 97.

Satellite lines in the XPS of Xe

[286] M. Y. Adam, F. Wuilleumier, N. Sandner, V. Schmidt, and G. Wendin, *J. Physique* **39** (1978), 129.

Satellite lines in the XPS of Ar

[287] M. Y. Adam, F. Wuilleumier, S. Krummacher, V. Schmidt, and W. Mehlhorn, *J. Phys. B* **11** (1978), L413.

Vibrational broadening of core electron lines, besides [35] and [36]

[288] U. Gelius, S. Svensson, H. Siegbahn, E. Basilier, A. Faxälv, and K. Siegbahn, *Chem. Phys. Lett.* 28 (1974), 1.

A theoretical study of vibrational fine structure in core ionization of HCN and C_2H_2

[289] D. T. Clark and L. Colling, *J. Electron Spectrosc.* 13 (1978), 317.

A theoretical investigation of vibrational excitations of $1b_1$ bands of H_2O

[290] H. Agren and J. Müller, *J. Electron Spectrosc.* 19 (1980), 285.

Photoelectron peak intensities and atom—ion overlaps

[291] M. Mehta, C. S. Fadley, and P. S. Bagus, *Chem. Phys. Lett.* 37 (1976), 454.

Multiplet broadening of $2p$ lines in the XPS of $3d$ transition metal compounds

[292] I. G. Main and J. F. Marshall, *J. Phys. C* 9 (1976), 1603.

Structure of $2p$ lines in XPS of low-spin $3d^6$ transition metal compounds

[293] I. G. Main, C. E. Johnson, J. F. Marshall, G. A. Robins, and G. Demazeau, *J. Phys. C* 10 (1977), L677.

Core-level ligand field splittings in XPS

[294] G. M. Bancroft and R. P. Gupta, *Chem. Phys. Lett.* 54 (1978), 226.

XPS of active species for the formation of C_2H_5Cl from C_2H_5OH

[295] E. Akiba, S. Naito, M. Soma, T. Onishi, and K. Tamaru, *Chem. Lett.* (1978), 483.

A study of 8-hydroxyquinolinates of Al, Ga, In, and Sc

[296] M. Thompson, *Analytica Chim. Acta* 98 (1978), 357.

XPS investigation of ambident ions and tautomerism; N-cyanobenzamides and benzohydroxamic acids

[297] B. Lindberg, A. Berndtsson, R. Nilsson, R. Nyholm, and O. Exner, *Acta Chem. Scand.* 32 (1978), 353.

XPS study of Fe (II) and Fe (III) fluorides

[298] M. Kasrai and D. S. Urch, *J. Chem. Soc. Faraday Trans. II* 75 (1979), 1522.

XPS of hexavalent Fe

[299] H. Konno and M. Nagayama, *J. Electron Spectrosc.* 18 (1980), 341.

XPS of dopant molecules for conducting polymers
[300]· P. Brant, M. J. Moran, and D. C. Weber, *Chem. Phys. Lett.* 76 (1980), 529.

XPS studies of metal complexes of sulfur-containing ligands
[301] R. A. Walton, *Coord. Chem. Rev.* 31 (1980), 183.

Brief notes on semiempirical molecular orbital methods follow. The neglect of differential overlap methods are based on [304] and HAM/3 on [391], in which further references are provided. For other semiempirical and general MO methods, see [303] through [310].

Slater type orbitals (STO). Each orbital is of the general form

$$\phi_{n,l,m,\zeta} = A(n,\zeta) r^{n-1} e^{-\zeta r} Y_l^m(\theta,\phi)$$

The exponential part has the form $e^{-\zeta r}$, the effective charge ζ is positive, and Y_l^m represents normalized spherical harmonics.

Gaussian type orbitals (GTO). Each orbital has the form

$$\phi_{n,l,m,\zeta} = A(n,\zeta) r^{n+1} e^{-\zeta r^2} Y_l^m(\theta,\phi)$$

exponential part has the form $e^{-\zeta r^2}$, ζ positive.

Slater determinant. The N-particle wave function in the product form can be written

$$\psi(1,2,\ldots,N) = \phi_1(1)\phi_2(2)\ldots\phi_N(N)$$

where ϕ's are single-particle wave functions. If indistinguishability of the particles is taken into account and all possible permutations in the orbitals are considered, then the antisymmetrized wave function can be written in a determinantal form:

$$\psi(1,2,\ldots N) = \begin{vmatrix} \phi_1(1) & \phi_1(2) & \ldots & \phi_1(N) \\ & \phi_2(2) & \ldots & \\ & & & \ldots \phi_N(N) \end{vmatrix}$$

The principal diagonal corresponds to the initial serial order; all possible permutations are involved.

The total Hamiltonian operator $\mathscr{H}_T$ of a system of electrons and nuclei that includes all possible electrostatic interactions between the particles as well as the particle kinetic energies is given by (in atomic units):

$$\mathscr{H}_T = \sum_{A<B} \frac{Z_A Z_B}{r_{AB}} - \sum_A \sum_i \frac{Z_A}{r_{Ai}} + \sum_{i<j} \frac{1}{r_{ij}}$$
$$- \sum_i \frac{h^2}{8\pi^2 m} \nabla_i^2 - \sum_A \frac{h^2}{8\pi^2 M_A} \nabla_A^2$$

Z_A and Z_B are nuclear charges, r_{AB} internuclear distances, r_{Ai} electron–nucleus distances, m electronic mass, M_A atomic masses, ∇_i^2 and ∇_A^2 are the kinetic energy operators for the electrons and the nuclei respectively. The first term in the above equation represents internuclear repulsion, the second electron–nucleus attraction, and the third interelectronic repulsion. In comparison with fast-moving electrons, the nuclei may be assumed stationary (Born–Oppenheimer approximation) and the internuclear repulsion may be treated separately. The second, third, and fourth terms constitute the electronic Hamiltonian $\mathscr{H}_e$, which contributes to the electronic energy of the system.

For a closed shell system, if each orbital is doubly occupied – by an electron of spin $\alpha(+\frac{1}{2})$ and another of spin $\beta(-\frac{1}{2})$, the wave function Ψ for an n-electron system can be expressed as a single Slater determinant:

$$\Psi = (n!)^{-1/2} |\psi_p^\alpha(1) \psi_p^\beta(2) \cdots \psi_z^\alpha(n-1) \psi_z^\beta(n)|$$

where $\psi_p^\alpha(1)$ denotes a one-electron molecular wave function of pth molecular orbital containing electron 1, and so on. The product shown in the determinant is the diagonal term. Each such ψ_p may be expressed as a linear combination of atomic orbitals (LCAO) ϕ_k (basis set) with coefficient c_k's, such as $\psi_p(i) = (1/\sqrt{N_p}) \Sigma_k c_k^p \phi_k(i)$, where N_p the normalization constant is given by $N_p = \Sigma_k \Sigma_l c_k^p c_l^p S_{kl}$, S_{kl} representing the overlap $\phi_k \phi_l$ between the k and l orbitals.

Electronic energy E of the system is given by the energy functional $\int \Psi^* \mathscr{H}_e \Psi d\tau / \int \Psi^* \Psi d\tau$. E is minimized variationally with respect to c_k^p, that is, $dE/dc_k^p = 0$ for each index k and p. If Ψ is represented by the determinant above, and the energy functional is constructed and minimized with respect to the coefficients c_k^p, one obtains equations of the type $dE/dc_k^p = \Sigma_l c_l^p (F_{kl} - E^p S_{kl}) = 0$ (Roothaan's equation) for each index k, which implies that the terms of the secular determinant $|F_{kl} - E^p S_{kl}| = 0$. The Fock matrix elements F_{kl} explicitly are as follows:

$$F_{kl} = \int \phi_k(i) \left[-\sum_i \left(\sum_A \frac{Z_A}{r_{Ai}} + \frac{h^2}{8\pi^2 m} \nabla_i^2 \right) \right] \phi_l(i) d\tau_i$$

resonance integral

$$+ \sum_m \sum_n 2 \sum_p (c_m^p c_n^p / N_p) \left[\int \phi_k(i) \phi_m(j) \left(\sum_{i<j} \frac{1}{r_{ij}} \right) \phi_l(i) \phi_n(j) d\tau_i d\tau_j \right.$$

Coulomb repulsion integral

$$\left. - \frac{1}{2} \int \phi_k(i) \phi_m(j) \left(\sum_{i<j} \frac{1}{r_{ij}} \right) \phi_l(j) \phi_n(i) d\tau_i d\tau_j \right]$$

exchange integral

The F_{kl} above may also be expressed in a simplified notation as

$$F_{kl} = H_{kl} + \sum_m \sum_n P_{mn} [\langle kl|mn\rangle - \tfrac{1}{2}\langle kn|lm\rangle]$$

$$= H_{kl} + \sum_m \sum_n P_{mn} [T]$$

The pth molecular orbital energy is $E^p = \Sigma_m \Sigma_n (c_m^p c_n^p/N_p) F_{mn}$, and the overlap integral $S_{kl} = \int \phi_k(i)\phi_l(i)d\tau_i$. The differential overlap δ_{kl} between the two atomic functions ϕ_k and ϕ_l is defined as $\phi_k(i)\phi_l(i)$; physically it means the probability of finding an electron i in a volume element common to ϕ_k and ϕ_l. Solution of the secular determinant $|F_{kl} - E^p S_{kl}| = 0$ yields E^p of each molecular orbital. The substitution of these energies into the energy minimization equation, the use of the normalization condition $\Sigma_k \Sigma_l c_k c_l S_{kl} = 1$, and the rules of orthogonality yield the coefficients c and hence the weightage of various atomic orbitals.

The solution of the secular equation, however, requires knowledge of F_{kl}, hence also the coefficient c's, as F_{kl}'s are functions of c's. The Hartree–Fock (HF) self-consistent field (SCF) procedure assumes an initial charge distribution, that is, an initial set of c's. F_{kl}'s are then set up and the determinantal equation solved. This yields an improved set of c values (hence E_p's and ψ's), which are then used to constitute an improved set of F_{kl}'s. The procedure is iterated until self-consistency is achieved. The best HF energy is still different from the true electronic energy. This is due to the electron correlation correction E_{corr} that arises because of dependence of the position and motion of an electron on the position and motion of other electrons in the system. It is also due to the relativistic correction E_{rel} when inner shell electrons have relativistic velocities and the electronic mass needs to be treated relativistically. Thus $E_{\text{true}} = E_{\text{HF}} + E_{\text{corr}} + E_{\text{rel}}$. Besides c's, setting up of F_{kl}'s requires H_{kl}'s and $\langle|\rangle$ integrals.

Hartree–Fock–Slater wave functions are the HF wave functions derived by using Slater exchange expression for the potential. (This is discussed in Chapter 6). The presence of configuration interaction means that the wave function of a state is described by a mixture of different configurations, including those of excited states. Such mixing causes a lowering of energy and hence constitutes a better description of the state. An alternative way of presenting the F_{kl} matrix elements above is in terms of the Fock operator $\mathscr{F}$. Appropriate terms of the Hamiltonian are incorporated into $\mathscr{F}$ in a manner such that $F_{kl} = \langle \phi_k | \mathscr{F} | \phi_l \rangle$. The H_{kl} and $\langle|\rangle$ terms can either be calculated as in *ab initio* methods or be approximated as follows.

Zero differential overlap (ZDO) approximation assumes $\delta_{kl} = 0$, unless $k = l$. As a result, in this approximation, overlap integral $S_{kl} = 1$ (if $k = l$) or $= 0$ (if $k \neq l$). Nuclear-electron attraction integrals $Z_A \int (\phi_k(i)\phi_l(i)/r_{Ai}) d\tau_i = 0$ unless $k = l$. The electron–electron repulsion (and exchange) integrals $\int (\phi_k(i)\phi_m(j)\phi_l(i)\phi_n(j)/r_{ij}) d\tau_i d\tau_j = \langle kl|mn\rangle = 0$ unless $k = l$ and $m = n$ (for exchange $\langle kn|lm\rangle = 0$ unless $k = n, l = m$). ZDO thus removes all three and four center integrals. Further, since

$$H_{kl} = -\sum_{A} Z_A^* \int \frac{\phi_k(i)\phi_l(i)}{r_{Ai}} d\tau_i - \frac{h^2}{8\pi^2 m} \int \phi_k(i) \nabla_i^2 \phi_l(i) d\tau_i$$

(Z_A^* denotes effective core with charge Z_A^*), if $k = l$, then no term cancels. Thus, if the attraction of the ith electron by all nuclei other than A is separated out as $-\Sigma_{B \neq A} V_{kk}^B$, and the attraction due to the nucleus A and the kinetic energy term are grouped together as U_{kk}^A, then $H_{kk} = U_{kk}^A - \Sigma_{B \neq A} V_{kk}^B$. if $k \neq l$, ZDO approximation cancels all nucleus–electron attraction terms; only the kinetic energy term remains, and H_{kl} is treated as an empirical parameter, proportional to the overlap integral $H_{kl} \approx S_{kl} \beta_{kl}$. The two-electron term $T = \langle kl|mn \rangle - \frac{1}{2} \langle kn|lm \rangle$ integrals cancel unless $k = l$ and $m = n$, or $k = n$ and $l = m$. If $k = l$, $m = n$, then $T = \langle kk|mm \rangle - \frac{1}{2}[0] = \Gamma_{km}$; if $k = n$, $l = m$, then $T = [0] - \frac{1}{2} \langle kk|mm \rangle = -\frac{1}{2} \Gamma_{km}$; and if $k = n = l = m$, $\Gamma = \langle kk|kk \rangle - \frac{1}{2} \langle kk|kk \rangle = \frac{1}{2} \Gamma_{kk}$.

CNDO/1. CNDO (complete neglect of differential overlap) is based on ZDO. It further assumes that repulsion between electrons in orbitals of the same atom depend only on the nature of the atom. Thus for any valence orbitals k and m of atom A, $\Gamma_{km} = \langle s_A s_A | s_A s_A \rangle = \Gamma_{AA}$ (s_A = spherically symmetric s orbitals). The two-center repulsion between electrons in the k orbital of A and the l orbital of B is taken as electron repulsion integral between Slater s orbitals, that is, $\Gamma_{kl} = \langle kk|ll \rangle = \langle s_A s_A | s_B s_B \rangle = \Gamma_{AB}$. $V_{kk}^B = Z_B^* \langle s_A s_A | B \rangle = V_A^B$ ($B = 1/r_{Bi}$). $H_{kl} = S_{kl} \beta_{kl} = S_{kl}(\beta_A^0 + \beta_B^0)/2$, where β_A^0, β_B^0 are characteristic parameters of atoms A and B determined empirically with reference to *ab initio* calculation. U_{kk}^A is obtained by comparing the theoretical ionization potential with the experimental ionization potential IP_k^A (loss of a k electron). The expression for U_{kk}^A turns out to be $U_{kk}^A = -IP_k^A - (Z_A^* - 1) \Gamma_{kk}$.

CNDO/2. The only differences from CNDO/1 are $V_{kk}^B = Z_B^* \Gamma_{AB} = Z_B^* \Gamma_{kk}$, that is, the two-center nuclear–electron interaction is assigned the same value in magnitude as the two-center electron–electron interaction, and U_{kk}^A is taken as $U_{kk}^A = -\frac{1}{2}(IP_k + EA_k) - (Z_A^* - \frac{1}{2}) \Gamma_{kk}$ where EA is the electron affinity.

In INDO (intermediate neglect of differential overlap) and MINDO (modified INDO) methods, the differential overlap is neglected only if it appears in a two-center integral. ZDO matrix elements are no longer applicable. The matrix elements obtained from $F_{kl} = H_{kl} + \Sigma\Sigma_{m\ n} P_{mn} (\langle kl|mn \rangle - \frac{1}{2} \langle kn|lm \rangle)$ for any set of electrons α or β are: $F_{kk}^u = U_{kk}^A + (P_{kk} - P_{kk}^u) \Gamma_{kk} + \Sigma_{m \neq k}^A (P_{mm} \Gamma_{km} - P_{mm}^u \Gamma_{km}^{ex}) + \Sigma_{B \neq A} (P_{BB} \Gamma_{AB} - V_A^B)$; $F_{km}^u = (2P_{km} - P_{km}^u) \Gamma_{km}^{ex} - P_{km}^u \Gamma_{km}$, and $F_{kl}^u = H_{kl} - P_{kl}^u \Gamma_{AB}$. Orbitals k and m both are on atom A, l and n on atom B. The u represents α- or β-spinned system; therefore, $P_{kk}^\alpha + P_{kk}^\beta = P_{kk}$. $\Gamma_{kk} = \langle kk|kk \rangle$, $\Gamma_{km} = \langle kk|mm \rangle$, and Γ_{km}^{ex} represents the exchange integral $\Gamma_{km}^{ex} = \langle km|km \rangle$. By sequential calculation of the α- and β-spinned sets of wave functions, self-consistency is achieved.

INDO. One-center electron–electron interactions in INDO and MINDO/1/2 are evaluated as follows: Γ_{kk}: $\langle ss|ss\rangle = \langle ss|p_x p_x\rangle = F^0 = \Gamma_{AA}$, Γ_{km}: $\langle p_x p_x | p_x p_x\rangle = F^0 + \frac{4}{25} F^2$, $\langle p_x p_x | p_y p_y\rangle = F^0 - \frac{2}{25} F^2$, Γ_{km}^{ex}: $\langle sp_x|sp_x\rangle = \frac{1}{3} G^1$, $\langle p_x p_y | p_x p_y\rangle = \frac{3}{25} F^2$. G^1 and F^2 are Slater–Condon parameters determined from UV spectra of corresponding atoms. The F^0 term is set equal to Γ_{AA} and calculated, as in CNDO, from Slater s orbitals.

$V_{kk}^B = Z_B^* \Gamma_{kl}$, $\Gamma_{kl} = \langle kk|ll\rangle$. $H_{kl} = S_{kl}(\beta_A^0 + \beta_B^0)/2$. $U_{kk}^B = -\frac{1}{2} (IP_k + EA_k) + ER$, where ER denotes total electron repulsion. Since one-center electron repulsion terms are modified by Slater–Condon parameters, ER would depend on the orbitals as well as nuclei. For example, for hydrogen $U_{ss}^H = -\frac{1}{2}(IP + EA) - \frac{1}{2}\Gamma_{AA}$, for boron to fluorine $U_{ss}^A = -\frac{1}{2}(IP_s + EA_s) - (Z_A^* - \frac{1}{2})\Gamma_{AA} + \frac{1}{6}(Z_A^* - \frac{3}{2}) G^1$, $U_{pp}^A = -\frac{1}{2}(IP_p + EA_p) - (Z_A^* - \frac{1}{2})\Gamma_{AA} + (1/3) G^1 + \frac{2}{25}(Z_A^* - \frac{5}{2}) F^2$. Further, $\Gamma_{kl} = \langle ss|ss\rangle$. The core repulsion correction CR to total energy is taken as $CR = Z_A^* Z_B^* \Gamma_{kl}$.

MINDO/1. The changes from INDO are $U_{kk}^A = IP_k - (Z_A^* - 1) ER$, $H_{kl} = S_{kl}(IP_k^A + IP_l^B) \beta'_{AB} + \beta''_{AB}/r_{AB}^2$ where β'_{AB} and β''_{AB} are adjustable parameters, and $\Gamma_{kl} = \{r_{AB}^2 + (1/4)[(1/\Gamma_{kk}) + (1/\Gamma_{ll})]^2\}^{-1/2}$.

MINDO/2. Same as MINDO/1 except $H_{kl} = \beta_{AB} S_{kl}(IP_k^A + IP_l^B)$ where β_{AB} is a parameter characteristic of the pair of atoms, and $CR = Z_A^* Z_B^* \{\Gamma_{AB} + [(1/r_{AB}) - \Gamma_{AB}] e^{-\alpha_{AB} r_{AB}}\}$, α_{AB} being an adjustable parameter that depends only on A and B.

MINDO/3. Changes from MINDO/2 are $H_{kl} = \beta_{AB} S'_{kl} (IP_k^A + IP_l^B)$ where in S'_{kl}, exponents of k and l orbital are used as parameters rather than usual Slater–Zener values, and the one–center integrals, rather than using parametrization with Slater–Condon parameters, are obtained from a best fit of known valence-state energies of several atoms.

SPINDO. Spectroscopic potential adjusted INDO is the same as MINDO except $H_{kl} = (IP_k^A + IP_l^B) S_{kl} f(R_{AB})$, where $f(R_{AB})$ has different values for different types of interactions, the magnitude of the latter obtained by the use of a small number of hydrocarbons as test molecules. For example, $1s:1sf(R) = 0.13647$, $1s:2sf(R) = 0.17832$, $1s:2p\sigma f(R) = 0.35100$, and so on.

HMO. In the HMO (Hückel molecular orbital) method applicable to π-electron systems, it is assumed that each electron moves in an average field of the core (nuclei, atomic core electrons, and σ electrons) and the other π-electrons. The one-electron molecular orbital is taken as a linear combination of atomic orbitals, $\psi = \sum_k c_k \phi_k$. For π electron profiles ϕ_k's are $2p_\pi$ ($\equiv 2p_z$) atomic orbitals centered on each atom, contributing electrons to the π-system. The expectation value for one-electron energy is given by $E = \Sigma_k \Sigma_l c_k c_l \int \phi_k \mathcal{H} \phi_l d\tau / \Sigma_k \Sigma_l c_k c_l \int \phi_k \phi_l \, d\tau$. If $k = l$, $H_{kk} = \int \phi_k \mathcal{H} \phi_k d\tau = \alpha_k$ (Coulomb integral); if $k \neq l$, $H_{kl} = \int \phi_k \mathcal{H} \phi_l d\tau = \beta_{kl}$ (resonance integral) and $S_{kl} = \int \phi_k \phi_l d\tau$ (overlap integral). E is then minimized variationally and the secular determinant solved for energy.

α_k, β_{kl}, the Hückel parameters, need to be evaluated empirically for each type of atom in the system.

Extended Hückel method. This method is based on HMO, but overlap is included: $S_{kl} \neq 0$ when $k \neq l$. H_{kk}, the diagonal matrix elements, are taken to be the same as the valence state ionization potentials. The Mulliken approximation is used to estimate the off-diagonal elements as $H_{kl} = 0.5\,k'(H_{kk} + H_{ll})S_{kl}$, where k' is an empirical constant. H $1s$ and C $2s$, $2p$ STO's are employed, the C $1s$ electrons are treated part of the core.

HAM/3. An improved version of the original HAM (hydrogen atom in molecules) method, which employs hydrogenic STO's with orbital exponents ζ's and screening constants σ's as parameters. With molecular orbital $\psi_p(i)$ constructed in terms of atomic orbitals as $\psi_p(i) = \Sigma_k c_k^p \phi_k(i)$, (charge) density matrix elements P'_{kl} may be defined as $P'_{kl} = \Sigma_p q_p c_k^p c_l^p$, where q_p represents the charge in orbital p. P'_{kl} is a function of the coefficients c_l^p, and the total energy E of the system is a function of P'_{kl}. If E is minimized variationally with respect to the coefficients c_l^p, then one obtains equations of the type $\Sigma_l(\partial E/\partial P'_{kl} - E^p S_{kl})c_l^p = 0$. Fock matrix elements thus correspond to $F_{kl} = \partial E/\partial P'_{kl}$. To obtain the Fock matrix elements, therefore, E needs to be constituted and its derivative with respect to the density matrix elements evaluated. $E = \Sigma_{kl} E_{kl}$, E_{kl} represents the energy of the electronic charge $P'_{kl}S_{kl}$ in the region kl.

E_{kk} and E_{kl} energies are separately estimated. E_{kk} is used as $E_{kk} = -P'_{kk}\zeta_k^2$, ζ_k being the orbital exponent in STO. $\zeta_k = (Z_A - s_k)/n_k$, where in a hydrogenic type of orbital Z_A is the nuclear charge, s_k the shielding, and n_k the principal quantum number. The off-diagonal matrix elements of energy E_{kl} (k orbital of atom A and l orbital of atom B) has the form $E_{kl} = -P'_{kl}S_{kl}[\frac{1}{2}(\zeta_k^2 + \zeta_l^2)]f_{kl}$; f_{kl} is an empirical factor which is 1.75 in extended Hückel theory and in SPINDO a complicated function $f(R_{AB})$. Generally, f_{kl} is a curved function of the internuclear distance R_{AB}.

Since in the shielding only repulsion between the electrons of the same atom is considered, a correction term $\Sigma_{A>B} Q_A Q_B \gamma_{AB}$ due to electrostatic interaction between the gross atomic charges Q_A and Q_B is added to the total energy. γ_{AB} is obtained by using the Ohno–Klopman equation (Γ_{kl} in MINDO/1) but calculated with specially determined parameters. By introducing the total effective number of electrons in ϕ_k as $T_k = P'_{kk} + \Sigma_{B \neq A}^B \Sigma_l^{\frac{1}{2}} (P'_{kl}S_{kl} + P'_{lk}S_{lk})f_{kl}$ and adding the $Q_A Q_B$ interaction, one obtains $E^{(1)} = -\Sigma_k T_k \zeta_k^2 + \Sigma_{A>B} Q_A Q_B \gamma_{AB}$, $F_{kk} = \partial E^{(1)}/\partial P'_{kk} \approx -\zeta_k^2 + \Sigma_l^A \sigma_{kl} (2/n_l)\zeta_l T_l - \Sigma_B Q_B \gamma_{AB}$, and $F_{kl} \approx \frac{1}{2} S_{kl} [F_{kk} + F_{ll} - (\zeta_k^2 + \zeta_l^2)(f_{kl} - 1)]$, where σ_{kl}'s are shielding constants which are functions of the nature of the atom. A function that represents σ_{kl} satisfactorily is $\sigma_{kl} = a_{kl} - (b_{kl} + c_{kl}Z_A)/\zeta_k$ where a, b, and c are constants that depend on kl and are determined from a consideration of total energies of a large number of atomic species. A better approximation to core repulsion, as is employed in MINDO/2, has been used. In order to obtain parameters for molecules that are built up of several atoms, one needs to simultaneously consider a large number of energies of occupied orbitals (such as ionization energies from photoelectron spectra) and energies of unoccupied orbitals (excitation data from electron spectroscopy). The method has been quite successful in predicting photoelectron band positions.

CNDO/1(1965), although somewhat successful in reproducing the orbitals and electron populations of some selected molecules, was unsuccessful in predicting ΔH_f and bond distances; CNDO/2(1966) at one time used quite extensively, gives only better bond distances and angles. INDO(1967) is appropriate when dealing with free radicals but otherwise comparable to CNDO/2. MINDO/1 (1969) provides ΔH_f within a few kcal but gives poor results for relative energies of rotational isomers. MINDO/2(1970) provides improved results on geometries and reasonable values of force constants. MINDO/3(1975) is the best NDO method to date that predicts satisfactorily ΔH_f, ionization potentials, dipole moments, $\Delta H_{reaction}$, and free radical stabilities. SPINDO(1972) and particularly HAM/3(1977) are the methods that predict photoelectron band positions most satisfactorily.

Slater—Condon parameters

[302] E. U. Condon and G. H. Shortley, *Theory of Atomic Spectra*, Cambridge University Press, Cambridge, 1970, p. 174.

Molecular orbital methods

[303] G. A. Segal, Ed., *Semiempirical Methods of Electronic Structure Calculations*, Parts A and B, Plenum, New York, 1977.

[304] G. Klopman and R. C. Evans, in [303], Part A, p. 29.

[305] A. Streitwieser, Jr., *Molecular Orbital Theory for Organic Chemists*, Wiley, New York, 1961.

[306] J. D. Roberts, *Notes on Molecular Orbital Calculations*, Benjamin, New York, 1962.

[307] M. J. S. Dewar, *The Molecular Orbital Theory of Organic Chemistry*, McGraw-Hill, New York, 1969.

[308] J. A. Pople and D. L. Beveridge, *Approximate Molecular Orbital Theory*, McGraw-Hill, New York, 1970.

[309] J. N. Murrell and A. J. Harget, *Semi-empirical Self-Consistent-Field Molecular Orbital Theory of Molecules*, Wiley-Interscience, New York, 1972.

[310] A. Streitwieser, Jr. and P. H. Owens, *Orbital and Electron Density Diagrams*, Macmillan, New York, 1973.

CHAPTER 4

Spectra of lone-pair electrons are discussed in considerable detail in [7]. The discussion on MO models and applications to organic molecules is based on [61].

Additivity rules

[311] A. W. Potts, T. A. Williams, and W. C. Price, *Faraday Disc. Chem. Soc.* **54** (1972), 104.

Sum rule

[312] K. Kimura, *Faraday Disc. Chem. Soc.* **54** (1972), 140 (Comments in general discussion).
[313] K. Kimura, S. Katsumata, Y. Achiba, H. Matsumoto, and S. Nagakura, *Bull. Chem. Soc. (Japan)* **46** (1973), 373.
[314] K. Kimura, S. Katsumata, T. Yamazaki, and W. Wakabayashi, *J. Electron Spectrosc.* **6** (1975), 41.

Spin–orbit splitting

[315] F. A. Grimm, *J. Electron Spectrosc.* **2** (1973), 475.
[316] J. L. Berkosky, F. O. Ellison, T. H. Lee, and J. W. Rabalais, *J. Chem. Phys.* **59** (1973), 5342.
[317] T. H. Lee and J. W. Rabalais, *J. Chem. Phys.* **60** (1974), 1172.

Multiplet splitting and band intensities

[318] J. W. Rabalais, L. O. Werme, T. Bergmark, L. Karlsson, M. Hussain, and K. Siegbahn, *J. Chem. Phys.* **57** (1972), 1185.
[319] P. A. Cox, S. Evans, and A. F. Orchard, *Chem. Phys. Lett.* **13** (1972), 386.

Jahn–Teller effect

[320] A. W. Potts and W. C. Price, *Proc. Roy. Soc. London* **A236** (1972), 175.

Perfluoro effect

[321] M. B. Robin, N. A. Kuebler, and C. R. Brundle, in [6], p. 351.
[322] C. R. Brundle, M. B. Robin, N. A. Kuebler, and H. Basch, *J. Amer. Chem. Soc.* **94** (1972), 1451.

Benzene

[323] E. Lindholm and B. O. Jonsson, *Chem. Phys. Lett.* **1** (1967), 501.
[324] A. D. Baker, C. R. Brundle, and D. W. Turner, *Int. J. Mass Spectrom. Ion Phys.* **1** (1968), 443.
[325] P. Natalis, J. E. Collin, and J. Momigny, *Int. J. Mass Spectrom. Ion Phys.* **1** (1968), 327.
[326] J. A. R. Samson, *Chem. Phys. Lett.* **4** (1969), 257.
[327] B. O. Jonsson and E. Lindholm, *Arkiv. Fysik*, **39** (1969), 65.
[328] L. Åsbrink, E. Lindholm, and O. Edqvist, *Chem. Phys. Lett.* **5** (1970), 609.
[329] A. W. Potts, W. C. Price, D. G. Streets, and T. A. Williams, *Faraday Disc. Chem. Soc.* **54** (1972), 168.
[330] W. C. Price, A. W. Potts, and T. A. Williams, *Chem. Phys. Lett.* **37** (1976), 17.

[331] L. Karlsson, L. Mattsson, R. Jadrny, T. Bergmark, and K. Siegbahn, *Physica Scripta* **14** (1976), 230.

Discussions in the text on spectral assignment are primarily based on [327], which also has a useful general discussion on how Rydberg transitions may affect photoelectron spectra. For a discussion on electron energy loss spectroscopy, see [45].

Acene

[332] P. A. Clark, F. Brogli, and E. Heilbronner, *Helv. Chim. Acta* **55** (1972), 1415.

Benzenoid hydrocarbons

[333] F. Brogli and E. Heilbronner, *Angew. Chem. Internat. Ed.* **11** (1972), 538.
[334] F. Brogli and E. Heilbronner, *Theoret. Chim. Acta* **26** (1972), 289.
[335] F. Brogli and E. Heilbronner, *Angew. Chem.* **84** (1972), 551.

For correlations in π-substituted systems, see [61].

Conformational studies

[336] E. Heilbronner and H. D. Martin, *Helv. Chim. Acta* **55** (1972), 1490.
[337] E. Heilbronner, *Israel J. Chem.* **10** (1972) 143.
[338] J. P. Maier and D. W. Turner, *Faraday Disc. Chem. Soc.* **54** (1972), 149.
[339] J. P. Maier and D. W. Turner, *J. Chem. Soc. Faraday Trans. II* **69** (1973), 196.
[340] A. D. Walsh, *Trans. Faraday Soc.* **45** (1949), 179.
[341] W. J. Le Noble, *Highlights of Organic Chemistry*, Marcel Dekker, New York, 1974, p. 201.
[342] I. Prins, J. W. Verhoeven, T. J. DeBoer, and C. Worrell, *Tetrahedron* **33** (1977), 127.

Pullman k-index

[343] B. Pullman and A. Pullman, *Quantum Biochemistry*, Wiley-Interscience, London, 1963.

DNA–RNA bases

[344] D. Dougherty and S. P. McGlynn, *J. Chem. Phys.* **67** (1977), 1289.

Biological piperazines

[345] S. D. Worley, S. H. Gerson, N. Bodor, and J. J. Kaminski, *Chem. Phys. Lett.* **60** (1978), 104.

Nicotinic acid, barbituric acid, xanthine, purine, pyrimidine, and other biological molecules

[346] D. Dougherty, E. S. Younathan, R. Voll, S. Abdulnur, and S. P. McGlynn, *J. Electron Spectrosc.* **13** (1978), 379.

Methylnitroimidazoles

[347] F. Kajfež, L. Klasinc, and V. Šunjic, *J. Heterocyclic Chem.* **16** (1979), 529.

Thalidomide

[348] L. Klasinc, N. Trinajstic, and J. V. Knop, *Int. J. Quant. Chem.* **7** (1980), 403.

Tautomeric equilibria

[349] A. Schweig, H. Vermeer, and U. Weidner, *Chem. Phys. Lett.* **26** (1974), 229.
[350] G. W. Mines and H. Thompson, *Proc. Roy. Soc. London.* **A342** (1975), 327.

Study of tautomeric equilibria: The first method is from [349], and the second is from [350].

General article on the application of electron spectroscopy to organic chemistry (also see [61] and [62])

[351] G. D. Mateescu and J. L. Riemenschneider, in [6], p. 661.

UPS of methane, thiophene, 2-bromothiophene, and 3-bromothiophene

[352] T. Bergmark, J. W. Rabalais, L. O. Werme, L. Karlsson, and K. Siegbahn, in [6], p. 413.

Photoelectron spectra of polycyclic aromatic hydrocarbons anthracene, phenanthrene, pentacene, perylene, chrysene, 1,2-benzanthracene, 1,2-benzpyrene, benzo[g, h, i]perylene, and ovalene

[353] R. Boschi, J. N. Murrell, and W. Schmidt, *Faraday Disc. Chem. Soc.* **54** (1972), 116.

Substituent effects in photoelectron spectra of monofunctional aliphatic compounds

[354] P. Carlier, R. Hernandez, P. Masclet, and G. Mouvier, *J. Electron Spectrosc.* **5** (1974), 1103.

HCNO, CH_2N_2, and N_3H

[355] J. Bastide and J. P. Maier, *Chem. Phys.* **12** (1976), 177.

Cyclopropanone

[356] P. C. Martino, P. B. Shevlin, and S. D. Worley, *J. Amer. Chem. Soc.* **99** (1977), 8003.

Alkyl cadmium compounds

[357] G. M. Bancroft, D. K. Creber, and H. Basch, *J. Chem. Phys.* **67** (1977), 4891.

Hexatriacontane (n–$C_{36}H_{74}$) polycrystal, a model compound of polyethylene

[358] K. Seki, S. Hashimoto, N. Sato, Y. Harada, K. Ishii, H. Inokuchi, and J. Kanbe, *J. Chem. Phys.* **66** (1977), 3644.

Heterocycles with bridgehead nitrogen

[359] M. H. Palmer, D. Leaver, J. D. Nisbet, R. W. Millar, and R. Egdell, *J. Mol. Struct.* **42** (1977), 85.

Carbonyls: 1,4-benzoquinones

[360] D. Dougherty and S. P. McGlynn, *J. Amer. Chem. Soc.* **99** (1977), 3234.

Chlorofluoromethanes

[361] T. Cvitaš, H. Güsten, and L. Klasinc, *J. Chem. Phys.* **67** (1977), 2687.

Orthosubstituted and metasubstituted benzonorbornadienes

[362] C. Santiago, E. J. McAlduff, K. N. Houk, R. A. Snow, and L. A. Paquette, *J. Amer. Chem. Soc.* **100** (1978), 6149.

Ferraboranes

[363] J. A. Ulman, E. L. Andersen, and T. P. Fehlner, *J. Amer. Chem. Soc.* **100** (1978), 456.

In (I) and Tl (I) cyclopentadienides

[364] R. G. Egdell, I. Fragala, and A. F. Orchard, *J. Electron Spectrosc.* **14** (1978), 467.

Substituted pyridines (pyridine, pentafluoro-, 2-fluoro-, 2-methyl, 2-cyano-, 2-aminopyridine, and pyrazine)

[365] H. Daamen and A. Oskam, *Inorg. Chim. Acta* **27** (1978), 209.

Aromatic molecules; aza derivatives of indole, benzofuran, and benzothiophen

[366] M. H. Palmer and S. M. F. Kennedy, *J. Mol. Struct.* **43** (1978), 203.

Small ring fused aromatics as probes of ring distortions

[367] C. Santiago, R. W. Gandour, K. N. Houk, W. Nutakul, W. E. Cravey, and R. P. Thummel, *J. Amer. Chem. Soc.* **100** (1978), 3730.

Methylenenortriquinacenes

[368] L. N. Domelsmith, K. N. Houk, C. R. Degenhardt, and L. A. Paquette, *J. Amer. Chem. Soc.* **100** (1978), 100.

Allylic alcohols and ethers

[369] R. S. Brown and R. W. Marcinko, *J. Amer. Chem. Soc.* **100** (1978), 5721.

Fulvenallene

[370] R. Botter, J. Jullien, J. M. Pechine, J. J. Piade, and D. Solgadi, *J. Electron Spectrosc.* **13** (1978), 141.

N-chlorosuccinimide and N-bromosuccinimide

[371] S. D. Worley, S. H. Gerson, N. Bodor, J. J. Kaminski, and T. W. Flechtner, *J. Chem. Phys.* **68** (1978), 1313.

Unbranched C_5-C_7 alkenes, aldehydes, and ketones

[372] F. S. Ashmore and A. R. Burgess, *J. Chem. Soc. Faraday Trans. II* **74** (1978), 734.

Diels–Alder reactions between *trans*-1-N-acylamino-1,3-dienes and methyl acrylate

[373] L. E. Overmann, G. F. Taylor, K. N. Houk, and L. N. Domelsmith, *J. Amer. Chem. Soc.* **100** (1978), 3182.

Hallucinogens

[374] L. N. Domelsmith and K. N. Houk, *"Quasar" Research Monograph-22*, G. Barnett, M. Trsic, and R. Willette, Eds., National Institute on Drug Abuse, 1978, p. 423.

Benzobicycloalkenes

[375] L. N. Domelsmith, P. D. Mollere, K. N. Houk, R. C. Hahn, and R. P. Johnson, *J. Amer. Chem. Soc.* **100** (1978), 2959.

Aliphatic anhydrides

[376] D. Colbourne, D. C. Frost, C. A. McDowell, and N. P. C. Westwood, *J. Electron Spectrosc.* **14** (1978), 391.

Substituted benzamides

[377] E. J. McAlduff, B. M. Lynch, and K. N. Houk, *Can. J. Chem.* **56** (1978), 495.

Dialkyl mercury compounds

[378] T. B. Fehlner, J. Ulman, W. A. Nugent, and J. K. Kochi, *Inorg. Chem.* **15** (1976), 2544.

The molecules in Fig. 4.16 are:
1-1 p-NO$_2$-, 1-2 p-CN-, 1-3 p-Cl-, 1-4 H-, 1-5 p-CH$_3$-, 1-6 p-CH$_3$O-benzamide. The imines in the complexes, for Cr, are 2-picoline, 4-t-butyl pyridine, 2,4-

lutidine, pyrazine, pyridazine, and, for both Cr and W, pyridine, 4-picoline, 4-Cl-pyridine, 4-Br-pyridine. 3-1 Pentadienoic acid, 3-2 $N(1,3$ butadienyl) trichloroacetamide, 3-3 $N(1,3$ butadienyl) ethyl carbamate, 3-4 $N(1,3$ butadienyl) phenyl carbamate, 3-5 $N(1,3$ butadienyl)N'-pyrollidino urea. 4-1 Benzobicyclo[4.2.1]nona-2,7-diene, 4-2 benzobicyclo[4.2.1]nona-3,7-diene, 4-3 benzobicyclo[4.2.1]nonene, 4-4 benzobicyclo[3.2.1]octene, 4-5 benzonorbornene, 4-6 benzoexotricyclo[3.2.1.0^{2,4}]octene, 4-7 benzobicyclo[3.2.1]octa-2,5-diene, 4-8 benzonorbornadiene. 5-1 $(CH_3)_2Hg$, 5-2 $CH_3HgC_2H_5$, 5-3 CH_3Hg (i-C_3H_7), 5-4 $CH_3Hg(t$-$C_4H_9)$. 6-1 $(CF_3CO)_2O$, 6-2 $(CClF_2CO)_2O$, 6-3 $(CCl_3CO)_2O$, 6-4 $(CH_3CO)_2O$. 7-1 Phenethylamine, 7-2 amphetamine, 7-3 4-methoxyphenethylamine, 7-4 3,4-dimethoxyphenethylamine, 7-5 mescaline, 7-6 5-methoxy-N,N,-dimethyltryptamine, 7-7 N,N-dimethyltryptamine, 7-8 tryptamine, 7-9 5-methyltryptamine, 7-10 lysergic acid diethylamide.

Cyclobutadiene and trimethylenemethane

[379] S. D. Worley, T. R. Webb, D. H. Gibson, and T. Ong, *J. Organometallic Chem.* **168** (1979), C16.

Gauche and trans conformers of 1,2-bromochloroethane

[380] F. Carnovale, T. H. Gan, and J. B. Peel, *J. Electron Spectrosc.* **16** (1979), 87.

Monomers and dimers of CH_3COOH and CF_3COOH

[381] F. Carnovale, T. H. Gan, and J. B. Peel, *J. Electron Spectrosc.* **20** (1980), 53.

4-Methylenethiacyclohexane derivatives

[382] R. Sarneel, C. W. Worrell, P. Pasman, J. W. Verhoeven, and G. F. Mes, *Tetrahedron* **36** (1980), 3241.

Photoelectron spectroscopy of carbonyls

[383] S. Chattopadhyay, J. L. Meeks, G. L. Findley, and S. P. McGlynn, *J. Phys. Chem.* **85** (1981), 968.

Polynuclear aromatics

[384] E. Clar, J. M. Robertson, R. Schlogl, and W. Schmidt, *J. Amer. Chem. Soc.* **103** (1981), 1320.

Ketene radical cation

[385] D. Hall, J. P. Maier, and P. Rosmus, *Chem. Phys.* **24** (1977), 373.

Cation radicals of aromatic hydrocarbons, aliphatic and aromatic amines

[386] T. Shida, Y. Nosaka, and T. Kato, *J. Phys. Chem.* **82** (1978), 695.

Singlet and triplet methylene

[387]   H. L. Hase, G. Lauer, K. W. Schulte, A. Schweig, and W. Thiel, *Chem. Phys. Lett.* **54** (1978), 494.

ClO radical

[388]   D. K. Bulgin, J. M. Dyke, N. Jonathan, and A. Morris, *J. Chem. Soc. Faraday Trans. II* **75** (1979), 456.

Methyl, ethyl, isopropyl, and t-butyl radicals

[389]   F. A. Houle and J. L. Beauchamp, *J. Amer. Chem. Soc.* **101** (1979), 4067.

FCO radical

[390]   J. M. Dyke, N. Jonathan, A. Morris, and M. J. Winter, *J. Chem. Soc. Faraday Trans. II* **77** (1981), 667.

HAM/3

[391]   L. Åsbrink, C. Fridh, and E. Lindholm, *Chem. Phys. Lett.* **52** (1977), 63.
[392]   ——, *Chem. Phys. Lett.* **52** (1977), 69.
[393]   ——, *Chem. Phys. Lett.* **52** (1977), 72.
[394]   ——, *Z. Naturforsch.* **33a** (1978), 172.

Spectra of furan, pyrrole, and cyclopentadiene, studied with HAM/3

[395]   L. Åsbrink, C. Fridh, and E. Lindholm, *J. Electron Spectrosc.* **16** (1979), 65.

Acetylacetone complexes, metal–oxygen bonding

[396]   S. Evans, A. Hamnett, A. F. Orchard, and D. R. Lloyd, *Faraday Disc. Chem. Soc.* **54** (1972), 227.

Photoelectron spectra of Se_2 and Te_2

[397]   D. G. Streets and J. Berkowitz, *J. Electron Spectrosc.* **9** (1976), 269.

Hund's coupling cases arise because of different modes of coupling of **L**, the resultant electronic orbital angular momentum, **S**, the resultant spin angular momentum, and **N**, the angular momentum of nuclear rotation in the molecule. **J** is the total angular momentum. If **L** and **S** have strong couplings with the internuclear axis, then Λ and Σ respectively are their components along the internuclear axis. The nature of fine structure in electronic bands depends on the nature of angular momentum couplings. The couplings in Hund's cases (a) (b), (c), and (d) are as follows. (a) The interaction of nuclear rotation with electronic motion is weak. Λ and Σ couple to form Ω, the latter couples with **N** to form **J**. (b) **S** does not couple with the internuclear axis; therefore, Ω is not defined. Λ and **N** form a resultant **K**; **K** and **S** couple to form **J**. (c) The interaction between **L** and **S** is stronger than interaction with the internuclear axis; therefore, Λ and Σ are not defined. L and S couple to form first J′, which couples with the internuclear axis with a component Ω along the latter. Ω then couples with N to form J. (d) If the interaction of L with the internuclear axis is very weak but the interaction with nuclear rotation N strong, then L and N couple to form K, which couples with S to form J.

For further details about Hund's coupling cases, see [88], p. 219, and for the λ-splitting (Λ-type doubling) that arises because of interaction between the rotation of the nucleus and L, see [88], p. 226.

UPS of group III, IV, V, and VI halides

[398] A. W. Potts, H. J. Lempka, D. G. Streets, and W. C. Price, *Phil. Trans. Roy. Soc. London* **A268** (1970), 59.

UPS of CH_4, SiH_4, and GeH_4

[399] A. W. Potts and W. C. Price, *Proc. Roy. Soc. London,* **A326** (1972), 165.

UPS of group V and VI hydrides

[400] A. W. Potts and W. C. Price, *Proc. Roy. Soc. London* **A326** (1972), 181.

UPS of halogeno, dimethylamino and methyl borane, BX_3 and BX_2Y

[401] G. H. King, S. S. Krishnamurthy, M. F. Lappert, and J. B. Pedley, *Faraday Disc. Chem. Soc.* **54** (1972), 70.

Dinitrogen tetroxide and dinitrogen pentoxide

[402] D. L. Ames and D. W. Turner, *Proc. Roy. Soc. London* **A348** (1976), 175.

[403] T. H. Gan, J. B. Peel, and G. D. Willett, *J. Chem. Soc. Faraday Trans. II* **73** (1977), 1459.

[404] D. C. Frost, C. A. McDowell, and N. P. C. Westwood, *J. Electron Spectrosc.* **10** (1977), 293.

[405] K. Nomoto, Y. Achiba, and K. Kimura, *Chem. Phys. Lett.* **63** (1979), 277.

Boranes and carboranes: five, six, and seven-atom frameworks

[406] J. A. Ulman and T. P. Fehlner, *J. Amer. Chem. Soc.* **100** (1978), 449.

Carbon subsulfide

[407] A. P. Ginsberg and C. R. Brundle, *J. Chem. Phys.* **68** (1978), 5231.

$M(CO)_5PX_3$ (M = Cr, Mo, W; X = F, Cl, Br)

[408] H. Daamen, G. Boxhoorn, and A. Oskam, *Inorg. Chim. Acta* **28** (1978), 263.

Monosubstituted chromium and tungsten hexacarbonyls $M(CO_5)L$ (L = amine, substituted pyridine, azine)

[409] H. Daamen and A. Oskam, *Inorg. Chim. Acta* **26** (1978), 81.

Chloramine and dichloramine

[410] M. K. Livett, E. Nagy-Felsobuki, J. B. Peel, and G. D. Willett, *Inorg. Chem.* **17** (1978), 1608.

NH_2Cl, $NHCl_2$, NCl_3, CH_3NHCl, CH_3NCl_2, $(CH_3)_2NCl$

[411] D. Colbourne, D. C. Frost, C. A. McDowell, and N. P. C. Westwood, *J. Chem. Phys.* **69** (1978), 1078.

Antimicrobial N-chloroamines

[412] S. H. Gerson, S. D. Worley, N. Bodor, and J. J. Kaminski, *J. Med. Chem.* **21** (1978), 686.

Bromamine

[413] E. Nagy-Felsobuki, J. B. Peel, and G. D. Willett, *J. Electron Spectrosc.* **13** (1978), 17.

High temperature vapors: alkali perrhenates

[414] D. O. Vick, D. G. Woodley, J. E. Bloor, J. D. Allen, Jr., T. C. Mui, and G. K. Schweitzer, *J. Electron Spectrosc.* **13** (1978), 247.

Transition metal dihalides

[415] J. Berkowitz, D. G. Streets, and A. Garritz, *J. Chem. Phys.* **70** (1979), 1305.

Alkali halide vapors

[416] J. Berkowitz, in *Alkali halide vapors: structure, spectra and reaction dynamics*, Academic, New York, 1979, p. 155.

Lithium halide monomers and dimers

[417] A. W. Potts and E. P. F. Lee, *J. Chem. Soc. Faraday Trans. II* **75** (1979), 941.

AgCl, AgBr, and AgI

[418] J. Berkowitz, C. H. Batson, and G. L. Goodman, *J. Chem. Phys.* **72** (1980), 5829.

Zn and Cd difluoride

[419] E. P. F. Lee, D. Law, and A. W. Potts, *J. Chem. Soc. Faraday Trans. II* **76** (1980), 1314.

Nb and Ta complexes

[420] H. van Dam, A. Terpstra, A. Oskam, and J. H. Teuben, *Z. Naturforsch.* **36b** (1981), 420.

General article on photoelectron spectroscopy and electronic structure of coordination compounds

[421] C. Furlani, *Coord. Chem. Rev.* **43** (1982), 355.

Photoelectron-photoion coincidence spectroscopy of gas phase clusters

[422] E. D. Poliakoff, P. M. Dehmer, J. L. Dehmer, and R. Stockbauer, *J. Chem. Phys.* **76** (1982), 5214.

Rotationally resolved study of H_2^+

[423] J. E. Pollard, D. J. Trevor, J. E. Reutt, Y. T. Lee, and D. A. Shirley, *Chem. Phys. Lett.* **88** (1982), 434.

Photoelectron spectrum and Rydberg transitions of CO

[424] L. Åsbrink, C. Fridh, E. Lindholm and K. Codling, *Physica Scripta* **10** (1974), 183.

Temperature-dependent studies on cyclopentadienone

[425] V. Eck, G. Lauer, A. Schweig, W. Thiel, and H. Vermeer, *Z. Naturforsch* **33a** (1978), 383 (in German).

Variable temperature studies on rotational isomerism in tetramethyl diphosphine

[426] A. Schweig, N. Thon, and H. Vermeer, *J. Electron Spectrosc.* **15** (1979), 65.

CHAPTER 5

Valence band spectra of Au

[427] J. Freeouf, M. Erbudak, and D. E. Eastman, *Solid State Commun.* **13** (1973), 771.

Na

[428] P. H. Citrin, *Phys. Rev.* **B8** (1973), 5545.

[429] S. P. Kowalczyk, L. Ley, F. R. McFeely, R. A. Pollak, and D. A. Shirley, *Phys. Rev.* **B8** (1973), 3583.

Pd, Ag, Cd, In, Sn, Sb, Te

[430] R. A. Pollak, S. P. Kowalczyk, L. Ley, and D. A. Shirley, *Phys. Rev. Lett.* **A29** (1972), 274.

Sm

[431] Y. Baer and G. Busch, *J. Electron Spectrosc.* **5** (1974), 611.

AuPd alloy

[432] J. A. Nicholson, J. D. Riley, R. C. G. Leckey, J. G. Jenkin, and J. Liesegang, *J. Electron Spectrosc.* **15** (1979), 95.

AgCl, AgBr, AgF, and AgI

[433] M. G. Mason, *J. Electron Spectrosc.* **5** (1974), 573.

GaAs

[434] G. Leonhardt, *J. Electron Spectrosc.* **5** (1974), 603.
[435] L. Ley, R. A. Pollak, F. R. McFeely, S. D. Kowalczyk, and D. A. Shirley, *Phys. Rev. B* **9** (1974), 600.
[436] J. Chelikowski, D. J. Chadi, and M. L. Cohen, *Phys. Rev. B* **8** (1973), 2786.
[437] D. E. Eastman, J. L. Freeouf, and M. Erbudak, Congrès du Centenaire de la Société Française de Physique, Vittel, France (1973).

$(SN)_x$

[438] R. P. Messmer and D. R. Salahub, *Chem. Phys. Lett.* **41** (1976), 73.
[439] D. R. Salahub and R. P. Messmer, *Phys. Rev. B* **14** (1976), 2592.
[440] V. V. Walatka Jr., M. M. Labes, and J. H. Perlstein, *Phys. Rev. Lett.* **31** (1973), 1139.
[441] P. Mengel, P. M. Grant, W. E. Rudge, B. H. Schechtman, and D. W. Rice, *Phys. Rev. Lett.* **35** (1975), 1803.
[442] L. Ley, *Phys. Rev. Lett.* **35** (1975), 1796.
[443] R. H. Findlay, M. H. Palmer, A. J. Downs, R. G. Egdell, and R. Evans, *Inorg. Chem.* **19** (1980), 1307.

Os

[444] R. Nilsson, A. Berndtsson, N. Mårtensson, R. Nyholm, and J. Hedman, *Phys. Stat. Sol. B* **75** (1976), 197.

Nonstoichiometric single crystals of TiC

[445] L. I. Johansson, A. L. Hagstrom, B. E. Jacobson, and S. B. M. Hagstrom, *J. Electron Spectrosc.* **10** (1977), 259.

In (I) and Tl (I) halides

[446] R. G. Egdell and A. F. Orchard, *J. Chem. Soc. Faraday Trans. II* **74** (1978), 1179.

GaS

[447] A. Balzarotti, R. Girlanda, V. Grasso, E. Doni, F. Antonangeli, and M. Piacentini, *Can. J. Phys.* **56** (1978), 700.

Solid NH_3, H_2O, CO_2, SO_2, and N_2O_4

[448] M. J. Campbell, J. Liesegang, J. D. Riley, and J. G. Jenkin, *J. Phys. C* **15** (1982), 2549.

Valence band relaxation energies of Si and Ge during photoelectron emission

[449] R. T. Poole, *Chem. Phys. Lett.* **54** (1978), 223.

Crystal-field and band broadening effects in Cd

[450] P. M. A. Sherwood and D. A. Shirley, *Chem. Phys. Lett.* **56** (1978), 404.

Core electrons in metals

[451] G. K. Wertheim, *Jpn. J. Appl. Phys.* **17** (1978), Suppl. 17-2, 33.
[452] W. E. Spicer, I. Lindau, C. Y. Su, P. W. Chye, and P. Pianetta, *Appl. Phys. Lett.* **33** (1978), 934.

Au Schottky-barrier formation on GaSb, GaAs, and InP

[453] P. W. Chye, I. Lindau, P. Pianetta, C. M. Garner, C. Y. Su, and W. E. Spicer, *Phys. Rev. B* **18** (1978), 5545.

Au, Cu, Al, Mg, Cr and Ti Schottky-barrier formation on GaAs

[454] S. P. Kowalczyk, J. R. Waldrop, and R. W. Grant, *Appl. Phys. Lett.* **38** (1981), 167.

LEED. Low energy electron diffraction method for studying crystal surfaces

[455] S. Andersson and J. B. Pendry, *J. Phys. C* **6** (1973), 601.
[456] A. Ignatiev, F. Jona, D. W. Jepsen, and P. M. Marcus, *J. Vac. Sci. Technol.* **12** (1975), 226.
[457] H. D. Shih, F. Jona, D. W. Jepsen, and P. M. Marcus, *Comm. Phys.* **1** (1976), 25.
[458] E. Zanazzi, F. Jona, D. W. Jepsen, and P. M. Marcus, *Phys. Rev. B* **14** (1976) 432.

Surface chemistry of Cu-, Co-, and Ni-ferrites

[459] D. L. Perry, D. W. Bonnell, G. D. Parks, and J. L. Margrave, *High Temp. Sci.* **9** (1977), 85.

XPS of localized levels near surfaces: scattering effects and relaxation shifts

[460] M. Šunjić, Ž. Crljen, and D. Šokčević, *Surf. Sci.* **68** (1977), 479.

GaAs surface states

[461] R. Ludeke and L. Ley, Proceedings of the International Conference on Physics of Semiconductors, Edinburgh 1978, p. 1069.

III—V surface—oxide interface

[462] W. E. Spicer, I. Lindau, P. Pianetta, P. W. Chye, and C. M. Garner, *Thin Solid Films* **56** (1979), 1.

Intermediate valence states in rare earth compounds

[463] G. K. Wertheim, *J. Electron Spectrosc.* **15** (1979), 5.

Amorphous hydrogenated Si

[464] B. von Roedern, L. Ley, and M. Cardona, *Solid State Commun.* **29** (1979), 415.
[465] B. von Roedern, L. Ley, and F. W. Smith, *Inst. Phys. Conf. Ser. No. 43* (1979), chapter 21, p. 701.

Temperature dependence of Ni and Pd d-band peak energy

[466] J. E. Rowe and J. C. Tracy, in [6], p. 551.

For a discussion on exchange splitting

[467] D. H. Martin, *Magnetism in Solids*, Iliffe, London, 1972.

Temperature dependence of the energy distribution curves of AgBr, Ge, LiI

[468] R. S. Bauer and W. E. Spicer, in [6], p. 569.

Curie temperature. Temperature above which the spontaneous magnetization vanishes. It separates the disordered paramagnetic phase at $T > T_c$ from the ordered ferromagnetic phase at $T < T_c$.

Ge photoelectron spectra with oxide overlayer

[469] M. Barber and P. Swift, *Methodes Physiques d'Analyse* **8** (1972), 43.

Depth resolution in XPS

[470] D. Briggs, in A. R. West, Ed., *Proceedings of the 6th Molecular Spectroscopy Conference*, Heyden, London, p. 468. See also [32].

Mean free paths of low-energy electrons in metals

[471] R. Payling and J. A. Ramsey, *Proceedings of the 7th International Vacuum Congress and 3rd International Conference on Solid Surfaces, Vienna*, 1977, p. 2451.

Photoelectron escape depths in polymers

[472] P. Cadman, G. Gossedge, and J. D. Scott, *J. Electron Spectrosc.* **13** (1978), 1.

Uniform depth profiling in XPS

[473] L. Bradley, Y. M. Bosworth, D. Briggs, V. A. Gibson, R. J. Oldman, A. C. Evans and J. Franks, *Appl. Spectrosc.* **32** (1978), 175.

Empirical mean free path curves for electron scattering in solids

[474] R. E. Ballard, *J. Electron Spectrosc.* **25** (1982), 75.

For further discussion and references on experimental determination of effective escape depth, see [27], p. 31.

Initial oxidation of Zn

[475] D. Briggs, *Faraday Disc. Chem. Soc.* **60** (1975), 81.
[476] C. R. Brundle, Comments in General Discussion, *Faraday Disc. Chem. Soc.* **60** (1975), 152.

Adsorption studies on Ni films from photoelectron spectra

[477] C. R. Bundle and A. F. Carley, *Faraday Disc. Chem. Soc.* **60** (1975), 51.

Adsorption on Pt(100) surfaces

[478] T. A. Clarke, I. D. Gay, B. Law, and R. Mason, *Faraday Disc. Chem. Soc.* **60** (1975), 119.
[479] I. Lindau, P. Pianetta, and W. E. Spicer, *Phys. Lett.* **57A** (1976), 225.
[480] G. Apai, P. S. Wehner, J. Stohr, R. S. Williams, and D. A. Shirley, *Solid State Commun.* **20** (1976), 1141.
[481] D. J. Kennedy and S. T. Manson, *Phys. Rev. A* **5** (1972), 227.

CO molecular orbitals: 1σ and 2σ are primarily O $1s$ and C $1s$ atomic orbitals. 3σ consists mostly of the oxygen $2s$, that is, a lone pair, but there is substantial σ bonding. 4σ has σ bonding but largely a lone-pair orbital on carbon and oxygen; O $2p_x$ orbital makes a significant contribution. In 1π, the electron density concentrates more towards electronegative oxygen. 5σ is strongly polarized towards carbon; there is some σ bonding character, but it is mostly a carbon lone-pair orbital. See [310].

Chemisorbed oxygen on metal surfaces

[482] E. L. Evans, J. M. Thomas, M. Barber and R. J. M. Griffiths, *Surf. Sci.* **38** (1973), 245.

Cu, CuO, Cu_2O, and Cu_2S thin films

[483] P. E. Larson, *J. Electron Spectrosc.* **4** (1974), 213.

UPS and XPS study of the adsorption and catalytic decomposition of HCO_2H by Cu, Ni and Au

[484] R. W. Joyner and M. W. Roberts, *Proc. Roy. Soc. London* **A350** (1976), 107.

Adsorption of NO by Fe surfaces

[485] K. Kishi and M. W. Roberts, *Proc. Roy. Soc. London* **A352** (1976), 289.

Rare earth metals and their oxides

[486] B. D. Padalia, W. C. Lang, P. R. Norris, L. M. Watson, and D. J. Fabian, *Proc. Roy. Soc. London* **A354** (1977), 269.

Adsorption of CO, O_2, and H_2 on Pt

[487] D. M. Collins and W. E. Spicer, *Surf. Sci.* **69** (1977), 114.

Oxidation and sulphidation of Pb(100) and (110) surfaces

[488] R. W. Joyner, K. Kishi, and M. W. Roberts, *Proc. Roy. Soc. London* **A358** (1977), 223.

Passive layer on Sn–Ni alloy

[489] J. H. Thomas III and S. P. Sharma, *J. Vac. Sci. Technol.* **14** (1977), 1168.

Chemical forms in oxidation of Ti

[490] L. Porte, M. Demosthenous, G. Hollinger, Y. Jugnet, P. Pertosa, and T. M. Duc, in [471], p. 923.

Surfaces complexes of CO, PF_3, and $CNCH_3$ on Fe(110)

[491] G. Ertl, J. Kuppers, F. Nitschke, and M. Weiss, *Chem. Phys. Lett.* **52** (1977), 309.

SiN films of varying refractive indices

[492] T. N. Wittberg, J. R. Hoenigman, W. E. Moddeman, C. R. Cothern, and M. R. Gulett, *J. Vac. Sci. Technol.* **15** (1978), 348.

(hcp-bcc) phase transformation on a $(10\bar{1}1)$ Ti surface

[493] Y. Fukuda, G. M. Lancaster, F. Honda, and J. W. Rabalais, *Phys. Rev. B* **18** (1978), 6191.

NiCO

[494] D. T. Clark, B. J. Cromarty, and A. Sgamellotti, *Chem. Phys. Lett.* **55** (1978), 482.

Oxygen adsorption and the surface electronic structure of GaAs (110) studied by synchrotron radiation

[495] I. Lindau, P. Pianetta, W. E. Spicer, P. E. Gregory, C. M. Garner, and P. W. Chye *J. Electron Spectrosc.* **13** (1978), 155.

Valence band studies of clean and oxygen-exposed GaAs(110) surfaces

[496] P. Pianetta, I. Lindau, P. E. Gregory, C. M. Garner, and W. E. Spicer, *Surf. Sci.* **72** (1978), 298.

Oxidation of Mo, Ti, and Co in the monolayer range

[497] A. Benninghoven, O. Ganschow, and L. Wiedmann, *J. Vac. Sci. Technol.* **15** (1978), 506.

Codeposited sulfur and oxygen layers on Ag(III)

[498] G. G. Tibbetts, and J. M. Burkstrand, *J. Vac. Sci. Technol.* **15** (1978), 497.

Ad–atom interactions with III–V semiconductor surfaces

[499] I. Lindau, W. E. Spicer, P. Pianetta, P. W. Chye, and C. M. Garner, *J. Electron Spectrosc.* **15** (1979), 197.

Adsorption of methanol on ZnO powder

[500] G. D. Parks and M. J. Dreiling, *J. Electron Spectrosc.* **16** (1979), 321.

Selection rules to photoemission from clear and adsorbate-covered metal single crystals

[501] N. V. Richardson, D. R. Lloyd, and C. M. Quinn, *J. Electron Spectrosc.* **15** (1979), 177.

Local atomic and electronic structure of oxide–GaAs and SiO_2–Si interfaces

[502] F. J. Grunthaner, P. J. Grunthaner, R. P. Vasquez, B. F. Lewis, J. Maserjian, and A. Madhukar, *J. Vac. Sci. Technol.* **16** (1979), 1443.

Chemisorption of NO and CO on NiO surfaces

[503] M. W. Roberts and R. St. C. Smart, *Surf. Sci.* **100** (1980), 590.

Chemisorption of NO on Ni(100)

[504] G. L. Price and B. G. Baker, *Surf. Sci.* **91** (1980), 571.

Chemisorption on Pt 6 (111) × (100)

[505] J. N. Miller, D. T. Ling, M. L. Shek, D. L. Weissman, P. M. Stefan, I. Lindau, and W. E. Spicer, *Surf. Sci.* **94** (1980), 16.

Chemisorption on clean Th and U surfaces

[506] W. McLean, C. A. Colmenares, R. L. Smith, and G. A. Somorjai, *Phys. Rev. B* **25** (1982), 8.

A recent application of the photoelectron spectroscopic technique has been in monitoring surfaces subjected to bombardment by ion beams, such as He^+, Ar^+, Xe^+, N_2^+, NO^+, in the energy range 30–3000 eV. Bombardment by low-energy ions causes ion penetration into the solid surface, thermalization, implantation of ions, surface alterations, and chemical reactions such as destruction of the original compounds and the formation of new compounds. For example, reactions of the N_2^+ ion beam with Si, Ge, and Sn produce nitrides; Ar^+ ion bombardment causes the reduction of iron oxides to Fe; and bombardment of N_2^+ and NO^+ induces chemical reactions of N and O atoms with surfaces of graphite, diamond, teflon, and graphite monofluoride.

Chemical reactions of N_2^+ ion beams with group IV elements and their oxides

[507] J. A. Taylor, G. M. Lancaster, and J. W. Rabalais, *J. Electron Spectrosc.* **13** (1978), 435.

Interaction of Ar^+ and Xe^+ beams with graphite, graphite monofluoride, and teflon

[508] J. A. Taylor, G. M. Lancaster, and J. W. Rabalais, *Appl. Surf. Sci.* **1** (1978), 503.

He^+ ion bombardment on Ta_2O_5

[509] G. C. Nelson, *J. Vac. Sci. Technol.* **15** (1978), 702.

Ion bombardment damage of Fe-S compounds

[510] T. Tsang, G. J. Coyle, I. Adler, and L. Yin, *J. Electron Spectrosc.* **16** (1979), 389.

Ion beam effects on surfaces

[511] S. Storp and R. Holm, *J. Electron Spectrosc.* **16** (1979), 183.

In electrochemical studies, XPS has been used to detect the nature of species formed on electrode surfaces, such as oxide layers. Nickel electrode after polarization in sulfuric acid shows the presence of NiO and $Ni(OH)_2$; different molybdenum species with oxidation states III, IV, and VI are indicated after polarization in NaOH solutions. Analysis of a stainless steel electrode after polarization in sulfuric acid shows oxide/hydroxide film, with Fe and Cr as major constituents in the film; Si is also observed.

Dissolution and passivation of Ni

[512] T. Dickinson, A. F. Povey, and P. M. A. Sherwood, *J. Chem. Soc. Faraday Trans. I* **73** (1977), 327.

Surface studies on acid–oxygen–carbon electrodes

[513] L. Y. Johansson, J. Mrha, M. Musilova, and R. Larsson, *J. Power Sources* **2** (1977–1978), 183.

Permethyl polysilanes chemically bound to electrode surface

[514] A. L. Allred, C. Bradley, and T. H. Newman, *J. Amer. Chem. Soc.* **100** (1978), 5081.

Stainless steel electrodes after polarization in the regions of transpassivity and secondary passivity

[515] A. F. Povey, R. O. Ansell, T. Dickinson, and P. M. A. Sherwood, *J. Electroanal. Chem.* **87** (1978), 189.

Thallium oxide electrode

[516] J. Knecht and G. Stork, *Fresenius Z. Anal. Chem.* **289** (1978), 206 (in German).

Fe-bearing clay minerals

[517] J. W. Stucki, C. B. Roth, and W. E. Baitinger, *Clays and Clay Minerals* **24** (1976), 289.

XPS study of vanadium-bearing aegerines

[518] I. Nakai, H. Ogawa, Y. Sugitani, Y. Niwa, and K. Nagashima, *Miner. J.* **8** (1976), 129.

Studies of lunar soil grains

[519] R. L. Baron, E. Bilson, T. Gold, R. J. Colton, B. Hapke, and M. A. Steggert, *Earth Planet. Sci. Lett.* **37** (1977), 263.

Cu minerals

[520] I. Nakai, Y. Sugitani, K. Nagashima, and Y. Niwa, *J. Inorg. Nucl. Chem.* **40** (1978), 789.

XPS study of zeolites

[521] J. Finster, P. Lorenz, and E. Angele, *Z. Phys. Chemie. Leipzig* **259** (1978), 113.

Co on the surface of kaolinite

[522] J. G. Dillard and M. H. Koppelman, *J. Colloid Interface Sci.* **87** (1982), 46.

Metal crystallite size dispersion in catalysis

[523] D. Briggs, *J. Electron Spectrosc.* **9** (1976), 487.

Hydrido phosphine complexes of Pt (II) chloride—dehydrogenation catalysts

[524] C. Andersson and R. Larsson, *Chemica Scripta* **11** (1977), 140.

XPS of catalysts obtained upon the interaction of $W(\pi\text{-}C_4H_7)_4$ and SiO_2

[525] Y. M. Shulga, A. A. Startsev, Y. I. Yermakov, B. N. Kuznetsov, and Y. G. Borod'ko, *React. Kinet. Catal. Lett.* **6** (1977), 377.

Surface characterization of catalysts

[526] T. E. Madey, C. D. Wagner, and A. Joshi, *J. Electron Spectrosc.* **10** (1977), 359.

$CoMo/Al_2O_3$ hydrodesulfurization catalysts

[527] R. I. Declerck-Grimee, P. Canesson, R. M. Friedman, and J. J. Fripiat, *J. Phys. Chem.* **82** (1978), 889.

Surface composition of Co–Mo binary oxide catalyst

[528] Y. Okamoto, T. Shimokawa, T. Imanaka, and S. Teranishi, *J. C. S. Chem. Commun.* (1978), 47.

Chemical state and reactivity of supported Pd

[529] F. Bozon-Verduraz, A. Omar, J. Escard, and B. Pontvianne, *J. Catalysis* **53** (1978), 126.

NiO/SiO_2 and $NiO-Al_2O_3/SiO_2$ catalysts

[530] P. Lorenz, J. Finster, G. Wendt, J. V. Salyn, E. K. Žumadilov, and V. I. Nefedov, *J. Electron Spectrosc.* **16** (1979), 267.

Characterization of fluoropolymer surfaces

[531] W. M. Riggs and D. W. Dwight, *J. Electron Spectrosc.* **5** (1974), 447.

Hexatriacontane ($n\text{-}C_{36}H_{74}$) polycrystal

[532] K. Seki, S. Hashimoto, N. Sato, Y. Harada, K. Ishii, H. Inokuchi, and J. Kanbe, *J. Chem. Phys.* **66** (1977), 3644.

Polyethylene–aluminum laminates

[533] D. Briggs, D. M. Brewis, and M. B. Konieczko, *Eur. Polym. J.* **14** (1978), 1.

Polymers: sample-charging phenomena

[534] D. T. Clark, A. Dilks, and H. R. Thomas, *J. Polym. Sci. Polym. Chem. Ed.* **16** (1978), 1461.

Electron irradiated $LiNO_3$ and Li_2SO_4

[535] T. Sasaki, R. S. Williams, J. S. Wong, and D. A. Shirley, *J. Chem. Phys.* **68** (1978), 2718.

Trapped oxygen species in irradiated $NaClO_3$

[536] R. G. Copperthwaite and J. Lloyd, *Nature* **271** (1978), 141.

Electron beam effects on hydrated $MgCl_2$

[537] F. Garbassi and L. Pozzi, *J. Electron Spectrosc.* **16** (1979), 199.

Texts on solid state physics

[538] C. Kittel, *Introduction to Solid State Physics*, Wiley, New York, 1971.
[539] L. V. Azároff, *Introduction to Solids*, McGraw-Hill, New York, 1960.

Brillouin zone

[540] W. Hume-Rothery, *Electrons, Atoms, Metals, and Alloys*, Iliffe, London, 1955.
[541] H. Jones, *Theory of Brillouin Zones and Electronic States in Crystals*, American Elsevier, New York, 1975.
[542] W. A. Wooster, *Tensor and Group Theory for the Physical Properties of Crystals*, Clarendon, Oxford, 1973.
[543] C. J. Bradley and A. P. Cracknell, *The Mathematical Theory of Symmetry in Solids*, Clarendon, Oxford, 1972.

Band theory of solids

[544] J. Callaway, *Energy Band Theory*, Academic, New York, 1964.

Brillouin zones: The momenta of electrons in a crystal along x, y, and z directions (p_x, p_y, and p_z respectively) can be represented by a point in a three-dimensional coordinate system, the momentum (p-) space. In this space, representation of all electrons in their lowest energy states would be bound by a surface called the Fermi surface. An alternate representation of momentum uses the wave number k: $k = 2\pi/\lambda = 2\pi p/h$ (λ is de Broglie wavelength, h Planck's constant). In this k-space, if we consider a straight line from the origin, which corresponds to a definite direction with respect to the crystal, then at certain values of k (i.e., at certain values of λ), when the Bragg reflection condition $n\lambda = 2d \sin\theta$ is satisfied, strong reflections occur. At each of these special points on the straight line an abrupt increase of energy occurs. For a given order, the locus of all these points in the three-dimensional k-space constitutes a surface, and the space enclosed is called the Brillouin zone. The first set of critical points closest to the origin encloses the first Brillouin zone, the second closest set encloses the second Brillouin zone, and so on. The shapes of Brillouin zones are characteristic of the crystal type.

In terms of lattice vectors, Brillouin zones can be defined as follows. Position vectors $\mathbf{r}$ and $\mathbf{r}'$ of two points from which positions of the atom in the crystal appear identical are related by $\mathbf{r} = \mathbf{r}' + m_1\boldsymbol{\tau}_1 + m_2\boldsymbol{\tau}_2 + m_3\boldsymbol{\tau}_3$, where m_1, m_2 and m_3 are arbitrary integers and $\boldsymbol{\tau}_1, \boldsymbol{\tau}_2, \boldsymbol{\tau}_3$ are called primitive lattice vectors. With combinations of $\boldsymbol{\tau}_1, \boldsymbol{\tau}_2$, and $\boldsymbol{\tau}_3$, and m_1, m_2, and m_3, all lattice points can be reached. If we define $\mathbf{k}_1 = 2\pi \boldsymbol{\tau}_1 \times \boldsymbol{\tau}_3/(\boldsymbol{\tau}_1 \cdot \boldsymbol{\tau}_2 \times \boldsymbol{\tau}_3)$, $\mathbf{k}_2 = 2\pi \boldsymbol{\tau}_3 \times \boldsymbol{\tau}_1/(\boldsymbol{\tau}_1 \cdot \boldsymbol{\tau}_2 \times \boldsymbol{\tau}_3)$, $\mathbf{k}_3 = 2\pi \boldsymbol{\tau}_1 \times \boldsymbol{\tau}_2/(\boldsymbol{\tau}_1 \cdot \boldsymbol{\tau}_2 \times \boldsymbol{\tau}_3)$, then it follows that $\boldsymbol{\tau}_1 \cdot \mathbf{k}_1 = \boldsymbol{\tau}_2 \cdot \mathbf{k}_2 = \boldsymbol{\tau}_3 \cdot \mathbf{k}_3 = 2\pi$, and $\mathbf{k}_1 \cdot \boldsymbol{\tau}_2 = \mathbf{k}_1 \cdot \boldsymbol{\tau}_3 = \mathbf{k}_2 \cdot \boldsymbol{\tau}_1 = \mathbf{k}_2 \cdot \boldsymbol{\tau}_3 = \mathbf{k}_3 \cdot \boldsymbol{\tau}_1 = \mathbf{k}_3 \cdot \boldsymbol{\tau}_2 = 0$. Any vector of the form $\mathbf{g} = n_1 \mathbf{k}_1 + n_2 \mathbf{k}_2 + n_3 \mathbf{k}_3$ is called a reciprocal lattice vector. Using combinations of n_1, n_2, n_3 which are integers, all reciprocal lattice points can be reached. In the reciprocal lattice, distances are reciprocal to those in the crystal. A reciprocal lattice can be constructed by drawing from a common origin a normal on each set of planes in the crystal and using, for the length of the vector, the reciprocal of distance between the planes. Perpendicular bisectors of the reciprocal lattice vectors that enclose the origin constitute the Brillouin zones. The first Brillouin zone is the smallest such volume enclosing the origin.

Alternative notations are also used, such as $\mathbf{a}$, $\mathbf{b}$, $\mathbf{c}$, $\mathbf{a}^*$, $\mathbf{b}^*$, and $\mathbf{c}^*$ for $\boldsymbol{\tau}_1, \boldsymbol{\tau}_2, \boldsymbol{\tau}_3, \mathbf{k}_1, \mathbf{k}_2$, and $\mathbf{k}_3$, and u, v, w, h, k, and l for m_1, m_2, m_3, n_1, n_2, and n_3.

Energy bands: For a free electron moving in an unbounded potential free region, the energy is given by

$$E = \frac{p^2}{2m} = \frac{h^2}{8\pi^2 m} k^2, \quad p = \frac{h}{\lambda} = \frac{h}{2\pi} \frac{2\pi}{\lambda} = \frac{h}{2\pi} k$$

where p is the electron momentum, λ the de Broglie wavelength, and k the corresponding wave vector. E therefore changes continuously with the square of k.

If a Schrödinger equation with one-dimensional periodic potential is used, the solutions have the form $\psi(x) = u_k(x) e^{\pm ikx}$, where $u_k(x)$ is periodic with periodicity of the lattice. In this case, $\psi(x)$ are called Bloch functions. The Kronig–Penney model of electrons in a solid assumes a square well periodic potential. This potential is zero near the nucleus and is equal to V_0 with width w halfway between adjacent nuclei that are separated by a distance a. Solutions of the Schrödinger equation using this potential are possible only when

$$\cos ka = P \frac{\sin \alpha a}{\alpha a} + \cos \alpha a \tag{5-1}$$

where

$$P = \frac{4\pi^2 ma}{h^2} V_0 w \quad \text{and} \quad \alpha = \frac{2\pi}{h} (2mE)^{1/2} \tag{5-2}$$

The right-hand side of Eq. (5–1) has, for small values of αa, the appearance of a damped cosine wave. It has a value > 1 near $\alpha a = 0$ and -1 at $\alpha a = 3\pi/2$, and gradually the oscillations merge to a cosine wave $\cos \alpha a$. The left-hand side of Eq. (5–1), however, puts a constraint on the range of values, which must now be between $+1$ and -1. Thus for those values of a, values of the right-hand side of Eq. (5–1) which are > 1 or < -1 are states that are not permissible. As a result, the energy spectrum is divided into allowed and forbidden zones. With an increase of energy, the widths of both the allowed and the forbidden zones increase. Further, since cos ka can have only one value for one E, and cos ka is an even periodic function of ka, E is also an even periodic function of ka. A plot of E against k shows that the energy overlaps with the free electron energy parabola ($\propto k^2$) except at and near the points where the discontinuities occur, that is, at $k = n\pi/a$, where $n = \pm 1, \pm 2, \ldots$. The energy therefore appears in bands. The first Brillouin zone width is of $2\pi/a$, that is, from $k = -\pi/a$ to $+\pi/a$. The plot of energy bands, as mentioned above, is called the extended zone representation. If energies of the second and higher Brillouin zones (from $k = \pi/a$ to $2\pi/a$, and from $-\pi/a$ to $-2\pi/a$) are also plotted against the first Brillouin zone k values, it is called the reduced zone representation.

In metals the core electrons are tightly bound and localized. The valence electron wave functions, however, overlap with those of adjacent atoms, and a series of continuous or valence bands are formed. In a metal the highest filled band (valence band) and the next higher energy band overlap, in a semiconductor there is a small gap (band gap) between these two bands, and in an insulator the band gap is large. The upper band is also called the conduction band.

Work function: The difference in energy between the vacuum level and the Fermi energy is called the work function. The potential just outside the solid, where there is no influence of the crystal, is called the vacuum level. In a metal at $0°K$, the highest energy electrons are at the Fermi energy; therefore, at this temperature the work function represents the minimum photon energy at which photoemission can occur. In a semiconductor or an insulator, the Fermi energy usually lies within the forbidden gap, the gap from the bottom of the conduction band to the top of the valence band.

For methods of determination of work function, see, for example, [22], Vol. 1, p. 17.

Binding energy references in solids. In ionization of a metallic sample, even if both the sample and the spectrometer are in electrical contact with

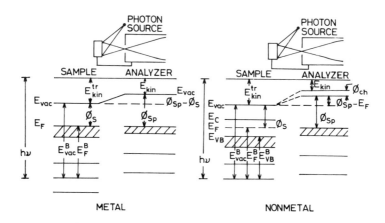

METAL NONMETAL

each other (they have the same Fermi level), there exists a small electrical field between them owing to the difference in the work functions of the sample and the spectrometer material. For example, if the spectrometer material has a higher work function and the vacuum level represents a state where the photoelectron is at the surface of the metal but beyond the influence of the solid, then owing to the work function difference and the fact that the spectrometer material and the sample metal have the same potential at the Fermi level, the surface of the spectrometer material will be at a higher potential than the surface of the sample metal. As a result, electrons leaving the sample will be decelerated by the field existing between the sample and the spectrometer. The magnitude of the field equals the work function difference $\phi_{sp} - \phi_s$, where ϕ_{sp} and ϕ_s are work functions of the spectrometer material and the sample respectively. The measured kinetic energy E_{kin} of photoelectrons thus will be lower than the true kinetic energy E_{kin}^{tr} by an amount $(\phi_{sp} - \phi_s)$. The binding energy with respect to vacuum level, after incorporating the correction, is therefore given by

$$E_{\text{vac}}^B = h\nu - E_{\text{kin}} - (\phi_{sp} - \phi_s)$$

where $h\nu$ is the incident photon energy. If binding energies are expressed with respect to the Fermi level of the metal sample, then we have

$$E_F^B = h\nu - E_{\text{kin}} - (\phi_{sp} - \phi_s) - \phi_s = h\nu - E_{\text{kin}} - \phi_{sp}$$

In insulators and semiconductors, the correction term equivalent to the one in the parentheses above would be $(\phi_{sp} - E_F)$ where E_F is the energy of the Fermi level of the sample with respect to the vacuum level. Besides this, another correction is needed. When the electrons leave a nonconducting sample, the sample acquires a positive charge and therefore decelerates the photoelectrons; in fact, it shifts the entire photoelectron spectrum to a lower kinetic energy by an amount ϕ_{ch}, equal to the amount of sample charging. If these two corrections are incorporated, the binding energy of an electron in a nonconducting sample is given by

$$E_{\text{vac}}^B = h\nu - E_{\text{kin}} - (\phi_{sp} - E_F) - \phi_{\text{ch}}$$

The determination of both E_F and ϕ in a nonconducting sample is difficult. To circumvent this, the best recourse is to refer all binding energies of such samples with respect to the top edge of the valence band E_{VB}.

CHAPTER 6

For experimental measurements of photoionization cross sections, see [2], p. 59, and the references below. A collection of experimental photoionization cross sections of atoms and molecules is given in [2].

[545] J. A. R. Samson, in D. R. Bates and I. Estermann, Eds., *Advances in Atomic and Molecular Physics*, Vol. 2, Academic, New York, 1966, p. 177.

[546] H. S. W. Massey, in H. S. W. Massey, E. H. S. Burhop, and H. B. Gilbody, Eds., *Electronic and Ionic Impact Phenomena*, Vol. 2, Clarendon, Oxford, 1969, p. 1078.
[547] A. L. Di Domenico, C. Graetzel, and J. L. Franklin, *Int. J. Mass. Spectrom. Ion Phys.* **33** (1980), 349.

The discussion on theoretical methods presented in Chapter 6 is based on material from [50] and the following.

[548] H. Hall, *Rev. Mod. Phys.* **8** (1936), 358.
[549] H. A. Bethe and E. E. Salpeter, *Quantum Mechanics of One- and Two-Electron Atoms*, Springer, Berlin, 1957.
[550] E. Merzbacher, *Quantum Mechanics*, Wiley, New York, 1961.
[551] L. I. Schiff, *Quantum Mechanics*, McGraw-Hill, New York, 1968.
[552] I. I. Sobel'man, *Introduction to the Theory of Atomic Spectra*, Pergamon, Oxford, 1972.
[553] J. C. Slater, *Quantum Theory of Atomic Structure*, 2 vols., McGraw-Hill, New York, 1960.
[554] J. T. J. Huang and J. W. Rabalais, in [21], Vol. 2, p. 197.
[555] J. T. J. Huang and J. W. Rabalais, in [21], Vol. 2, p. 225.

Units and dimensions

[556] A. G. Chertov, *Units of Measurement of Physical Quantities*, Hayden, New York, 1964.

The equivalence of Eqs. (6.14), (6.15), and (6.16) in the text is shown below. Here f is used to represent both the final-state wave function and the index (under the summation sign) for various final states, and i represents variously the initial-state wave function, the index for the various initial states, and the imaginary quantity $\sqrt{-1}$ (such as in the definition of $\mathbf{P}_\mu = -i\hbar \nabla_\mu$). We start with Eq. (6.15).

$$|M_{if}|^2 = \sum_{i,f} \left| \left\langle f \left| \sum_\mu \mathbf{r}_\mu \right| i \right\rangle \right|^2$$

using

$$\frac{i\hbar \mathbf{P}_\mu}{m} = (E_f - E_i)\mathbf{r}_\mu$$

[see Eq. (6–13) below for proof] we get

$$|M_{if}|^2 = \frac{\hbar^2}{m^2(E_f - E_i)^2} \sum_{i,f} \left| \left\langle f \left| \sum_\mu i\mathbf{P}_\mu \right| i \right\rangle \right|^2 \qquad (6-1)$$

$$= \frac{\hbar^2}{m^2(E_f - E_i)^2} \sum_{i,f} \left| \left\langle i \left| \sum_\mu (i\mathbf{P}_\mu)^* \right| f \right\rangle^* \right|^2 \qquad (6-2)$$

(see [551], p. 164; * denotes complex conjugate)

$$= \frac{\hbar^2}{m^2(E_f - E_i)^2} \sum_{i,f} \left| \left\langle i \left| \sum_\mu i^* \mathbf{P}_\mu^* \right| f \right\rangle \right|^2 \quad (6\text{-}3)$$

$i^* = -i$, $(i^*)^2 = 1$. $\mathbf{P}_\mu$ and $\mathbf{P}_\mu^*$ differ only in sign as $\mathbf{P}_\mu = -i\hbar \boldsymbol{\nabla}_\mu$. Since the integral will be squared, one may write $\mathbf{P}_\mu$ in place of $\mathbf{P}_\mu^*$:

$$|M_{if}|^2 = \frac{\hbar^2}{m^2(E_f - E_i)^2} \sum_{i,f} \left| \left\langle i \left| \sum_\mu \mathbf{P}_\mu \right| f \right\rangle \right|^2 \quad (6\text{-}4)$$

Dimensional compatibility of the left- and right-hand sides of Eqs. (6.15) and (6–4) can be checked with f in $\text{cm}^{-3/2}\,\text{cm}^{-1}\,\text{energy}^{-1/2}$ (product of the core photoion wave function $\text{cm}^{-3/2}$ and the wave function of the continuum photoelectron $\text{cm}^{-1}\,\text{energy}^{-1/2}$; see [549], p. 33, where the continuum wave function varies as $\sim (kr)^{-1}$ with $k = (2\epsilon)^{1/2}$), r in cm, i in $\text{cm}^{-3/2}$, and the volume element $d\tau$ in cm^3, that is, an overall dimension of energy^{-1}, as it would be in Ry^{-1} for the left side of the equation. By putting $\mathbf{P}_\mu = -i\hbar \boldsymbol{\nabla}_\mu$ in Eq. (6–1), we obtain

$$|M_{if}|^2 = \frac{\hbar^4}{m^2(E_f - E_i)^2} \sum_{i,f} \left| \left\langle f \left| \sum_\mu \boldsymbol{\nabla}_\mu \right| i \right\rangle \right|^2 \quad (6\text{-}5)$$

We now express Eq. (6–5) in atomic units. In atomic units, the unit of action is $\hbar$, and the unit of mass is electronic mass m. Therefore, $\hbar$ and m in Eq. (6–5) are replaced by unity. If energy was expressed in atomic units ($e^2/a_0 = 27.2\,\text{eV}$), we would have $(E_f - E_i)^2$ a.u. in the denominator. But since we are expressing energy in Rydberg units ($e^2/2a_0 = 13.6\,\text{eV}$), we have $(E_f - E_i)^2/4$ instead, as 1 a.u. of energy is equal to 2 Rydberg units of energy. Thus, Eq. (6–5) becomes

$$|M_{if}|^2 = \frac{4}{(E_f - E_i)^2} \sum_{i,f} \left| \left\langle f \left| \sum_\mu \boldsymbol{\nabla}_\mu \right| i \right\rangle \right|^2$$

which is Eq. (6.14) as $(E_f - E_i) = I_{ij} + \epsilon$.

To show the equivalence of Eq. (6.16), we proceed as follows. Potential V of a general atomic system can be written

$$V = -Z \sum_\mu \frac{e^2}{r_\mu} + \sum_{\mu < \mu'} \frac{e^2}{|\mathbf{r}_\mu - \mathbf{r}_{\mu'}|} \quad (6\text{-}6)$$

Since the Hamiltonian is $\mathscr{H} = \sum_\mu (\mathbf{P}_\mu^2/2m) + V$, we have

$$\langle i|[\mathbf{P}_\mu, \mathscr{H}]|f\rangle = \langle i|[\mathbf{P}_\mu, V]|f\rangle = -i\hbar \langle i|\boldsymbol{\nabla}_\mu V|f\rangle$$

$$= -i\hbar e^2 Z \left\langle i \left| \frac{\mathbf{r}_\mu}{r_\mu^3} \right| f \right\rangle \tag{6-7}$$

(as the gradient of the second term in V is zero)

But we also have

$$\langle i|[\mathbf{P}_\mu, \mathscr{H}]|f\rangle = (E_f - E_i)\langle i|\mathbf{P}_\mu|f\rangle \tag{6-8}$$

derived in a manner similar to Eq. (6–11) below. Combining Eqs. (6–7) and (6–8), we have

$$\langle i|\mathbf{P}_\mu|f\rangle = \frac{-i\hbar e^2 Z}{(E_f - E_i)} \left\langle i \left| \frac{\mathbf{r}_\mu}{r_\mu^3} \right| f \right\rangle \tag{6-9}$$

But from Eq. (6–4):

$$|M_{if}|^2 = \frac{\hbar^2}{m^2(E_f - E_i)^2} \sum_{i,f} \left| \left\langle i \left| \sum_\mu \mathbf{P}_\mu \right| f \right\rangle \right|^2$$

Hence, using Eq. (6–9),

$$|M_{if}|^2 = \frac{\hbar^4 e^4 Z^2}{m^2(E_f - E_i)^4} \sum_{i,f} \left| \left\langle i \left| \sum_\mu \frac{\mathbf{r}_\mu}{r_\mu^3} \right| f \right\rangle \right|^2 \tag{6-10}$$

Expressing energy in Rydbergs and other quantities in atomic units,

$$|M_{if}|^2 = \frac{16 Z^2}{(E_f - E_i)^4} \sum_{i,f} \left| \left\langle i \left| \sum_\mu \frac{\mathbf{r}_\mu}{r_\mu^3} \right| f \right\rangle \right|^2$$

which is Eq. (6.16).

$[x, \mathscr{H}] = x\mathscr{H} - \mathscr{H}x$. $\langle i|[x, \mathscr{H}]|f\rangle = \langle i|x\mathscr{H} - \mathscr{H}x|f\rangle = \langle i|x\mathscr{H}|f\rangle - \langle i|\mathscr{H}x|f\rangle = \langle i|xE_f|f\rangle - E_i\langle i|x|f\rangle = (E_f - E_i)\langle i|x|f\rangle$. Therefore, $(E_f - E_i)x = [x, \mathscr{H}]$. In three dimensions,

$$(E_f - E_i)\mathbf{r} = [\mathbf{r}, \mathscr{H}] \tag{6-11}$$

$[x, -i\hbar(d/dx)]|\phi\rangle = i\hbar|\phi\rangle$. Therefore $[x, -i\hbar(d/dx)] = i\hbar$. Now, $[x_\mu, \mathscr{H}] = [x_\mu, \sum_\mu(P_{x\mu}^2/2m) + V] = [x_\mu, (-i\hbar(d/dx_\mu))(-i\hbar(d/dx_\mu))(1/2m)]$ (since V commutes with x_μ) $= i\hbar P_{x\mu}/m$. In three dimensions,

$$[\mathbf{r}_\mu, \mathscr{H}] = \frac{i\hbar \mathbf{P}_\mu}{m} \tag{6-12}$$

Combining Eqs. (6–11) and (6–12),

$$\frac{i\hbar \mathbf{P}_\mu}{m} = (E_f - E_i)\mathbf{r}_\mu \qquad (6-13)$$

In the electric dipole approximation one affirms $e^{i\mathbf{k}\cdot\mathbf{r}} \simeq 1$. The distance r of electrons in atoms is on the order of 10^{-8} cm, and for visible and uv radiation $k_\nu = 2\pi/\lambda$ is $\ll 10^8$ cm^{-1} (at $\lambda = 58.4$ nm, $k_\nu \simeq 10^6$ cm^{-1}). Therefore $\mathbf{k}\cdot\mathbf{r}$ can be neglected and the exponential term replaced by unity. This approximation fails for high nuclear charge Z ($\mathbf{k}\cdot\mathbf{r}$ increases with Z) and also for transitions to very high energy states.

Using classical electrodynamics, Eq. (6.17) is derived in

[557]   H. Kramers, *Phil. Mag.* **46** (1923), 836.

See also an approximate formula for the continuous radiative absorption cross section of the lighter neutral atoms and positive and negative ions.

[558]   D. R. Bates, *Mon. Not. Roy. Astr. Soc.* **106** (1946), 423.

Calculation of cross section of neutral atoms and positive and negative ions

[559]   D. R. Bates, *Mon. Not. Roy. Astr. Soc.* **106** (1946), 432.

Besides bound–free Gaunt factors, there are bound–bound and free–free Gaunt factors

[560]   D. H. Menzel and C. L. Pekeris, in D. H. Menzel, Ed., *Selected papers on Physical Processes in Ionized Plasmas*, Dover, New York, 1962.

Central field calculations: Ne, Ar, and Kr

[561]   J. W. Cooper, *Phys. Rev.* **128** (1962), 681.

Many-body theory

[562]   J. Goldstone, *Proc. Roy. Soc. London* **239** (1957), 267.

[563]   D. J. Thouless, *The Quantum Mechanics of Many-Body Systems*, Academic, New York, 1972.

Cross section characteristics and nodal properties of orbitals: see [29], and

[564]   W. C. Price, A. W. Potts, and D. G. Streets, in [6], p. 187.

An XPS intensity model based on plane wave approximation

[565]   J. T. J. Huang and F. O. Ellison, *J. Electron Spectrosc.* **4** (1974), 233.

A model for approximate photoionization cross section

[566] U. Gelius, *J. Electron Spectrosc.* 5 (1974), 985.
[567] U. Gelius, in [6], p. 311.

Photoionization of atomic nitrogen and oxygen

[568] P. S. Ganas, *Phys. Rev.* 7A (1973), 928.

Photoionization cross section data of He

[569] G. L. Weissler, *Handbuch der Physik*, Vol. 21, Springer Verlag, Berlin, 1956.

Photoionization cross section data, Ar and Ne

[570] J. A. R. Samson, *J. Opt. Soc. Amer.* 54 (1964), 420.
[571] J. A. R. Samson, *J. Opt. Soc. Amer.* 55 (1965), 935.

Multiple photoionization in rare gases

[572] J. A. R. Samson and G. N. Haddad, *Phys. Rev. Lett.* 33 (1974), 875.

Photoionization cross section of rare gas outer s-subshell electrons

[573] J. A. R. Samson and J. L. Gardner, *Phys. Rev. Lett.* 33 (1974), 671.

Cross section of Ar and Xe

[574] M. Y. Adam, F. Wuilleumier, N. Sandner, S. Krummacher, V. Schmidt, and W. Mehlhorn, *Jpn. J. Appl. Phys.* 17, Suppl. 2 (1978), 170.

Cross section of Ne

[575] F. Wuilleumier and M. O. Krause, *J. Electron Spectrosc.* 15 (1979), 15.

Intensity ratios of Xe $4d_{5/2}/4d_{3/2}$ photoelectrons

[576] M. S. Banna, M. O. Krause, and F. Wuilleumier, *J. Phys. B* 12 (1979), L125.

Photoionization cross sections of Ne and Ar

[577] J. A. R. Samson, private communication, 1981.

Photoabsorption cross sections for atomic ions with $Z \leq 30$

[578] R. F. Reilman and S. T. Manson, *Astrophys. J. Suppl. Ser.* 40 (1979), 815.

Photoionization cross sections of atoms from H–Zn for Au M_α line

[579] V. G. Yarzhemsky, V. I. Nefedov, I. M. Band, and M. B. Trzhaskovskaya, *J. Electron Spectrosc.* 18 (1980), 173.

Cross section of Fe^{24+}

[580] R. F. Reilman and S. T. Manson, *Phys. Rev. A* **18** (1978), 2124.

Cross section and spin polarization of photoelectrons from Tl

[581] N. A. Cherepkov, *J. Phys. B* **10** (1977), L653.

Pb $5d$ core level cross sections using synchrotron radiation

[582] G. M. Bancroft, W. Gudat, and D. E. Eastman, *Phys. Rev. B* **17** (1978), 4499.

Elemental sensitivity factors and cross section for elements for use with SiK_α source

[583] J. E. Castle and R. H. West, *J. Electron Spectrosc.* **19** (1980), 409.

Photoionization cross section for H_2

[584] G. R. Cook and P. H. Metzger, *J. Opt. Soc. Amer.* **54** (1964), 968.

Variation of relative cross sections of molecules from He I to MgK_α limits

[585] D. A. Allison and R. G. Cavell, *J. Chem. Phys.* **68** (1978) 593.

Dipole oscillator strengths for the photoabsorption photoionization and fragmentation of molecular oxygen

[586] C. E. Brion, K. H. Tan, M. J. van der Wiel, and Ph. E. van der Leeuw, *J. Electron Spectrosc.* **17** (1979), 101.

Photoionization cross sections using HeI, HeII, YMζ, and ZrMζ lines

[587] V. G. Yarzhemsky, V. I. Nefedov, M. Ya. Amusia, N. A. Cherepkov, and L. V. Chernysheva, *J. Electron Spectrosc.* **19** (1980), 123.

Cross sections of CH_4, NH_3, and H_2O using the He I line

[588] Y. Achiba, T. Yamazaki, and K. Kimura, *J. Electron Spectrosc.* **22** (1981), 187.

Franck–Condon factors in H_2, HD, and D_2 photoionization

[589] J. Berkowitz and R. Spohr, *J. Electron Spectrosc.* **2** (1973), 143.

Rotational band shapes on HF and DF

[590] T. E. H. Walker, P. M. Dehmer, and J. Berkowitz, *J. Chem. Phys.* **59** (1973), 4292.

Vibrational transition probabilities H_2 and N_2

[591] T. H. Lee and J. W. Rabalais, *J. Chem. Phys.* **61** (1974), 2747.

Vibrational intensities of H_2

[592]   J. L. Gardner and J. A. R. Samson, *J. Electron Spectrosc.* **8** (1976), 123.

CO partial photoionization cross sections

[593]   J. A. R. Samson and J. L. Gardner, *J. Electron Spectrosc.* **8** (1976), 35.

Vibrational intensities of O_2

[594]   P. Natalis, *J. Phys. Chem.* **80** (1976), 2829.

Franck–Condon factors for H_2O, D_2O, and HDO

[595]   R. Botter and J. Carlier, *J. Electron Spectrosc.* **12** (1977), 55 (in French).

Total and partial photoionization cross sections of O_2 in the 10–80 nm range

[596]   J. A. R. Samson, J. L. Gardner, and G. N. Haddad, *J. Electron Spectrosc.* **12** (1977), 281.

Vibrational intensities of NO

[597]   P. Natalis, J. Delwiche, J. E. Collin, G. Caprace, and M.-Th. Praet, *J. Electron Spectrosc.* **11** (1977), 417.

Vibrational intensities of O_2, N_2, and CO

[598]   J. L. Gardner and J. A. R. Samson, *J. Electron Spectrosc.* **13** (1978), 7.

Effects of shape resonance on vibrational intensities in molecular photoionization

[599]   R. Stockbauer, B. E. Cole, D. L. Ederer, J. B. West, A. C. Parr, and J. L. Dehmer, *Phys. Rev. Lett.* **43** (1979), 757.

Branching ratio cross section in 3s subshell of Ar

[600]   K. H. Tan and C. E. Brion, *J. Electron Spectrosc.* **13** (1978), 77.

Branching ratio in the 5p subshell of Xe

[601]   W. Ong and S. T. Manson, *J. Phys. B* **11** (1978), L163.

Partial oscillator strengths for N_2O and CO_2

[602]   C. E. Brion and K. H. Tan, *Chem. Phys.* **34** (1978), 141.

Partial cross sections for CO_2 and N_2O

[603]   C. E. Brion and K. H. Tan, *J. Electron Spectrosc.* **15** (1979), 241.

Photoionization cross sections from 1 to 1500 keV

[604] J. H. Scofield, *Lawrence Livermore Laboratory Report* UCRL-51326 (1973).

Subshell photoionization cross sections at 1254 and 1487 eV

[605] J. H. Scofield, *J. Electron Spectrosc.* **8** (1976), 129.

Application of XPS to analytical chemistry

[606] C. D. Wagner, *Anal. Chem.* **44** (1972), 1050.
[607] W. J. Carter, G. K. Schweitzer, and T. A. Carlson, *J. Electron Spectrosc.* **5** (1974), 827.
[608] V. I. Nefedov, N. P. Sergushin, I. M. Band, and M. B. Trzhaskovskaya, *J. Electron Spectrosc.* **2** (1973), 383.

Quantitative analysis by XPS

[609] C. J. Powell, *Surf. Sci.* **44** (1974), 29.
[610] D. R. Penn, *J. Electron Spectrosc.* **9** (1976), 29.

Quantitative analysis without standards

[611] K. Hirokawa and M. Oku, *Z. Anal. Chem.* **285** (1977), 192.

Application of photoionization cross section and mean free paths to quantitative surface analysis by XPS

[612] K. Hirokawa, M. Oku, and F. Honda, *Z. Anal. Chem.* **286** (1977), 41.

Electron mean free path and quantitative analysis

[613] J. Szajman, J. G. Jenkin, R. C. G. Leckey, and J. Liesegang, *J. Electron Spectrosc.* **19** (1980), 393.

The absorption edge jump S_{qj} in Eq. (6.53) (for element j) is defined as the ratio of the photoabsorption coefficient just above and just below the experimental threshold frequency for photoionization from a shell (specified by q). Above the q ionization threshold, ionization takes place from one more (inner) shell q in addition to all those Σp having lower ionization potentials. Therefore $S_{qj} = (\zeta_q + \zeta_{\Sigma p})/\zeta_{\Sigma p}$ where ζ denotes the absorption coefficient. The factor $(S_{qj} - 1)/S_{qj} = \zeta_q/(\zeta_q + \zeta_{\Sigma p})$ hence represents the fraction of incident radiation absorbed for q ionization.

Eq. (6.53) and those which immediately follow it are based on [614]; for uniformity, terminologies used in other papers such as [616] and [617] for SI, RSI$_\infty$ are changed to conform to that of [614]. For interconversion between sublevel schemes $K, L_\text{I}, L_\text{II}, \ldots (K, L_1, L_2, \ldots)$ and $1s_{1/2}, 2s_{1/2}, 2p_{1/2}, \ldots$, see Table 3.1.

XPS model derived from X-ray fluorescence analysis

[614] H. Ebel and M. F. Ebel, *X-ray Spectrom.* **2** (1973), 19.
[615] M. F. Ebel, *J. Electron Spectrosc.* **5** (1974), 837.

Elemental sensitivities of metals in XPS

[616] M. Janghorbani, M. Vulli, and K. Starke, *Anal. Chem.* **47** (1975), 2200.

Intraelemental photoelectron line intensities

[617] M. Vulli, M. Janghorbani, and K. Starke, *Anal. Chim. Acta* **82** (1976), 121.

XPS: quantitative surface analysis with variable take-off angle

[618] H. Ebel, in [471], p. 2251.

Relations between calculated cross section and measured intensities of photoelectron lines

[619] M. Vulli and K. Starke, *J. Microsc. Spectrosc. Electron.* **3** (1978), 45.
[620] M. Vulli and K. Starke, *J. Microsc. Spectrosc. Electron.* **3** (1978), 57.

Quantitative surface measurement of elemental species on metal oxide powder surface

[621] M. J. Dreiling, *Surf. Sci.* **71** (1978), 231.

Surface analaysis of HTR fuel particles

[622] G. C. Allen, P. M. Tucker, and R. K. Wild, *J. Nucl. Mat.* **71** (1978), 345.

Quantitative analysis of GaAs(III), (111) and (110) single crystals

[623] M. Kudo, Y. Nihei, and H. Kamada, *Jpn. J. Appl. Phys.* **17** (1978), 797.

Evaluation of XPS-data of oxide layers

[624] M. F. Ebel and W. Liebl, *J. Electron Spectrosc.* **16** (1979), 463.

Quantitative surface analysis and angular distribution of photoelectrons

[625] M. Vulli, *Surf. Interface Anal.* **3** (1981), 67.

Trace analysis in solution, extraction onto solid surface, and matrix dilution technique

[626] D. M. Hercules, L. E. Cox, S. Onisick, G. D. Nichols, and J. C. Carver, *Anal. Chem.* **45** (1973), 1973.
[627] D. M. Hercules, *J. Electron Spectrosc.* **5** (1974), 811.
[628] D. M. Hercules, *Physica Scripta* **16** (1977), 169.

Simultaneous analytical detection of Sn and Pb at ppb level

[629] L. Sabbatini, T. Dickinson, and P. M. A. Scherwood, *Ann. Chim.* (1980), 137.

Quantitative analysis of polymers

[630] D. T. Clark, in [18], p. 211.

Other quantitative analytical applications

[631] J. S. Brinen and J. E. McClure, *Anal. Lett.* **5** (1972), 737.
[632] R. D. Giauque, F. S. Goulding, J. M. Jaklevic, and R. H. Pehl, *Anal. Chem.* **45** (1973), 671.
[633] J. S. Brinen and J. E. McClure, *J. Electron Spectrosc.* **4** (1974), 243.
[634] G. M. Bancroft, J. R. Brown, and W. S. Fyfe, *Anal. Chem.* **49** (1977), 1044.
[635] M. B. Carvalho and D. M. Hercules, *Anal. Chem.* **50** (1978), 2030.
[636] K. Hirokawa and M. Oku, *Talanta* **25** (1978), 539.

Analysis of trace elements in water

[637] D. Briggs, V. A. Gibson, and J. K. Becconsall, *J. Electron Spectrosc.* **11** (1977), 343.

Application to archaeological chemistry

[638] D. Briggs, M. Thompson, and E. T. Leventhal, *Analyt. Lett.* **10** (1977), 153.

Reference [638] involves an XPS study of resin and decorative pigments in a sample of Egyptian mummy cartonnage, stylistically of the 12th dynasty (1000–725 B.C.). Cleaning and depth profiling are done by Ar^+ etching followed by XPS scan of various regions of the surface for multielement analysis. The pigment layer shows a complex composition containing O, Na, Mg, Al, Si, P, S, Ca, and the transition metals Fe and Cu.

Besides XPS, the major methods of solid and surface analysis discussed in Chapter 6 are AES, SIMS, LEED, and ELS. Brief descriptions follow.

AES (Auger Electron Spectroscopy). In common Auger spectrometers, electron beams of energy in the range 5–10 keV are used. In the 0.1–3 keV range, KLL, LMM, and MNN series are observed. The cross section of the Auger process for any shell becomes maximum when the incident electron energy is four times that of the binding energy E_B of the shell. Under this condition the cross section is given by: $\sigma(K \text{ shell}) = 2 \times 10^{-14} E_B^{-2}$ cm^2; $\sigma(L \text{ shell}) = 5 \times 10^{-15} E_B^{-1.56}$ cm^2. The Auger chemical shift is much larger than the XPS shift. This is indicated for metal/metal-oxide systems: Zn 3.7, Ga 4.2, Ge 3.5, As 3.4 eV. The kinetic energy of the sharpest Auger line minus the kinetic energy of the most intense photoelectron line [1486.6 − B.E. (photoelectron)] is defined as the *Auger parameter* (if $E_{\text{Auger}}^{\text{kin}} - E_{\text{photoelectron}}^{\text{kin}}$ is negative then 1000 or 2000 is added to make it a positive number between 0 and 999). It is independent of static charge referencing and is characteristic of a molecular compound or solid state. AES can be used for surface analysis (1–2% of a monolayer or even less). Scanning Auger microscopy (SAM) involves fast elemental mapping using raster scanning of the surface with a highly collimated electron beam followed by Auger electron analysis. This, along with surface etching (such as with an Ar^+ beam), can be used for fast depth profiling. The electron beam size in AES is of the order of microns, and data collection time is in milliseconds. See [18], p. 249.

SIMS (Secondary Ion Mass Spectrometry). In SIMS, a beam of primary ions of high energy (several keV) is used to bombard the sample and the secondary ions produced by sputtering of target particles are analyzed mass spectrometrically. SIMS is capable of qualitative and semiquantitative analysis of all elements with an average detection limit of about 1 ppm and information depth of 0.3–2 nm. As SIMS is a destructive method, the minimum detectable concentration is related to material consumption. For example, Si bombarded with O_2^+(5.5 keV) having a primary ion current density 4×10^{-4} A/cm^2, bombardment area 7.0×10^{-4} cm^2, erosion rate 0.8 nm/sec, sputtered-volume/sec of 6.0×10^{-11} cm^3/sec, gives, on the basis of minimum detectable ion current of 3×10^{-18} A, an extrapolated minimum detectable concentration for Si of 1 ppm. It may be used for detection of surface contamination or for analysis of thin films, semiconductors, ceramic materials, diffusion profiles, or for local analysis of spots having dimensions of a few microns, or for the structure of organic compounds. See [23], p. 324.

LEED (Low Energy Electron Diffraction). In LEED, low energy electrons (15–1000 eV) that are capable of penetrating only a few atomic layers into a solid are made to fall on a solid surface, and the diffraction pattern is observed on a hemispherical fluorescent screen maintained at a positive potential of 5 kV with respect to the specimen. A fine wire grid maintained at the same potential as the specimen is mounted close in front of the screen to ensure that the trajectories of the diffracted electrons are not distorted by potential gradients in the space between the specimen and the screen. In an automatic scanning diffractometer, arrangement is provided for rotation to vary the azimuth ϕ of the crystal with respect to the incident electron beam. The diffracted beam produces a current in a Faraday cup that can move in an equatorial plane, permitting detection of electrons at various values of θ, the angle between the incident beam and the diffracted beam. The intensity of the diffracted beam can also be measured as a function of the energy of incident electrons. LEED phenomena are influenced by the nature of the surface of the crystal, such as the presence of adsorbates, and hence may be used for studying surface properties. The LEED pattern shows arrangements of atoms on the surface layer as well as on planes in the interior of the crystal near the surface region. See [12].

ELS (Electron Energy Loss Spectroscopy). In ELS inelastic scattering of low energy ($<$ 1 keV) electrons from a solid surface are studied by measuring the energy spectrum of backscattered electrons. If $E_0(\mathbf{k})$ is the energy of the primary electrons with momentum $\hbar k$ and $E_s(\mathbf{k}_s)$ is the energy of the reflected electron, then the energy loss $\hbar\omega$ of the electron due to scattering is given by $E_s(\mathbf{k}_s) = E_0(\mathbf{k}) - \hbar\omega$. Keeping $\mathbf{k}$, E_0 and the angle of incidence of electrons to the solid surface as constants, an energy spectrum of the scattered electrons leads to information on the characteristic $\hbar\omega$ spectrum. ELS is capable of yielding information on microscopic structure of adsorbates and on the type of adsorption, whether dissociative or molecular. It also provides information on binding potential and the effects of coadsorption; the kinetics of adsorption can be studied by intensity versus coverage measurements. Effects due to structural anisotropies can also be studied by ELS. Because of low incident electron energy and low values of current density ($<10^{-12}$ A cm^{-2}) ELS is a nondestructive method of analysis. See [19], p. 205.

CHAPTER 7

Angular distribution of valence shell photoelectrons

[639]   J. Berkowitz and H. Ehrhardt, *Phys. Lett.* **21** (1966), 531.

Photoelectron angular distributions in one-electron and many-electron problems

[640]   J. Cooper and R. N. Zare, in S. Geltman, K. Mahanthappa, and W. Brittin, Eds., *Lectures in Theoretical Physics,* Gordon and Breach, Vol. XI-C, New York, 1969, p. 317.

In the scattering problem involving a spherically symmetric potential, the method of *partial waves* for calculating the differential cross section involves connecting the asymptotic form of the solution [$\psi(r, \theta, \phi) \rightarrow A(e^{ikz} + r^{-1}f(\theta, \phi)e^{ikr})$ as $r \rightarrow \infty$] obtained from general physical considerations with that derived from solving the Schrödinger equation that is separable. The wave function is written as $\Sigma_{l=0}^{\infty}(2l+1)i^l R_l(r) P_l(\cos\theta)$ — an infinite sum of terms (*partial waves*) — each being a product of a radial function and a Legendre function with index l. The asymptotic form comes as $\sim r^{-1} \sin(kr + \delta_l)$ where δ_l is a phase angle. The difference in phases between the asymptotic forms of the actual radial function $R_l(r)$ and the radial function $j_l(kr)$ in the absence of a scattering potential is called the *phase shift* of the lth partial wave. The phase shifts completely determine $f(\theta)$ and hence the scattering cross section $f(\theta)^2$ [551]. In the calculation of photoionization cross section for dipole transition, the presence of $3-j$ symbols (from angular momentum algebra) in $I(\theta)$ makes the matrix element non-zero only when $l' = l \pm 1$, which constitutes the selection rule [640].

Angular distribution of photoelectrons; calculation of intensity expressions

[641]   J. A. R. Samson, *J. Opt. Soc. Amer.* **59** (1969), 356.

Angular distribution of photoelectrons: subshell structure

[642] J. W. Cooper and S. T. Manson, *Phys. Rev.* **177** (1969), 157.

Angular distribution in photoionization of N_2

[643] B. Schneider and R. S. Berry, *Phys. Rev.* **182** (1969), 141.

Review of angular distribution work up to 1971

[644] T. A. Carlson, G. E. McGuire, A. E. Jonas, K. L. Cheng, C. P. Anderson, C. C. Lu, and B. P. Pullen, in [6], p. 207.

Angular distribution of photoelectrons in the soft X-ray region

[645] P. Wuilleumier and M. O. Krause, *J. Electron Spectrosc.* **5** (1974), 921.

Photoelectron angular distribution with partially polarized light

[646] J. A. R. Samson, *Phil. Trans. Roy. Soc. London* **A268** (1970), 141.
[647] J. A. R. Samson and A. F. Starace, *J. Phys.* **B 8** (1975), 1806.

Angular distribution expression for partially polarized radiation:

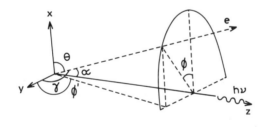

$$\frac{d\sigma}{d\Omega} = I(\gamma, \theta) = k \left[1 - \frac{\beta}{2} + \frac{3}{2} \frac{\beta}{g+1} (g \cos^2 \gamma + \cos^2 \theta) \right] \quad (7-1)$$

where g is the ratio of the intensities of polarization components in the x and y directions, $g = I_y/I_x$ [646]. Eq. (7.3) in the text is from [647].
Integration of the above equation over γ and θ, with $\cos \gamma = \sin \theta \cos \phi'$, gives

$$\sigma_{tot} = k \iint \left[1 - \frac{\beta}{2} + \frac{3}{2} \frac{\beta}{g+1} (g \cos^2 \gamma + \cos^2 \theta) \right] \sin \theta \, d\theta \, d\phi' = 4\pi k$$

that is,

$$k = \sigma_{tot}/4\pi$$

For polarized radiation, $g = 0$, and one obtains:

$$\frac{d\sigma}{d\Omega} = \frac{\sigma_{tot}}{4\pi}\left[1 + \frac{\beta}{2}(3\cos^2\theta - 1)\right] \quad (7-2)$$

For unpolarized radiation, Eq. (7–2) can be written as follows, since $\cos\theta = \sin\alpha\cos\phi$,

$$\frac{d\sigma}{d\Omega} = \frac{\sigma_{tot}}{4\pi}\left[1 + \frac{\beta}{2}(3\cos^2\phi\sin^2\alpha - 1)\right]$$

Integration over all ϕ and averaging give

$$\frac{d\sigma}{d\Omega} = \frac{\sigma_{tot}}{4\pi}\left[1 + \frac{\beta}{2}\left(\frac{3}{2}\sin^2\alpha - 1\right)\right] = \frac{\sigma_{tot}}{4\pi}\left[1 - \frac{\beta}{4}(3\cos^2\alpha - 1)\right]$$

β calculations

[648]   C. N. Yang, *Phys. Rev.* **74** (1948), 764.
[649]   V. L. Jacobs and P. G. Burke, *J. Phys. B* **5** (1972), 2257.
[650]   U. Fano and D. Dill, *Phys. Rev. A* **6** (1972), 185.
[651]   D. Dill, A. F. Starace, and S. T. Manson, *Phys. Rev. A* **11** (1975), 1596.

Photoelectron angular distribution in autoionization resonances

[652]   D. Dill, *Phys. Rev. A.* **7** (1973), 1976.

Determination of atomic pseudopotentials from photoemission angular distributions

[653]   D. L. Miller and J. D. Dow, *Phys. Lett.* **60 A** (1977), 16.

Angular distribution of photoelectrons in photodetachment from C^-, O^-, F^-, I^-

[654]   J. Cooper and R. N. Zare, *J. Chem. Phys.* **48** (1968), 942.

Angular distribution of photoelectrons in Kr photoionization between 300 and 1500 eV

[655]   M. O. Krause, *Phys. Rev.* **177** (1969), 151.

Photoelectron angular distribution measurements near threshold: Ar, Kr, Xe, H_2, N_2, O_2

[656]   J. W. McGowan, D. A. Vroom, and A. R. Comeaux, *J. Chem. Phys.* **51** (1969), 5626.

Calculations on photoelectron angular distribution in noble gas photoionization

[657] S. T. Manson and J. W. Cooper, *Phys. Rev. A* **2** (1970), 2170.

Angular distribution of photoelectrons in Cd and Zn atomic beam photoionization

[658] H. Harrison, *J. Chem. Phys.* **52** (1970), 901.

Angular distribution of photoelectrons from atomic O, N, and C

[659] D. J. Kennedy and S. T. Manson, *Planet. Space Sci.* **19** (1972), 621.
[660] S. T. Manson, D. J. Kennedy, A. F. Starace, and D. Dill, *Planet. Space Sci.* **22** (1974), 1535.

Angular distribution as a probe of anisotropic electron–ion interactions

[661] D. Dill, S. T. Manson, and A. F. Starace, *Phys. Rev. Lett.* **32** (1974), 971.

Angular distribution of photoelectrons from atomic O

[662] A. F. Starace, S. T. Manson, and D. J. Kennedy, *Phys. Rev. A* **9** (1974), 2453.

Atomic O photoelectron angular distributions

[663] E. R. Smith, *Phys. Rev. A* **13** (1976), 1058.

Photoelectron angular distributions of s electrons in open-shell atoms

[664] A. F. Starace, R. H. Rast, and S. T. Manson, *Phys. Rev. Lett.* **38** (1977), 1522.

Photoelectron angular distribution and β parameters for Ne and Ar

[665] K. T. Taylor, *J. Phys. B* **10** (1977), L699.

β for Kr $4p$ ionization as a function of energy by synchrotron radiation

[666] D. L. Miller, J. D. Dow, R. G. Houlgate, G. V. Marr, and J. B. West, *J. Phys. B* **10** (1977), 3205.

Angular distribution of photoelectrons from atomic O at 73.6 nm and 58.4 nm

[667] J. A. R. Samson and W. H. Hancock, *Phys. Lett.* **61A** (1977), 380.

Anisotropic effects in the angular distribution of photoelectrons from Cs $6s$

[668] W. Ong and S. T. Manson, *Phys. Lett.* **66A** (1978), 17.

Angular distribution and spin polarization of Xe $5s - \epsilon p$ photoelectrons

[669] N. A. Cherepkov, *Phys. Lett.* **66A** (1978), 204.

A comparison of theories of angular distributions of photoelectrons emitted from rare gas atoms

[670] C. A. Swarts, D. L. Miller, and J. D. Dow, *Phys. Rev. A* **19** (1979), 734.

Angular distribution for the $4d^{10}$ subshell of Cd

[671] C. E. Theodosiou, *J. Phys. B* **12** (1979), L673.

Angular distribution for the outer shell of alkali-metal atoms

[672] W. Ong and S. T. Manson, *Phys. Rev. A* **20** (1979), 2364.

Branching ratios and angular distributions in the outer p shells of noble gases

[673] W. Ong and S. T. Manson, *Phys. Rev. A* **21** (1980), 842.

Angular distribution from the subshells of high Z elements

[674] Y. S. Kim, R. H. Pratt, A. Ron, and H. K. Tseng, *Phys. Rev. A* **22** (1980), 567.

Photoelectron angular distribution in small molecules

[675] J. C. Tully, R. S. Berry, and B. J. Dalton, *Phys. Rev.* **176** (1968), 95.
[676] A. D. Buckingham, B. J. Orr, and J. M. Sichel, *Phil. Trans. Roy. Soc. London* **A268** (1970), 147.

Angular distribution of photoelectrons in photoionization of benzene

[677] T. A. Carlson and C. P. Anderson, *Chem. Phys. Lett.* **10** (1971), 561.

Photoelectron angular distribution in halomethanes

[678] T. A. Carlson and R. M. White, *Faraday Disc. Chem. Soc.* **54** (1972), 285.

Using time-dependent perturbation theory, Eq. (7.42) is derived in [551], p. 285 and [550], p. 470. The Hamiltonian $\mathcal{H}$ of the system is expressed as $\mathcal{H} = \mathcal{H}_0 + \mathcal{H}'$. $\mathcal{H}'$, the perturbation to $\mathcal{H}_0$, is small and time-dependent, and its effect is to cause transitions between eigenstates of $\mathcal{H}_0$. Equation (7.42) is called Fermi's "golden rule no. 2."

The Hamiltonian for a charged particle in an electromagnetic field is given by ([551] pp. 176, 398):

$$\mathcal{H} = \frac{\mathbf{P}^2}{2m} + V - \frac{e}{2mc}(\mathbf{P}\cdot\mathbf{A} - \mathbf{A}\cdot\mathbf{P}) + \frac{e^2}{2mc^2}\mathbf{A}^2 + e\phi \qquad (7\text{--}3)$$

where V is the atomic potential felt by the electron. The vector potential $\mathbf{A}$ and the scalar potential ϕ are related to electric and magnetic field strengths by

$$\mathbf{E} = -\frac{1}{c}\frac{\partial \mathbf{A}}{\partial t} - \nabla\phi \quad \text{and} \quad \mathbf{H} = \nabla \times \mathbf{A} \qquad (7\text{--}4)$$

where e is the electronic charge and c is the velocity of light. From [551] p.178, we get:

$$[A, P_x] = A_x \cdot P_x - P_x \cdot A_x = i\hbar \frac{\partial A_x}{\partial x}$$

In three dimensions,

$$[A, P] = A \cdot P - P \cdot A = i\hbar \nabla \cdot A \tag{7-5}$$

With the help of Eq. (7–5), now Eq. (7–3) can be written

$$\mathcal{H} = \frac{P^2}{2m} + V - \frac{e}{mc} A \cdot P + \frac{ie\hbar}{2mc} \nabla \cdot A + \frac{e^2}{2mc^2} A^2 + e\phi \tag{7-6}$$

(Note that in Eq. (7–6), $\nabla \cdot A$ is simply a multiplication term, the gradient of the vector potential). If the radiation field is weak, then the fifth term can be neglected. It is possible to choose conditions such that $\nabla \cdot A = 0$ and $\phi = 0$ (Coulomb gauge; see [551], p. 522). Under this condition, $\mathcal{H}'$, the perturbation to the Hamiltonian $\mathcal{H}_0$ is equal to the term $-(e/mc) A \cdot P$. Equation (7.43) is then derived by substituting in Eq. (7.42) $\mathcal{H}' = -(e/mc) A \cdot P$ and $A = A_0 u$, where u is the unit vector in the direction of polarization, and A_0 is the amplitude of the vector potential.

Solution of the wave equation gives for the *real* part of the vector potential ([551], p. 400):

$$A(r, t) = A_0 e^{i(k \cdot r - \omega t)} + \text{its complex conjugate}$$

The constant complex vector $A_0 (A_0 = A_0' + iA_0'')$ is defined as $|A_0| e^{i\alpha}$, where $|A_0|$ is a real vector. Then, the sum of the two terms for $A(r, t)$ above gives

$$A(r, t) = 2|A_0| \cos(k \cdot r - \omega t + \alpha)$$

Using the definitions of electric and magnetic field strengths E and H given above:

$$E = -2k|A_0| \sin(k \cdot r - \omega t + \alpha) \qquad H = -2k \times |A_0| \sin(k \cdot r - \omega t + \alpha)$$

The Poynting vector S expresses the rate of energy flow per unit area in a plane electromagnetic wave; it is defined by $S = (c/4\pi) E \times H$. The direction of S (i.e., the direction of $E \times H$), is the direction of movement of energy, here the direction of k. Its magnitude, averaged over a period $2\pi/\omega$ of the oscillation, which is the intensity associated with the plane wave, is $(\omega^2/2\pi c)|A_0|^2$ where $|A_0|^2 = |A_0| \cdot |A_0| = A_0 \cdot A_0^*$.

Calculations on photoelectron angular distribution of molecules

[679] L. L. Lohr, Jr., in [6], p. 245.

[680] B. Ritchie, *J. Chem. Phys.* **61** (1974), 3291.
[681] B. Ritchie, *J. Chem. Phys.* **60** (1974), 898.
[682] B. Ritchie, *J. Chem. Phys.* **61** (1974), 3279.
[683] B. Ritchie, *Phys. Rev. A* **14** (1976), 359.
[684] J. W. Rabalais and T. P. Debies, *J. Electron Spectrosc.* **5** (1974), 847.

High energy approximations for angular distribution calculations

[685] B. Ritchie, *J. Chem. Phys.* **64** (1976), 3050.

Angular distribution of photoelectrons from cyclopropane

[686] F. J. Leng and G. L. Nyberg, *J. Electron Spectrosc.* **11** (1977), 293.

Angular distribution of photoelectrons from allene

[687] F. J. Leng and G. L. Nyberg, *J. Chem. Soc. Faraday Trans. II* **73** (1977), 1719.

Angular distribution of photoelectrons from π-orbital states of 1,4-cyclohexadiene

[688] M. H. Kibel, M. K. Livett, and G. L. Nyberg, *J. Electron Spectrosc.* **14** (1978), 155.
[689] F. J. Leng and G. L. Nyberg, *J. Phys. E* **10** (1977), 686.

Dynamical and structural information from angular distributions of oriented or rotationally resolved unoriented molecular samples

[690] B. Ritchie and B. R. Tambe, *J. Chem. Phys.* **68** (1978), 755.

Angular distribution of photoelectrons from unsaturated hydrocarbons: ethylene, 1,3-butadiene, benzene, cyclohexene, norbornadiene, 1,3- and 1,4-cyclohexadiene

[691] M. H. Kibel, K. K. Livett, and G. L. Nyberg, *J. Electron Spectrosc.* **15** (1979), 275.

Angular distribution of photoelectrons from H_2, N_2, O_2, CO, CO_2, Ar, and Xe

[692] M. H. Kibel, F. J. Leng, and G. L. Nyberg, *J. Electron Spectrosc.* **15** (1979), 281.

Angular distribution of photoelectrons from CO

[693] J. A. Sell, A. Kuppermann, and D. M. Mintz, *J. Electron Spectrosc.* **16** (1979), 127.

Angular distribution of photoelectrons from H_2

[694] S. Katsumata, Y. Achiba, and K. Kimura, *Chem. Phys. Lett.* **63** (1979), 281.

Angular distribution of photoelectrons from NO

[695] M. H. Kibel and G. L. Nyberg, *J. Electron Spectrosc.* **17** (1979), 1.

Angular distribution of photoelectrons for C_2H_2

[696] J. Kreile and A. Schweig, *Chem. Phys. Lett.* **69** (1980), 71.

Calculation of β for N_2, CO, CO_2, COS, and CS_2

[697] F. A. Grimm, T. A. Carlson, W. B. Dress, P. Agron, J. O. Thomson, and J. W. Davenport, *J. Chem. Phys.* **72** (1980), 3041.

Photoelectron angular distribution of HBr and HI

[698] M. G. White, S. H. Southworth, P. Kobrin, and D. A. Shirley, *J. Electron Spectrosc.* **19** (1980), 115.

Photoelectron angular distribution from 3d orbitals of Kr, HBr, Br_2, and CH_3Br using synchrotron radiation: Molecular effects on the angular distribution parameter

[699] T. A. Carlson, M. O. Krause, F. A. Grimm, P. R. Keller, and J. W. Taylor, *Chem. Phys. Lett.* **87** (1982), 552.

Effects of rotational autoionization in photoelectron angular distribution from H_2

[700] D. Dill, *Phys. Rev. A* **6** (1972), 160.

Calculation for individual rotational transitions

[701] J. M. Sichel, *Mol. Phys.* **18** (1970), 95.

Photoelectron angular distribution in rotational transitions $H_2 \rightarrow H_2^+$

[702] A. Niehaus and M. W. Ruf, *Chem. Phys. Lett.* **11** (1971), 55.

Angular dependence of photoelectrons in vibrational structure, N_2 and O_2

[703] T. A. Carlson, *Chem. Phys. Lett.* **9** (1971), 23.
[704] T. A. Carlson and A. E. Jonas, *J. Chem. Phys.* **55** (1971), 4913.

Angular dependence of vibrational structure

[705] O. F. Kalman, *J. Electron Spectrosc.* **8** (1976), 335.

Vibrational–rotational structure and intensity in angular distribution of photoelectrons from diatomic molecules: General theory and application to H_2

[706] Y. Itikawa, *Chem. Phys.* **37** (1979), 401.
[707] Y. Itikawa, *Chem. Phys. Lett.* **62** (1979), 261.

Bloch's theorem states that for a periodic potential the eigenfunctions (Bloch functions) of the wave equation are of the form $\psi_k(r) = e^{i\mathbf{k}\cdot\mathbf{r}} u_k(r)$, where $u_k(r)$ is periodic in the crystal lattice. Under a crystal lattice translation that takes $\mathbf{r}$ to $\mathbf{r} + \mathbf{l}$, where $\mathbf{l}$ is a real lattice vector, one may write

$$\psi_k(\mathbf{r} + \mathbf{l}) = e^{i\mathbf{k}\cdot(\mathbf{r}+\mathbf{l})} u_k(\mathbf{r} + \mathbf{l}) = e^{i\mathbf{k}\cdot\mathbf{l}} e^{i\mathbf{k}\cdot\mathbf{r}} u_k(r) = e^{i\mathbf{k}\cdot\mathbf{l}} \psi_k(r)$$

since $u_k(\mathbf{r} + \mathbf{l}) = u_k(r)$. If the wave vector happens to be $\mathbf{g}$, which is a reciprocal lattice vector, then from above one can write

$$\psi_g(\mathbf{r} + \mathbf{l}) = e^{i\mathbf{g}\cdot\mathbf{l}} \psi_g(\mathbf{r}) = \psi_g(\mathbf{r})$$

since $e^{i\mathbf{g}\cdot\mathbf{l}} = e^{2\pi i} = 1$ for all $\mathbf{l}$. If the wave vector $\mathbf{k}$ of the state ψ_k can be represented as $\mathbf{k} = \mathbf{k}' + \mathbf{g}$ where $\mathbf{g}$ is a reciprocal lattice vector (e.g., any vector in $\mathbf{k}$ space can be written as a sum of a vector $\mathbf{k}'$ in the first Brillouin zone and some reciprocal lattice vector $\mathbf{g}$), then from above

$$\psi_k(\mathbf{r} + \mathbf{l}) = e^{i(\mathbf{g}+\mathbf{k}')\cdot\mathbf{l}} \psi_k(\mathbf{r}) = e^{i\mathbf{g}\cdot\mathbf{l}} e^{i\mathbf{k}'\cdot\mathbf{l}} \psi_k(\mathbf{r}) = e^{i\mathbf{k}'\cdot\mathbf{l}} \psi_k(\mathbf{r})$$

This means that ψ_k satisfies Bloch's theorem as if it had the wave vector $\mathbf{k}'$. The label $\mathbf{k}$ is thus not unique. Every state has a set of possible wave vectors; one differs from another by reciprocal lattice vectors, and it is always possible to write $\mathbf{k} = \mathbf{k}' + \mathbf{g}$. (See [708] p. 20.)

In the matrix element $\langle \psi_k^f | \mathbf{A} \cdot \nabla | \psi_k \rangle$, if Bloch functions are used for ψ_k^f and ψ_k, the matrix element assumes the explicit form

$$\int e^{i(\mathbf{k}-\mathbf{k}^f)\cdot\mathbf{r}} [u_{k^f}^*(\mathbf{r})(\mathbf{A}\cdot\nabla) u_k(\mathbf{r})] d^3r$$

Expanding $u_k(r)$ in a Fourier series in the reciprocal lattice vectors $\mathbf{g}$, $u_k(\mathbf{r}) = \sum_g a_g e^{i\mathbf{g}\cdot\mathbf{r}}$ (see theorem, [709], p. 2), one obtains for the factor $-$ (say, $f(\mathbf{r})$) $-$ within the square brackets as $i\Sigma(\mathbf{A}\cdot\mathbf{g}) e^{i\mathbf{g}\cdot\mathbf{r}} u_{k^f}^*(\mathbf{r})$, which has the periodicity of the lattice. Under this periodicity of $f(\mathbf{r})$, the matrix element integral is zero unless $\mathbf{k} - \mathbf{k}^f = -\mathbf{g}$, a reciprocal lattice vector (see theorem, [709], p. 3). This means that for nonzero transition probability, $\mathbf{k}^f = \mathbf{k} + \mathbf{g}$.

[708] J. M. Ziman, *Principles of the Theory of Solids*, Cambridge University Press, Cambridge, 1965.

[709] C. Kittel, *Quantum Theory of Solids*, Wiley, New York, 1963.

For symmetry selection rules, see also [63] and [78]; for single crystal effect, see [63].

Surface energy band structure for Cl adsorbed on Si (111)

[710] P. K. Larsen, N. V. Smith, M. Schlüter, H. H. Farrell, K. M. Ho, and M. L. Cohen, *Phys. Rev. B* 17 (1978), 2612.

Photoelectron intensities from oriented molecules

[711] J. W. Davenport, *Phys. Rev. Lett.* **36** (1976), 945.
[712] J. W. Davenport, Thesis, University of Pennsylvania, 1976.

Parity considerations in angle-resolved photoemission experiments

[713] J. Hermanson, *Solid State Commun.* **22** (1977), 9.
[714] G. W. Gobeli, F. G. Allen, and E. O. Kane, *Phys. Rev. Lett.* **12** (1964), 94.

Photoelectron diffraction: Theory of photoemission from localized adsorbate levels

[715] A. Liebsch, *Phys. Rev. B* **13** (1976), 544.

Diffraction of photoelectrons emitted from core levels of Te and Na atoms adsorbed on Ni(001)

[716] D. P. Woodruff, D. Norman, B. W. Holland, N. V. Smith, H. H. Farrell, and M. M. Traum, *Phys. Rev. Lett.* **41** (1978), 1130.

Azimuthal anisotropy in deep core level X-ray photoemission from an adsorbed atom: oxygen on Cu (001)

[717] S. Kono, C. S. Fadley, N. F. T. Hall, and Z. Hussain, *Phys. Rev. Lett.* **41** (1978), 117.

Angle-resolved core level XPS from Cu single crystals

[718] R. N. Lindsay, C. G. Kinniburgh, and J. B. Pendry, *J. Electron Spectrosc.* **15** (1979), 157.

Structural sensitivity of photoelectron diffraction azimuthal patterns

[719] B. W. Holland, M. S. Woolfson, D. P. Woodruff, P. D. Johnson, D. Norman, H. H. Farrell, M. M. Traum, and N. V. Smith, *Solid State Commun.* **35** (1980), 225.

Photoelectron diffraction effects in GaAs(110) and Ge(110)

[720] M. Owari, M. Kudo, Y. Nihei, and H. Kamada, *J. Electron Spectrosc.* **22** (1981), 131.

Photoelectron diffraction from the atomic sites in the layer compound Sb_2Te_2Se

[721] M. R. Thuler, R. L. Benbow, and Z. Hurych, *Solid State Commun.* **42** (1982), 803.

Azimuthal anisotropy in core level X-ray photoemission from c(2 × 2) oxygen on Cu(001)

[722] S. Kono, S. M. Goldberg, N. F. T. Hall, and C. S. Fadley, *Phys. Rev. Lett.* **41** (1978), 1831.

Angle resolved UPS studies of the chemisorption of ethylene and acetylene on Ni(100)

[723] K. Horn, A. M. Bradshaw, and K. Jacobi, *J. Vac. Sci. Technol.* **15** (1978), 575.

Angle resolved photoemission from surfaces and adsorbates

[724] N. V. Smith, *J. Physique* **39** (1978), 4–161.

Inelastic electron scattering from adsorbate vibrations

[725] C. H. Li, S. Y. Tong, and D. L. Mills. *Phys. Rev. B* **21** (1980), 3057.

Angular distribution in metal photoemission

[726] G. D. Mahan, *Phys. Rev. Lett.* **24** (1970), 1068.

Angular distribution of photoelectrons from metal single crystal

[727] C. S. Fadley and S. A. L. Bergstrom, *Phys. Lett.* **35A** (1971), 375.

Angular distribution of photoelectrons from metal crystals

[728] C. S. Fadley and S. A. L. Bergstrom, in [6], p. 233.

XPS photoelectron angular distribution from solids

[729] C. S. Fadley, *J. Electron Spectrosc.* **5** (1974), 725.

Angular distribution of extreme ultraviolet photoelectrons

[730] W. McMahon and L. Heroux, *Applied Optics* **13** (1974), 438.

Angular dependence of X-ray photoelectrons from solids

[731] J. Brunner and H. Zogg, *J. Electron Spectrosc.* **5** (1974), 911.

Spatial symmetries of valence band structures by angle-resolved XPS

[732] R. H. Williams, P. C. Kemeny, and L. Ley, *Solid State Commun.* **19** (1976), 495.

Spin–orbital splitting in the valence bands of PbSe from angle-resolved UV photoemission

[733] T. Grandke, L. Ley, M. Cardona, and H. Preier, *Solid State Commun.* **24** (1977), 287.

Angle-resolved photoemission from PbS(100) for 16.85 eV and 21.22 eV excitation energy

[734] T. Grandke, L. Ley, and M. Cardona, *Solid State Commun.* **23** (1977), 897.

Surface d-band narrowing in Cu from angle-resolved XPS

[735] M. Mehta and C. S. Fadley, *Phys. Rev. Lett.* **39** (1977), 1569.

Angular resolved UPS of the $c(2 \times 2)$ overlayer of Se on Ni(001) using polarized and unpolarized He I radiation

[736] K. Jacobi and C. V. Muschwitz, *Solid State Commun.* **26** (1978), 477.

Fixed molecule high-energy photoelectron angular distribution

[737] W. Domcke and L. S. Cederbaum, *Surface Sci.* **72** (1978), 223.

Differential photoelectric cross sections for fixed-orientation p and d orbitals

[738] S. M. Goldberg, C. S. Fadley, and S. Kono, *Solid State Commun.* **28** (1978), 459.

Angle-dependent studies on some prototype vertically and laterally inhomogeneous samples

[739] D. T. Clark, A. Dilks, D. Shuttleworth, and H. R. Thomas, *J. Electron Spectrosc.* **14** (1978), 247.

Symmetry method for the absolute determination of energy-band dispersions $E(k)$ using angle-resolved photoelectron spectroscopy

[740] E. Dietz and D. E. Eastman, *Phys. Rev. Lett.* **41** (1978), 1674.

Angle-resolved photoemission and band structure of solids

[741] L. F. Ley, *J. Electron Spectrosc.* **15** (1979), 329.

Effect of elastic scattering on angular distribution

[742] O. A. Baschenko and V. I. Nefedov, *J. Electron Spectrosc.* **17** (1979), 405.

Strong temperature dependence and direct transition effects from the valence bands of single crystal W

[743] Z. Hussain, S. Kono, R. E. Connelly, and C. S. Fadley, *Phys. Rev. Lett.* **44** (1980), 895.

Angular distribution in multiphoton transitions

[744] S. N. Dixit and P. Lambropoulos, *Phys. Rev. Lett.* **46** (1981), 1278.

Resonant enhanced photoemission from CO chemisorbed on Pt(111) surface

[745] G. Loubriel, T. Gustafsson, L. I. Johansson, and S. J. Oh, *Phys. Rev. Lett.* **49** (1982), 571.

Magnetic surface states from angle-resolved photoelectron spectroscopy

[746] E. W. Plummer, *J. Appl. Phys.* **53** (1982), 2002.

Complementary to the angle-resolved photoelectron spectroscopy (ARPES) method which provides $E(\mathbf{k}_\parallel)$ information on occupied states, there is now an experimental technique which involves the *inverse* of the photoemission process. The sample surface is bombarded with a beam of electrons of variable energy and the photon flux emitted due to radiative transitions is measured. This inverse photoelectron spectroscopic technique (IPES) permits investigation of the empty states between the Fermi level and the vacuum level which are inaccessible in ordinary photoelectron spectroscopy. Some of the first experiments are described in the following references.

[747] G. Denninger, V. Dose, and H. Scheidt, *Appl. Phys.* **18** (1979), 375.

Position of the lowest empty orbitals of CO and O chemisorbed on Ni(111) by inverse ultraviolet photoemission

[748] F. J. Himpsel and Th. Fauster, *Phys. Rev. Lett.* **49** (1982), 1583.

k-resolved inverse photoelectron spectroscopy (KRIPES) of an unoccupied surface state on Pd(111)

[749] P. D. Johnson and N. V. Smith, *Phys. Rev. Lett.* **49** (1982), 290.

Angle-resolved spin-polarized inverse photoelectron spectroscopy (SPIPES)

[750] J. Unguris, A. Seiler, R. J. Celotta, D. T. Pierce, P. D. Johnson, and N. V. Smith, *Phys. Rev. Lett.* **49** (1982), 1047.

INDEX

[Numbers in *italics* indicate pages where only references are provided.]

Absolute electron donor capabilities, 125
Absolute line intensities, 222
Absorption edge jump, 215, 354
Acenes, 109, *325*
Acetylacetone, 126, 131
Acetylacetone complexes, 131, *330*
Acylaminobutadienes, 129
Adiabatic ionization potential, 3
Additivity rules, 99, *324*
Adsorption studies, 163–178, 277–282,
    337–339, 366–368, 370
  on GaAs(110) surface, 174
  on Ni, polycrystalline, 166
  on Pt(100) and 6(111) × (100) surface,
    171, 177
  on Si(111) surface, 278
  on W(001) surface, 281
  on Zn, polycrystalline, 163
AES, *see* Auger electron spectroscopy
Ag, 50, 143, 148, 276, *333*
Ag-Au, 208, 219
AgBr, 145, 147, 159, 160, *334, 336*
AgCl, 145, 147, 149, *334*
AgF, 145, 147, *334*
AgI, 148, *334*
Al, 47, 143, 154
Aliphatic anhydrides, 129, *328*
Alkyl mercury compounds, 129, *328*
Allene, 242, *364*
Al_2O_3, 154, 211
Analyzer, electron energy:
  Bessel box, 35, *298*
  calibration of, 49, *303*
  cylindrical condenser, 31, *297*
  cylindrical grid, 26, *297*
  cylindrical mirror, 34, *297*
  differential retarding field, 30, *297*
  dispersive, 23
  low-pass mirror high-pass filter, 29, *297*

  magnetic, 31
  retarding potential, 23
  spherical condenser, 32, *297*
  spherical grid, 27, *297*
Angle-resolved photoelectron spectroscopy,
    233–284, *290, 292, 358–370*
  gases, 234–270, *358–365*
    asymmetry parameter, 20, 234. *See also*
      β
    autoionization, effect of, 241, 250, *360,
      365*
    differential cross section, 234–270,
      358–365
    molecular orbital parameters, correlations
      with, 266–270
    organic molecules, 242, *364, 365*
    with partially polarized light, 235, *359*
    rotational structure, 240, *365*
    vibrational structure, 240, 241, 260–266,
      365
  solids, 270–284, *288, 366–370*
    absorption studies, 277, *366–370*
    band narrowing effect, 274, *369*
    band structure, 277, *366–370*
    emission from oriented adsorbed
      molecules, 279, *367*
    parity considerations, 281, *367*
    surface emission enhancement, 273
    surface states, 277, *366, 370*
    symmetry selection rule, 274, 366
Anthracene, 109
Applications to:
  adsorption studies, 163–178, 277–282, *288,
    289, 337–339, 366–368, 370*
  aqueous solutions, *299*
  archaeological chemistry, 356
  band structure, 140–152, 155–160,
    277–279, *333–335, 366, 368, 369*

biological chemistry, 125, 126, *289, 294, 328*
catalysis, *341, 342.* See also Adsorption studies
corrosion, *288*
depth profiling of solids, 160–163, *336*
electrochemistry, 340, *341*
electron beam bombardment on surfaces, *342, 343*
frozen solutions, 43, *299*
geology, *341*
high temperature vapors, *332*
inorganic chemistry, 76–80, 131–138, *288, 314–317, 331–333*
ion beam bombardment of surfaces, 340
liquid samples, 42, 43, *299*
organic chemistry, 68–76, 93–131, *289, 313–316, 324–330, 333*
polymers, 230, 231, *317, 342*
psychotropic drugs, 22, 131, *295, 328*
quantitative analysis, 202–232, *354–356*
surface studies, *see* Adsorption studies; Applications to, band structure
teaching photoelectron spectroscopy, 305
transient species, 294, *329–330*
Ar, 7, 50, 180, 190
Argon ion etching, 162
ARPES, *see* Angle-resolved photoelectron spectroscopy
Arylcyclopropanes, 122
As, 174, 229
Asymmetry parameter, 20, 234. *See also* β
Atomic charge determination, Pauling method, 61
Au, 50, 51, 140, 143, 145, 216, 272, 276, 277, 282, *333*
Au-Ag, 208, 219
Au-Cu, 208
Auger electron spectroscopy, 4, 6, 152, 162, 164, 215, 230, *286, 287, 290, 292, 296*, 356
Au-Pd, 145, *334*
Autoionization, 12–15, 241, *288, 289, 296*

B, 10
β, 20, 234, *358–362*
  from experiment, 235–243, *358–362*
  from theory, 243–270, *347, 358–362*
$BaSO_4$, 230
Be, 10

Benzamides, 129, *328*
1,2 Benzanthracene, 113
Benzene, 97, 102, 109, 201, 242, *324, 325*
Benzenoid hydrocarbons, 111, *325*
Benzobicycloalkenes, 129, *328*
3,4 Benzophenanthrene, 113
Bessel box analyzer, 35, *298*
Biguanide complexes, 77, *314*
Binding energies, characteristic, of elements, 54, 308
Binding energy references in solids, 345
Binding energy shifts, *see* Chemical shifts
Biomolecules, 125, *325, 326*
Biphenyl, 119
BLE, 232
Bloch functions, 344, 365, 366
Bombardment-induced light emission spectroscopy, 232
Born-Oppenheimer approximation, 240, 261
Born theory, 267
Brillouin zone, 146, 149, 158, 278, 343
Bromomethanes, 61
Bulk states, 142
1,3 Butadiene, 242

C, 10, 173, 206, 248
C 1s binding energies, 55, 62
Ca, 187, 195, 224
Calibration of electron energy analyzers, 49, *303*
Catalysis, *341.* See also Adsorption studies
CBr_4, 19
CCl_4, 85
Cd, 143, *333*
CF_4, 50, 85, 201
CH_4, 19, 41, 84, 100, 200, 259, 268
C_2H_4, 172
C_3H_6, 172
Channel electron multiplier, 37
Charge potential model, 60, *310*
CH_2Br_2, 98
CH_2Cl_2, 97
Chemical shifts, 11, 12, 55, 75–80, *307–314*
  applications, 75–80
  thermodynamic estimates of, 66
CH_2I_2, 98
$C_{18}H_{12}$ isomers, 113
Chlorophyll, 21
C_4H_4O, 201
Chrysene, 113

INDEX

C_4H_4S, 201
Cl, 255, 278
CMA, 34, 297
CNDO, see Complete neglect of differential overlap
CO, 11, 13, 15, 79, 85, 167, 169, 173, 177, 240, 279, 337
CO_2, 11, 50, 167, 170
C_3O_2, 201
Collective resonance, 86
Complete neglect of differential overlap, 61, 64, 95, 97, 320, 323
Compton scattering, 295
Conduction band, 140
Conformational studies, 119, 122, *325*
Continuum wave functions, 186, 189, 194, 198, 246, 247, 249, 257, 348
CoO-MgO, 208
Cooper minimum, 178, 255
Core binding energies, 10, 54, *307, 308*
Core electrons, 3
  in solids, 152
Core relaxation energies, 57
Corrosion studies, *288*
Coster-Kronig transition, 83, 90, 206
Cross section, 293
  photoionization, 179–202, *351–354*
    applications, 202–226
    differential, see Angle-resolved photoelectron spectroscopy
    theoretical calculations:
      approximate model, 200, *351*
      central field calculations, 188, *350*
      close coupling scheme, 195
      Hartree-Fock and other methods, 193
      hydrogenic systems, 186
      many-body perturbation theory, 195, *350*
      for molecules, 197
      nodal properties of orbitals, correlations with, 199, *350*
      quantum defect method, 187
CS_2, 226
Cu, 50, 274, 276, 277
Cu-Au, 208
Cu-Ni, 208, 219
CuO-MgO, 208
Curie temperature, 157, 336
Cyclohexadiene, 242
Cyclohexene, 242

Cylindrical condenser analyzer, 31, *297*
Cylindrical grid analyzer, 26, *297*
Cylindrical mirror analyzer, 34, *297*

D_2, 27
Deconvolution, *289, 303, 305*
  Fourier transform technique, using, 305
Density of states, 143
Depth profiling, 160–162, *336*
Dialkyl mercury compounds, 129, *328*
Diels-Alder reaction, 130, *328*
Differential cross section, see Angle-resolved photoelectron spectroscopy
Differential retarding field analyzer, 30, *297*
Dimethyl acetylacetone, 126
Dipole approximation, 186, 350
Direct transition, 275
Direct transition model, 276
Dispersion, 24
Dispersive analyzers, 23
Dithiole compounds, 71, *313*
DNA-RNA bases, 125, *325*
D_2O, 92
DOS, 143
Double filter retarding potential analyzer, 29, *297*
Dy, 226

Early experiments, UPS-XPS, *292*
EDC, 141
Electron detectors, 39
Electron energy analyzers, see Analyzer, electron energy
Electron energy loss spectra, 107, 152, *288, 358*
Electron impact spectroscopy, *289*
Electron-ion interaction effects, anisotropic, 255, *361*
Electron irradiation on surfaces, effects of, *342, 343*
Electron multipliers, 40
Electron relaxation, 57
Electronic autoionization, 14
Elemental sensitivities in XPS, *355*
ELS, see Electron energy loss spectroscopy
Energies of Auger transitions, 6
Energy bands in solids, 344
Energy distribution curve, 141
Equalization of electronegativities, 59
Equivalent cores approximation, 66

ESCA, 4
Escape depth, 142, 160, *336, 337*
Étendue, 24, *296*
Ethyl chloroformate, 11
Ethylene, 97, 171, 242
Ethyl trifluoroacetate, 12
EXAFS, 283
Exchange potential, 189
Exchange splitting, 157, *336*
Extended Hückel theory, 64, 322
Extended x-ray absorption fine structure, 283

F, 10, 55, 69, 173
F 1s binding energies, 57
Fe, 74
$Fe(CO)_5$, 79, *314*
Fe_2O_3, 19
Fermi level, 140
 determination, photoelectron spectroscopic, 155
Ferrites, *335*
Feshbach resonance, 241
Fluorescence transition probability, 215
Fluorobenzene, 75
Fluoromethanes, 69, 238, *313*
Fluorophenyl sulfur compounds, *313*
Fourier transforms:
 deconvolution, application to, 305
 orbitals, of, 277
Franck-Condon excitation of lattice, 153
Franck-Condon factors, 3, 8, 15, 53, 84, 95, 240, 260, 293
Franck-Condon principle, 293
Free radicals, 294, *329, 330*
Frozen solutions, 43, *299*
FWHM, 23

Ga, 149, 174, 175
GaAs, 149, 174, 279, *334, 335*
GaSe, 38, 49
$GaSe_2$, 276, 277
Gaunt factors, 186, *350*
Ge, 159, 162, 277, *336*
GeH_4, 19
Graphite, 50, 62
Group electronegativities, 65
Group shifts, 64, *310*

H, 195, 281
H_2, 9, 15, 27, 35, 46, 180, 198, 238, 263

Hallucinogens, 129, *328*
Halomethanes, 68, 237, *313*
HAM, *see* Hydrogen atom in molecules method
Hammett σ constants, 75, 116, 129, 314
HBr, 9, *293*
HCl, 102, 238
$HCONH_2$, 42
$(HCO)_2O$, 131
He, 26, 180, 187
HEIS, 232
Hexacarbonyls, W and Cr, 129, *331*
Hexafluoroacetylacetone complexes, 131, *330*
Hexafluorobenzene, 97
High energy ion scattering, 232
HMO, *see* Hückel molecular orbital theory
H_2O, 91, 167, 171, 198, 199, 200, 238, 259
H_2S, 167, 238
Hückel molecular orbital theory, 109, 112, 117, 120, 321
Hund's coupling, 136, 330
Hydrogen atom in molecules method, 95, 322, 323, *330*
Hydrogenic systems, 186
8-Hydroxyquinoline complexes, 228

In, 143, *333*
INDO, *see* Intermediate neglect of differential overlap
Ielastic mean free path, 160
Inelastic scattering cross section, 208
Inorganic complexes, 76–79, 131, *315–317, 332, 333*
Inorganic halide vapors, *332*
Intensities, line:
 absolute, 222
 intraelemental, 222
 relative, 206
Intermediate neglect of differential overlap, 64, 95, 262, 320, 323
Inverse photoelectron spectroscopy, 370
Ion absorption spectroscopy, 9, *293*
Ion cyclotron resonance, 44, *299*
Ionization efficiency curve, 1. *See also* Retarding potential difference method
Ion-molecule reactions, 67
IPES, *see* Inverse photoelectron spectroscopy
7-Isopropylidene norbornadiene, 118

Jahn-Teller effect, 19, *324*

K, 153, 187
k-conserving transitions, 275
Kikuchi bands, 282
Koopmans's approximation, 15, 57, 59, 115, 293
Kr, 50, 142, 190
k-resolved inverse photoelectron spectroscopy, 370
KRIPES, 370

Langmuir, 164
LEED, see Low energy electron diffraction
LFER, 115, 314
LEIS, 232
Li, 10
Lifetime broadening, 83
LiI, 159, *336*
Linear free energy relationships, 115, 314
Linear inelastic scattering coefficient, 216
Linear total attenuation coefficient, 215
Line broadening, core, 83
Linewidth, ionizing radiation, 25
Liquid beam, 42
Liquids, UPS-XPS, 42, *299*
Lone pair peaks, 96, *323*
Low energy electron diffraction, 165, 175, 283, *286, 335*, 357
Low energy ion scattering, 232
Luminosity, 24, 30, 51

Magic angle, 235
Magnetic electron energy analyzer, 31
Magnetic surface states, *370*
Many-body effects, 17, *288*
Many-body perturbation theory, 195, *350*
Mass attenuation coefficients, 215
Matrix dilution technique, 229
Matrix elements, 185, 188, 241, 276, 347–350, 366
    dipole acceleration, length, velocity forms, 186
Mean free path, 204, 209, *336, 337, 354*
    estimates from angle-resolved photoelectron spectroscopy, 270
Metalloporphyrins, 21
Metalloproteins, 21
Metal phthalocyanine complexes, 21
Methyl acetylacetone, 126
Methylene halides, 97

Methyl methanes, 100
Mg, 47, 154, 187, 226, 247, 3
MgO, 154
Microcomputer automation, sample probe, *304*
MINDO, see Modified INDO
Mixed oxides, 208
Modified INDO, 64, 95, 126, 321, 323
Molecular beams, 45, *300*
Molecular orbital theory, semiempirical models, 317, *323*
Monosubstituted methanes, 69, *313*
MoO_3, 229
MoS_2, 276, 277
Mössbauer-XPS chemical shift correlations, 71, *314*
Multidetector system, 24, 40, 41
Multiphoton ionization spectroscopy, *289*
Multiplet splitting, 18, 153, *324*

N, 10, 64, 66, 248
N 1s binding energies, 55
N_2, 50, 66, 85, 89, 167, 170, 198, 199, 238, 240, 259, 263, 279
Na, 143, 187, 227, 283
NaCl, 145, 282
Naphthalene, 109
Ne, 26, 50, 81, 180, 190, 200, 238
NH_3, 66
Ni, 157, 165, 166, 178, 272, 280, 283
$Ni(CO)_4$, 168
NMR-XPS chemical shift correlations, 68, *313*
NO, 18, 66, 85, 167
N_2O, 83, 86, 167, 170, 201, 202
Nonbonded orbital peaks, 96
Nondirect transitions, 275
Norbornadiene, 117, 242

O, 10, 55, 64, 66, 165, 248
O 1s binding energies, 55
O_2, 15, 18, 77, 90, 135, 137, 184, 238
OPW, see Orthogonalized plane wave
Orbital-β correlations, 20, 237
Organometal phenyl sulfides, 69, *313*
Oriented molecules, 279, *367*
Orthogonalized coulomb wave, 198
Orthogonalized plane wave, 197, 256, 261
Oxidation of:
    GaAs(110) surface, 174
    Zn, polycrystalline, 163, *337*

Partial waves, 246, 247, 266, 270, 358
Partial photoionization cross section, *see* Cross section, photoionization
Pb, 230
Pb(CH$_3$)$_4$, 19
PbCl$_2$, 229, 230
PbI$_2$, 229, 230
PbSO$_4$, 229, 230
Pd, 143, 145, 157, *333, 336*
p-deutero phenyl cyclopropane, 122
pentacene, 109
Perfluoro effect, 96, 133, *324*
Perfluoroethylene, 97
PESIS, 4
PESOS, 4
p-fluorophenyl sulfur compounds, 69, *313*
Phase shift, 189, 249, 266, 358
Photoabsorption coefficient, 179
Photoabsorption cross section, 179
Photoelectron angular distribution, *see* Angle-resolved photoelectron spectroscopy
Photoelectron diffraction, 282, *367*
Photoelectron-photoion coincidence spectrometry, *305, 333*
Photoelectron spectromicroscopy, 45, *300*
Photoemission partial yield spectroscopy, 152
Photoionization cross section, *see* Cross section, photoionization
Photon sources, 4, 301
Phthalocyanine, 21, *295*
Plane mirror analyzer, 37, *298*
Plane wave, 197, 257, 261
Plane wave matrix element model, 276
Plasmon peaks, 154
Plasmons, 141
Polymers, 230, 231, *317, 336, 342, 356*
Post monochromator, 29
Poynting vector, 256, 363
Preretardation, 31
Pressure effect, 306
Propene, 171, 173
Psychotropic drugs, 22, 129, *295, 328*
$\overline{\pi}$-systems, substituted, 115, *325*
Pt, 171, 173, 177, 276, *337*
p-type bands, 99
Pullman *k* index, 125, *325*
PW, *see* Plane wave
Pyridine, 21, 78

Quadrupole mass spectrometry, 46, *300*

Quantitative analysis, 202–232, 289, 354–356
  constraints, 202
  polymers, 230, 231
  trace analysis, 226–230
  two models, 204, 214
Quantum defect method, 187
Quick-frozen solutions, 43, *299*

Radiation lines, characteristic, UPS-XPS, 4, 301
Radical cations, *329*
Rb, 153
Relative sensitivity index, 221
Relaxation energies, 16, 17, 43, 58, 309
Relaxation shifts, *335*
Renner-Teller splitting, 19
Resolution, 23
Resolving power, 23
Resonance light sources, 46, *300–301*
  line broadening, 25
  wavelengths, characteristic, 301
Retarding potential analyzers, 23, *297*
Retarding potential difference method, 291
Rotational autoionization, 14
Rotational isomerism studies, *333*
Ru, 280
Rydberg atomic units, 185
Rydberg states, 13, 14, 107, 198, 252, 294

S, 255
S 2p binding energies, 57, 62
S$_2$, 135
Salazopyrin, 21, 295
Satellite lines, 18, 81, 153
Sb, 143, 144, *333*
Se$_2$, 135–138, *330*
Secondary ion mass spectrometry, 232, *289, 290*, 357
SF$_6$, 85, 201
Semiempirical models, 95, 317–323
Sensitivity index, 220
Shake-off process, 81, *315*
Shake-up process, 81, 153, *315*
Shape resonance, 241
Schottky barrier formation, *335*
Si, 151, 152, 155, 187, 274, 277, 279, *336*
SiH$_4$, 19
SIMS, *see* Secondary ion mass spectrometry
Slater determinant, 188, 317
Sm, 144, 226, *333*

Sn, 143, *333*
(SN)$_x$, 149, *334*
SnSe$_2$, 276
SO, 137
SO$_2$, 167
Spectroscopic potential adjusted INDO, 95, 321, 323
Spherical condenser analyzer, 32, *297*
Spherical grid analyzer, 27, *297*
Spherical wave, 198
SPINDO, see Spectroscopic potential adjusted INDO
Spin-orbit splitting, 7, 18, 149, *324*
Spin-polarized inverse photoelectron spectroscopy, 370
Spin-polarized photoelectrons, *296*
SPIPES, 370
s-type bands, 99
Substituted benzenes, 314
Sum rule, 97, *324*
Super Coster-Kronig transition, 87
Surface energy band sructure, 151, 277
Surface photoemission intensity enhancement:
  grazing photoelectron exit angles, at, 273
  grazing X-ray incidence angles, at, 273
Surface states, 140, 142, 151, 277, *335*, 370
Symmetry selection rule, 274, 366
Synchrotron radiation, *288, 289, 302*
  applications, 39, 47, 174–178, 235, 283, *338, 352, 361, 365*
  properties, 47
  facilities, location of, 302

TaS$_2$, 278
TaSe$_2$, 278
Tautomeric equilibria, 126, *316, 326*
Te, 143, 283, *333*
Te$_2$, 135–138, *330*
Teichoic acid, 21, 295
Temperature dependent photoemission:
  organic compounds, 128, *333*
  solids, 157, *336*
Tetracene, 109, 113
Tetracyanoethylene, 129
Three-step model, 276
Threshold photoelectrons, 164, 303
Threshold photoelectron spectroscopy, 303
by electron attachment, 304
Total linear absorption coefficient, 214
TPES, 303
TPSA, 304
Trace analysis, *355, 356*
  extraction onto a solid surface, by, 226, *355*
  matrix dilution technique, 229, *355*
  volatilization technique, 229, *356*
Transmission factor, 24
Trifluoropropene, 173
Triphenylene, 113

UPS, 4

V, 76
Vacuum level, 140
Valence band, 140
Valence band photoemission, 140
Valance band spectra, 143–150, *333–335*
Valence electrons, 3
Valence electron spectra, 89–138
Valence potential model, 63, *312*
Vanadium complexes, 76, 134, *314*
Vertical ionization potential, 3
Vibrational autoionization, 14
Vibrational broadening, core lines, 83, *315–316*
Vibrational intensities, *352–353*
Vibrational structure, 35, 108, 240, *316, 365*
Vinyl chloride, 171
Vinyl fluoride, 171

W, 152, 281
Walsh type orbitals, 122
Work function, 141, 164, 167, 345

χ^2-, F-tests, 80, *315*
Xe, 14, 50, 57, 67, 86, 250
Xe 3d binding energies, 57, 67
XPS, 4
XPS-Mössbauer chemical shift correlations, 71, *314*
XPS-NMR chemical shift correlations, 68, *313*
X-ray absorption coefficient, 224
X-ray fluorescence, 4, 214

Zn, 163
Zn oxidation of, 163, *337*